# Monitoring and Visualizing Membrane-Based Processes

*Edited by*
*Carme Güell, Montserrat Ferrando, and Francisco López*

## *Further Reading*

K.-V. Peinemann, S. Pereira Nunes (Eds.)

**Membrane Technology**
**6 Volume Set**

ISBN: 978-3-527-31479-9

S. Pereira Nunes, K.-V. Peinemann (Eds.)

**Membrane Technology**
**in the Chemical Industry**

2006
ISBN: 978-3-527-31316-7

A.F. Sammells, M.V. Mundschau (Eds.)

**Nonporous Inorganic Membranes**
**for Chemical Processing**

2006
ISBN: 978-3-527-31342-6

B.D. Freeman

**Materials Science of Membranes**
**for Gas and Vapor Separation**

2006
ISBN: 978-0-470-85345-0

K. Ohlrogge, K. Ebert (Eds.)

**Membranen**
**Grundlagen, Verfahren und industrielle Anwendungen**

2006
ISBN: 978-3-527-30979-5

# Monitoring and Visualizing Membrane-Based Processes

*Edited by*
*Carme Güell, Montserrat Ferrando, and Francisco López*

WILEY-VCH Verlag GmbH & Co. KGaA

**The Editors**

***Dr. Carme Güell***
University Rovira i Virgili
Department d'Enginyeria Química
Campus Sescelades
Avda Països Catalans 26
45007 Tarragona
Spain

***Dr. Montserrat Ferrando***
University Rovira i Virgili
Department d'Enginyeria Química
Campus Sescelades
Avda Països Catalans 26
45007 Tarragona
Spain

***Dr. Francisco López***
University Rovira i Virgili
Department d'Enginyeria Química
Campus Sescelades
Avda Països Catalans 26
45007 Tarragona
Spain

**Cover illustration**
The membrane module image was kindly provided by Alfa Laval Corporate AB.



**Library of Congress Card No.:** applied for

**British Library Cataloguing-in-Publication Data**
A catalogue record for this book is available from the British Library.

**Bibliographic information published by the Deutsche Nationalbibliothek**
Die Deutsche Nationalbibliothek lists this publication in the Deutsche Nationalbibliografie; detailed bibliographic data are available in the Internet at http://dnb.d-nb.de



Printed in the Federal Republic of Germany
Printed on acid-free paper

**Cover design** Adam Design, Weinheim
**Typesetting** Macmillan Publishing Solutions, Bangalore, India
**Printing** Strauss GmbH, Mörlenbach
**Bookbinding** Litges & Dopf GmbH, Heppenheim

**ISBN:** 978-3-527-32006-6

# Contents

**Preface** *XIII*

**List of Contributors** *XV*

**1** **Introduction: Opportunities and Challenges of Real Time Monitoring on Membrane Processes** *1*
*Marianne Nyström and Mika Mänttäri*
1.1 Introduction *1*
1.1.1 Monitoring from Permeate and Concentrate Properties *1*
1.2 Microscopic Techniques in Membrane Characterization *2*
1.3 Electrical, Laser, Magnetic and Acoustic Techniques in Membrane or Membrane Process Characterization *3*
1.4 Process Oriented Monitoring Techniques *4*
1.5 Future Scope of Sensors in Membrane Process Characterization *5*
References *7*

**Part I** **Optical and Electronic Microscopic Techniques on Membrane Process Characterization** *9*

**2** **Direct Visual Observation of Microfiltration Membrane Fouling and Cleaning** *11*
*Jeffrey S. Knutsen and Robert H. Davis*
2.1 Introduction *11*
2.2 Particle Deposition *12*
2.2.1 Observations Through the Sidewalls *12*
2.2.2 Observations Through the Membrane *14*
2.2.3 Observations of the Membrane Surface *16*
2.3 Particle Removal *24*
2.4 Concluding Remarks *31*
References *31*

*Monitoring and Visualizing Membrane-Based Processes*
Edited by Carme Güell, Montserrat Ferrando, and Francisco López

ISBN: 978-3-527-32006-6

**3 Microscopy Techniques for the Characterization of Membrane Morphology** *33*
*Carles Torras, Tània Gumí, and Ricard Garcia-Valls*
3.1 Introduction *33*
3.2 Membrane Characterization Morphology Parameters *34*
3.3 Microscopy Techniques for Membrane Morphology Characterization *35*
3.3.1 Electron Microscopy *35*
3.3.2 Atomic Force Microscopy *36*
3.3.3 Confocal Scanning Laser Microscopy *37*
3.3.4 Analysis of SEM Images *38*
3.4 Case Studies *39*
3.4.1 Flat Membranes *39*
3.4.1.1 Synthesis Considerations *39*
3.4.1.2 Morphological Characterization *41*
3.4.1.3 Morphological Characterization Versus Membrane Performance *45*
3.4.2 Microcapsules *48*
3.4.2.1 Synthesis Considerations *48*
3.4.2.2 Morphological Characterization *49*
3.5 Final Remarks *52*
References *53*

**4 Confocal Scanning Laser Microscopy: Fundamentals and Uses on Membrane Fouling Characterization and Opportunities for Online Monitoring** *55*
*Montserrat Ferrando, Maria Zator, Francisco López, and Carme Güell*
4.1 Introduction *55*
4.2 Fundamentals of Confocal Scanning Laser Microscopy *56*
4.2.1 General Principle *56*
4.2.2 Sample Preparation and Fluorescent Labeling *59*
4.2.3 Image Analysis *60*
4.2.3.1 Surface Porosity Measurements *60*
4.2.3.2 Cake Characterization *61*
4.3 Applications of CSLM to Membranes and Membrane Processes *61*
4.3.1 Membrane Characterization *62*
4.3.2 Membrane Fouling Characterization by CSLM *63*
4.3.3 Characterization of Bio-fouling Layers in Membrane Bioreactors by CSLM *70*
4.3.4 Online Monitoring of Membrane Processes by CSLM *72*
4.4 Evaluating Limits and Prospects of CSLM in the Characterization of Membranes and Membrane Processes *73*
References *74*

**5 Scanning Probe Microscopy Techniques in the Investigation of Homogeneous and Heterogeneous Dense Membranes: the Case for Gas Separtion Membranes** *77*
*Antonio Hernández, Petro Prádanos, Laura Palacio, Roberto Recio, Ángel Marcos-Fernández, and Ángel Emilio Lozano*
5.1 Introduction *77*
5.2 Microscopic Techniques *80*
5.2.1 Electron Microscopy: SEM and TEM *81*
5.2.2 Scanning Probe Microscopy: STM and AFM *82*
5.2.3 Computerized Image Analysis *83*
5.2.4 Roughness and Fractal Dimension *84*
5.3 Gas Separation Membranes *85*
5.4 Case Studies *87*
5.4.1 Phase Segregated Membranes *87*
5.4.2 Solvent Evaporation *89*
5.4.3 Asymmetric Polymeric Membranes *94*
5.4.4 Mixed Matrix Membranes *96*
References *101*

**6 Atomic Force Microscopy Investigations of Membranes and Membrane Processes** *105*
*W. Richard Bowen and Nidal Hilal*
6.1 Introduction *105*
6.2 The Range of Possibilities *106*
6.3 Correspondence Between Surface Pore Dimensions from AFM and MWCO *111*
6.4 Imaging in Liquid and the Determination of Surface Electrical Properties *114*
6.5 Effects of Surface Roughness on Interactions with Particles *118*
6.6 "Visualization" of the Rejection of a Colloid by a Membrane Pore and Critical Flux *120*
6.7 The Use of AFM in Membrane Development *122*
6.8 Conclusions *125*
References *125*

**7 Confocal Raman Microscopy for Membrane Content Visualization** *127*
*Philippe Sistat, Patrice Huguet, and Stefano Deabate*
7.1 Introduction *127*
7.2 The Raman Effect *127*
7.2.1 Partial Quantum Mechanical Treatment of the Raman Effect *128*
7.3 Raman Microspectrometry *130*
7.4 Confocal Raman Microscopy *131*
7.5 Specific Data Processing *133*

7.5.1 Baseline and Fluorescence *133*
7.5.2 Raman Lines Assignment *134*
7.5.3 Quantitative Processing *134*
7.5.3.1 Intensity Ratio *134*
7.5.3.2 Nonlinear Least Squares Fitting *135*
7.5.3.3 Principal Component Analysis and Related Methods *137*
7.6 Visualization Using Raman Spectroscopy *138*
7.6.1 Line Scanning *138*
7.6.2 Plane *140*
7.6.3 Volume *140*
7.7 Membrane Systems Applications *140*
7.7.1 Cell Design *140*
7.7.2 Visualization of Ionic Species Distribution *141*
7.7.3 Visualization of Solvent Distribution *144*
7.7.4 Gas Permeation *146*
7.7.5 Membrane Synthesis *146*
7.7.6 Biological Media *147*
7.8 Conclusion *147*
References *148*

**8 In Situ Characterization of Membrane Fouling and Cleaning Using a Multiphoton Microscope** *151*
*Robert Field, David Hughes, Zhanfeng Cui, and Uday Tirlapur*
8.1 Overview *151*
8.2 Optical Techniques for Characterization of Membrane Fouling *151*
8.2.1 Introduction *151*
8.2.2 Principles of Multiphoton Microscopy *153*
8.3 Partial Review of Fouling and Cleaning *155*
8.4 Materials and Methods *156*
8.4.1 Outline of the Application of MPM to Membrane Filtration *156*
8.4.2 Module Design *157*
8.4.3 Filtration Circuit *158*
8.4.4 Image Analysis *158*
8.5 Imaging Cake Fouling *160*
8.5.1 Introduction *160*
8.5.2 Effect of Fluorophore Labeling *161*
8.5.3 Insights Into Cake Development *163*
8.6 Imaging of Protein Fouling *166*
8.6.1 Introduction *166*
8.6.2 Effect of Fluorophore Labeling *166*
8.6.3 Using Images to Cross-check Fouling Models *167*
8.6.4 Deposition of Protein Mixtures *168*
8.6.5 Chemical Cleaning of Protein Fouled Membranes *169*
8.7 In Situ Characterization of Cell–Protein Fouling *170*
8.7.1 Introduction *170*

8.7.2 Filtering Protein Solutions Through Preformed Yeast Cakes *171*
8.8 Conclusions *172*
References *173*

**Part II Electrical, Laser and Acoustic Techniques for Membrane Process Characterization** *175*

**9 Electrical Characterization of Membranes** *177*
*Juana Benavente*
9.1 Introduction *177*
9.2 Electrical Measurements *178*
9.2.1 Streaming Potential *178*
9.2.2 Membrane Potential *180*
9.2.3 Impedance Spectroscopy *181*
9.3 Experimental *185*
9.3.1 Membranes and Solutions *185*
9.3.2 Streaming Potential Measurements *186*
9.3.3 Membrane Potential and Impedance Spectroscopy Measurements *187*
9.4 Results of Electrical Measurements for Different Types of Membranes *188*
9.4.1 Streaming Potential *188*
9.4.2 Membrane Potential *192*
9.4.3 Impedance Spectroscopy *197*
9.5 Conclusions *203*
References *206*

**10 X-ray Tomography Application to 3D Characterization of Membranes** *209*
*Jean-Christophe Remigy*
10.1 Principles of X-ray Tomography *209*
10.2 Application of X-ray Tomography in the Membrane Field *212*
10.2.1 X-ray Tomography and Membranes: the Limits and Design of Experiments *212*
10.2.2 Reported Examples of the Application of Tomography on Membranes and Membrane Processes *215*
10.2.2.1 Structural Characterization of Membranes *216*
10.2.2.2 Concentration, Fouling or Deposit Characterization *225*
10.3 Conclusions *226*
References *227*

11 **Optical and Acoustic Methods for in situ Characterization of Membrane Fouling** *229*
*J. Mendret, C. Guigui, P. Schmitz, and C. Cabassud*
11.1 Introduction *229*
11.2 Approach *230*
11.3 In Situ Deposit Characterization with an Optical Method Using a Laser Sheet at Grazing Incidence *230*
11.3.1 Optical Methods Using a Laser for Fouling Characterization *230*
11.3.1.1 Optical Laser Sensor *231*
11.3.1.2 Laser Triangulometry *231*
11.3.2 Filtration Set-up *232*
11.3.3 Principle of the Optical Method *232*
11.3.4 Measurement of Cake Thickness Growth *233*
11.3.5 Image Analysis *235*
11.3.6 Capabilities and Limitations of the LSGI Technique *236*
11.3.7 Typical Results *237*
11.3.8 Verification of the Order of Magnitude *240*
11.3.9 Conclusion *242*
11.4 In Situ Deposit Characterization with an Acoustic Method *242*
11.4.1 Acoustic Methods for Fouling Characterization *243*
11.4.2 Development of an Acoustic Method for Fouling Characterization *244*
11.4.3 Example of the Development of an Acoustic Method for the Characterization of Dead-end UF Deposits *245*
11.4.4 In Situ Application of the Method *247*
11.4.5 Combined Use of the Two Methods: Approach *248*
11.5 Conclusion *249*
References *251*

**Part III Process-oriented Monitoring Techniques** *253*

12 **Monitoring of Membrane Processes Using Fluorescence Techniques: Advances and Limitations** *255*
*Carla A. M. Portugal and João G. Crespo*
12.1 Introduction: Why Use Natural Fluorescence as a Monitoring Technique? *255*
12.2 Natural Fluorescence Techniques *256*
12.2.1 Steady-state Fluorescence *257*
12.2.2 Time-resolved Fluorescence *258*
12.2.3 Steady-state Fluorescence Anisotropy *261*
12.2.4 Time-resolved Fluorescence Anisotropy *263*
12.2.5 Steady-state Fluorescence versus Time-resolved Fluorescence Techniques *264*
12.3 Monitoring of Membrane Processes *264*

12.3.1 Steady-state Fluorescence for the Monitoring of Membrane Water Treatment Processes and Membrane Bioreactors *264*
12.3.2 Natural Fluorescence Techniques for Monitoring the Membrane Processing of Biological Molecules *271*
12.4 Concluding Remarks *280*
References *281*

**13 Membrane Emulsification Processes and Characterization Methods** *283*
*Gabriela G. Badolato, Barbara Freudig, Ping Idda, Uwe Lambrich, Helmar Schubert, and Heike P. Schuchmann*
13.1 Introduction *283*
13.2 Emulsification Technology *284*
13.2.1 Emulsions *284*
13.3 Membrane Emulsification Processes *286*
13.3.1 Membranes *286*
13.3.2 Conventional or Direct Membrane Emulsification *287*
13.3.3 Membrane Emulsification in the Jetting Regime *289*
13.3.4 Premix Membrane Emulsification *290*
13.3.4.1 Process Principles *290*
13.3.4.2 Influence of Process Parameters *291*
13.3.4.3 Process Flux *295*
13.3.4.4 Inline Measurements *295*
13.3.5 Microchannel Emulsification *298*
13.4 Summary and Conclusions *300*
References *302*

**14 Towards Fouling Monitoring and Visualization in Membrane Bioreactors** *305*
*Yulita Marselina, Pierre Le-Clech, Richard M. Stuetz, and Vicki Chen*
14.1 Introduction *305*
14.2 Factors Affecting Fouling in MBR *306*
14.2.1 Nature of the Feed *306*
14.2.2 Membrane Properties *307*
14.2.3 Operating Conditions *307*
14.3 Fouling by Biological Material *308*
14.4 Fouling Characterization Without Visualization Techniques *309*
14.4.1 Permeate Flux and Transmembrane Pressure *309*
14.4.2 Empirical Fouling Models *310*
14.4.3 Rejection Performance *310*
14.5 Invasive Methods for Fouling Observation *311*
14.5.1 Electron Microscopy *311*
14.5.1.1 Scanning Electron Microscopy *311*
14.5.1.2 Field Emission Scanning Electron Microscopy *311*

14.5.1.3 Environmental Scanning Electron Microscopy *312*
14.5.1.4 Transmission Electron Microscopy *312*
14.5.2 Atomic Force Microscopy *313*
14.5.3 Confocal Laser Scanning Microscopy *314*
14.6 Non-invasive Observation Methods *315*
14.6.1 Projector Technique *316*
14.6.2 Microscope Observation *316*
14.6.2.1 Direct Observation Through Membrane *316*
14.6.2.2 Direct Observation on Hollow Fiber Membrane *317*
14.6.2.3 Membrane Fouling Simulator *319*
14.6.3 Laser Applications *320*
14.6.3.1 Laser Beam Excitation *320*
14.6.3.2 Laser Excitation Near Infrared Region *320*
14.6.4 Ultrasonic Time Domain Reflectometry *320*
14.6.5 Electrochemical Shear Probe *321*
14.6.6 Photo-interrupt Sensor *322*
14.6.7 Nuclear Magnetic Resonance *322*
14.6.8 Particle Image Velocimetry *323*
14.7 Conclusions *323*
References *325*

**15 Monitoring Technique for Water Treatment Membrane Processes** *329*
*Kuo-Lun Tung*
15.1 Introduction *329*
15.2 Development of Fouling Monitoring Techniques *330*
15.2.1 Requirements for a Successful Fouling Monitoring Technique *330*
15.2.2 Classification of Fouling Monitoring Techniques *330*
15.2.3 Installation of a Process-oriented Fouling Monitoring System *333*
15.3 Dynamic Analysis of Online Fouling Monitoring *334*
15.3.1 Online Measurement of Fouling Layer Thickness *334*
15.3.1.1 Optical Method *335*
15.3.1.2 Acoustic Method *337*
15.3.2 Dynamic Analysis of Fouling Layer Structure *338*
15.3.2.1 Formation of the Surface Fouling Layer *339*
15.3.2.2 Compression of a Fouling Layer *340*
15.3.2.3 Resistance Estimation of a Fouling Layer *341*
15.3.2.4 Procedures for Analyzing the Fouling Layer Structure During a Membrane Filtration Process *342*
15.3.2.5 Case Analysis *343*
15.3.3 Monitoring of Water Quality in a Membrane Filtration Process *348*
15.4 Conclusions and Future Perspectives *351*
References *353*

**Index** *355*

# Preface

Increasing interest in the use of membranes and membrane-based processes in many kinds of industries, and the interesting prospects that they offer for the near future, has resulted in a need to improve our knowledge, both on the processes themselves and also on the main "driving force" of those technologies, which are the membranes. During three days in October 2006, researchers from all over the world met in Tarragona (Spain) to discuss and present their experiences and expertise, within the frame of a workshop on the characterization and use of monitoring techniques applied to membrane processes. The workshop served as the initiation site for this book. All presenters, and some other researchers who could not participate at that moment, have collaborated in the preparation of the book. We would like to have a very special memory for Dr. Frank Reineke, who engaged in the project with great enthusiasm, making a great contribution during the workshop and who unfortunately passed away a few months after.

This book follows the same structure proposed for the workshop, with a first part devoted to microscopic techniques, a second part for electrical, laser and acoustic techniques and a third part on some examples of monitoring techniques applied to membrane processes, with a special emphasis on membrane bioreactors, which has become a very relevant application. Each chapter of the book offers the reader a basic introduction to the specific characterization and/or monitoring technique, followed by selected examples showing its capabilities, strengths and weaknesses, if any. This book has been constructed as a link between the books devoted to commercial applications of membrane processes and the more specific literature on materials characterization. Despite the availability of a vast literature on membrane processes, we feel this book fills an important gap, since it offers a broad overview of the different characterization methodologies currently available or under development and it also gives a thorough state of the art for each of them. Going through the different chapters of the book, any scientist, whether already in the membrane field or a beginner, will find helpful information to select the appropriate methodology for off- and/or online monitoring of the process.

Tarragona (Spain), December 2008

*Carme Güell*
*Montserrat Ferrando*
*Francisco López*

*Monitoring and Visualizing Membrane-Based Processes*
Edited by Carme Güell, Montserrat Ferrando, and Francisco López

ISBN: 978-3-527-32006-6

# List of Contributors

***Gabriela G. Badolato***
Universität Karlsruhe
Institut für Bio- und Lebensmittel-technik
Lebensmittelverfahrenstechnik
Kaiserstrasse 12
76128 Karlsruhe
Germany

***Juana Benavente***
Universidad de Málaga
Facultad de Ciencias
Departamento de Física Aplicada I
Grupo de Caracterización
Electrocinética y Electroquímica de
Membranas e Interfases
Avda Cervantes 2
29071 Malaga
Spain

***W. Richard Bowen***
i-Newton Wales
54 Llwyn y mor, Caswell
Swansea SA3 4RD
United Kingdom

***C. Cabassud***
Université de Toulouse
Laboratoire d'Ingénierie des Systèmes
Biologiques et Procédés
INSA/CNRS/INRA
135 Avenue de Rangueil
31077 Toulouse Cedex
France

***Vicki Chen***
University of New South Wales
School of Chemical Sciences and
Engineering
UNESCO Centre for Membrane
Science and Technology
High Street, Kensington
2052 Sydney
Australia

***João G. Crespo***
Universidade Nova de Lisboa
Faculdade de Ciências e Tecnologia
Departamento de Química
Requimte-CQFB
Campus da Caparica
2829–516 Caparica
Portugal

***Zhanfeng Cui***
University of Oxford
Department of Engineering Science
Parks Road
Oxford OX1 3PJ
United Kingdom

***Robert H. Davis***
University of Colorado
Department of Chemical and
Biological Engineering
424 UCB
Boulder, CO 80309–0424
USA

*Monitoring and Visualizing Membrane-Based Processes*
Edited by Carme Güell, Montserrat Ferrando, and Francisco López

ISBN: 978-3-527-32006-6

***Stefano Deabate***
Université Montpellier II
Institut Européen des Membranes
CC 047, Place Eugène Bataillon
34095 Montpellier Cedex 5
France

***Montserrat Ferrando***
University Rovira i Virgili
Department d'Enginyeria Química
Campus Sescelades
Avda Països Catalans 26
45007 Tarragona
Spain

***Robert Field***
University of Oxford
Department of Engineering Science
Parks Road
Oxford OX1 3PJ
United Kingdom

***Barbara Freudig***
Universität Karlsruhe
Institut für Bio- und Lebensmitteltechnik
Lebensmittelverfahrenstechnik
Kaiserstrasse 12
76128 Karlsruhe
Germany

***Ricard Garcia-Valls***
University Rovira i Virgili
Department d'Enginyeria Química
Avda Països Catalans 26
43983 Tarragona
Spain

***Carme Güell***
University Rovira i Virgili
Department d'Enginyeria Química
Campus Sescelades
Avda Països Catalans 26
45007 Tarragona
Spain

***C. Guigui***
Université de Toulouse
Laboratoire d'Ingénierie des Systèmes Biologiques et Procédés
INSA/CNRS/INRA
135 Avenue de Rangueil
31077 Toulouse Cedex
France

***Tània Gumí***
University Rovira i Virgili
Department d'Enginyeria Química
Avda Països Catalans 26
43983 Tarragona
Spain

***Antonio Hernández***
Universidad de Valladolid
Facultad de Ciencias
Departamento Física Aplicada
Real de Burgos s/n
47071 Valladolid
Spain
and
Surface and Porous Materials (SMAP)
UA-CSIC-UVA
Edificio de I+D
Belen s/n
Campus Miguel Delibes
47071 Valladolid
Spain

***Nidal Hilal***
University of Nottingham
School of Chemical and Environmental Engineering
Nottingham, NG7 2RD
United Kingdom

***David Hughes***
University of Oxford
Department of Engineering Science
Parks Road
Oxford OX1 3PJ
United Kingdom

***Patrice Huguet***
Université Montpellier II
Institut Européen des Membranes
CC 047, Place Eugène Bataillon
34095 Montpellier Cedex 5
France

***Ping Idda***
Universität Karlsruhe
Institut für Bio- und Lebensmittel-technik
Lebensmittelverfahrenstechnik
Kaiserstrasse 12
76128 Karlsruhe
Germany

***Jeffrey S. Knutsen***
University of Colorado
Department of Chemical and Biological Engineering
424 UCB
Boulder, CO 80309–0424
USA

***Uwe Lambrich***
Universität Karlsruhe
Institut für Bio- und Lebensmittel-technik
Lebensmittelverfahrenstechnik
Kaiserstrasse 12
76128 Karlsruhe
Germany

***Pierre Le-Clech***
University of New South Wales
School of Chemical Sciences and Engineering
UNESCO Centre for Membrane Science and Technology
High Street, Kensington
2052 Sydney
Australia

***Francisco López***
University Rovira i Virgili
Department d'Enginyeria Química
Campus Sescelades
Avda Països Catalans 26
45007 Tarragona
Spain

***Ángel Emilio Lozano***
CSIC
Instituto de Ciencia y Tecnología de Polímeros
Juan de la Cierva 3
28006 Madrid
Spain
and
Surface and Porous Materials (SMAP)
UA-CSIC-UVA
Edificio de I+D
Belen s/n
Campus Miguel Delibes
47071 Valladolid
Spain

***Mika Mänttäri***
Lappeenranta University of Technology
Department of Chemical Technology
Laboratory of Membrane Technology and Technical Polymer Chemistry
Skinnarilankatu 34
53851 Lappeenranta
Finland

***Ángel Marcos-Fernández***
CSIC
Instituto de Ciencia y Tecnología
de Polímeros
Juan de la Cierva 3
28006 Madrid
Spain
and
Surface and Porous Materials (SMAP)
UA-CSIC-UVA
Edificio de I+D
Belen s/n
Campus Miguel Delibes
47071 Valladolid
Spain

***Yulita Marselina***
University of New South Wales
School of Chemical Sciences and
Engineering
UNESCO Centre for Membrane
Science and Technology
High Street, Kensington
2052 Sydney
Australia

***J. Mendret***
Université de Toulouse
Laboratoire d'Ingénierie des Systèmes
Biologiques et Procédés
INSA/CNRS/INRA
135 Avenue de Rangueil
31077 Toulouse Cedex
France

***Marianne Nyström***
Lappeenranta University of
Technology
Department of Chemical Technology
Laboratory of Membrane Technology
and Technical Polymer Chemistry
Skinnarilankatu 34
53851 Lappeenranta
Finland

***Laura Palacio***
Universidad de Valladolid
Facultad de Ciencias
Departamento Física Aplicada
Real de Burgos s/n
47071 Valladolid
Spain
and
Surface and Porous Materials (SMAP)
UA-CSIC-UVA
Edificio de I+D
Belen s/n
Campus Miguel Delibes
47071 Valladolid
Spain

***Carla A. M. Portugal***
Universidade Nova de Lisboa
Faculdade de Ciências e Tecnologia
Departamento de Química
Requimte-CQFB
Campus da Caparica
2829–516 Caparica
Portugal

***Pedro Prádanos***
Universidad de Valladolid
Facultad de Ciencias
Departamento Física Aplicada
Real de Burgos s/n
47071 Valladolid
Spain
and
Surface and Porous Materials (SMAP)
UA-CSIC-UVA
Edificio de I+D
Belen s/n
Campus Miguel Delibes
47071 Valladolid
Spain

***Roberto Recio***
Universidad de Valladolid
Facultad de Ciencias
Departamento Física Aplicada
Real de Burgos s/n
47071 Valladolid
Spain
and
Surface and Porous Materials (SMAP)
UA-CSIC-UVA
Edificio de I+D
Belen s/n
Campus Miguel Delibes
47071 Valladolid
Spain

***Jean-Christophe Remigy***
Université Paul Sabatier
Laboratoire de Génie Chimique
CNRS UMR 5503
118 Route de Narbonne
31062 Toulouse Cedex 9
France

***P. Schmitz***
Université de Toulouse
Laboratoire d'Ingénierie des Systèmes Biologiques et Procédés
INSA/CNRS/INRA
135 Avenue de Rangueil
31077 Toulouse Cedex
France

***Helmar Schubert***
Universität Karlsruhe
Institut für Bio- und Lebensmitteltechnik
Lebensmittelverfahrenstechnik
Kaiserstrasse 12
76128 Karlsruhe
Germany

***Heike P. Schuchmann***
Universität Karlsruhe
Institut für Bio- und Lebensmitteltechnik
Lebensmittelverfahrenstechnik
Kaiserstrasse 12
76128 Karlsruhe
Germany

***Philippe Sistat***
Université Montpellier II
Institut Européen des Membranes
CC 047, Place Eugène Bataillon
34095 Montpellier Cedex 5
France

***Richard M. Stuetz***
University of New South Wales
School of Civil and Environmental Engineering
UNSW Water Research Centre
High Street, Kensington
2052 Sydney
Australia

***Uday Tirlapur***
Department of Engineering Science
University of Oxford
Parks Road
Oxford OX1 3PJ
United Kingdom

***Carles Torras***
University Rovira i Virgili
Department d'Enginyeria Química
Campus Sescelades
Avda Països Catalans 26
43983 Tarragona
Spain

***Kuo-Lun Tung***
Chung Yuan Christian University
R&D Center for Membrane Technology
Department of Chemical Engineering
200 Chung-Pei Road, Chungli
320 Taoyuan
Taiwan

***Maria Zator***
University Rovira i Virgili
Department d'Enginyeria Química
Campus Sescelades
Avda Països Catalans 26
45007 Tarragona
Spain

# 1
# Introduction: Opportunities and Challenges of Real Time Monitoring on Membrane Processes

*Marianne Nyström and Mika Mänttäri*

## 1.1
## Introduction

Membrane technology applications today focus considerable interest on the continuous performance of a membrane. It is well known that fouling reduces the flux and changes the retention properties of membranes, which makes the process less economic. Characterization of protein fouling has been reviewed by Chan and Chen [1], Chen et al. [2, 3]. The goal would be a non-fouling process with a steady flux and retention.

The possible tools today to make this come true are: cleaning the membrane and running the process at constant flux; and, in order to avoid fouling, running at a subcritical flux [4]. To find the best long-term conditions some online monitoring is needed. It has been noticed that in many discussions with industry one of the most important items on their "wish list" is to monitor the membrane process and characterize the membrane, possibly on-/inline.

The membrane properties of interest would be: flux, retention (pore size and distribution), charge, wearing, pinholes, fouling and hydrophilicity, just to mention a few. Membrane stability during the process would contribute to a sustainable process. The best monitoring processes would be in real time, non-invasive, in situ/in vivo, at a molecular scale and using pattern recognition approaches.

### 1.1.1
### Monitoring from Permeate and Concentrate Properties

Of the properties mentioned above, flux may be the easiest to monitor continuously. The monitoring could be done online by having a measurement device that could be connected to any membrane element. This device could also contain information on whether the flux is within acceptable limits or whether the module should go for cleaning. This type of device could also detect too high a flux and

*Monitoring and Visualizing Membrane-Based Processes*
Edited by Carme Güell, Montserrat Ferrando, and Francisco López

ISBN: 978-3-527-32006-6

thus maybe any wear or pinholes in the membrane. In the long run, too frequent cleaning does not increase the sustainability of a membrane.

One of the most difficult characteristics to handle is retention. In cases where the pore is much smaller than the retained molecules (ultrafiltration, UF; or microfiltration, MF,) the problem is not so delicate, but when the question comes to nanofiltration (NF), passage through the membrane can very easily be destroyed by a foulant (for instance calcium sulfate), making the fractionation in question impossible. The typical causes of changes in retention are fouling or other modifications of the membrane.

The cut-off of the membrane can be monitored with a standardized procedure, as developed in the FP4 CHARMME project for UF and NF and modified in the FP6 NanoMemPro project. Monitoring online faces the problem that the module in most cases should be taken out of use during the time needed for measurement. This could of course be done in connection with the cleaning of the module. Another way is to use tracers which could be analyzed from the normal permeate [5].

The following introduces the main methods available, online or not; and then they are further studied in the separate chapters of this book.

## 1.2 Microscopic Techniques in Membrane Characterization

In most microscopic techniques, characterization has to be done on dried samples and thus an online study is not possible. Normal *optical microscopy* might be done by direct observation if the sample of the membrane has big pores through which one can look at (for instance) fouling [6], using a camera (see Chapter 2). In this case you can only distinguish particles which are around a micrometer or larger. The method can be used to study bacterial or particle fouling or when looking at bubbles in the feed. With *confocal microscopy* (see Chapters 4 and 7) one can also look at the membrane in situ without having to dry it, since characterization can be performed on wet samples.

*Multi-photon microscopy* for in situ characterization is a fluorescence laser scanning technique (see Chapter 8). This technique can produce 3D images, which can be stacked to get time series for different interesting phenomena in UF and MF. Fouling inset can be characterized and thus also the critical flux can be measured. Fouling of membranes, especially by proteins, has been studied with this technique. Because the target substance needs to be fluorescent, not all kinds of filtration processes can be monitored in this way.

When attempting to see membrane structures below one micrometer the possible microscopic techniques are *electron microscopy* (EM), which can be supplied by an additional *elemental scanning* (EDS) device (see Chapter 3). In this case and also in *transmission electron microscopy* (TEM) [7] the samples need to be dry. It is often quite difficult to study polymer membranes with these microscopic techniques because the densities of most materials are the same. Therefore, marker systems should be used like radioactive tracers, fluorescent staining or dendrimeric staining. Today, in *environmental scanning electron microscopy* (ESEM) one can work also with wet samples.

*Atomic force microscopy* (AFM) is a technique where, in principle, you can also look at samples under water, but the images are usually sharper when done on dry samples. AFM is mostly used for UF membranes to look at pore size, but according to Bowen (see Chapter 6) one can actually see nanopores using AFM. When using AFM it is also possible to make scans on pore size distribution and on the roughness of samples [8]. Both of these characteristics would be important for online measurements because both fouling and wear of membranes could be observed as a function of time, which is of great value in industrial applications. Using AFM equipment, characterization of membranes used in gas separation processes can be performed (see Chapter 5) and also charge analysis of the surface can be made using its force balance analysis scheme, where repulsion or attraction between the tip of the instrument and the membrane can be analyzed and pictured as functions of interaction distance (see Chapter 6).

One of the new techniques to scan membranes and determine pore size distributions is *positron annihilation spectroscopy* (PAS). With this method also the free spaces in nanofiltration membranes can be determined. For instance in a study by Boussu et al. [9] it was found that some much-studied NF membranes (like Desal-5 DL, NTR7450) have two sizes of spaces, one size about 0.12–0.15 nm and the other between 3 nm and 4 nm. It could be speculated that diffusive transport would happen through the smaller spaces (the size of a water molecule), depending on the hydrophilicity properties of the membrane, and convective transport through the larger spaces, depending on size and charge conditions.

## 1.3 Electrical, Laser, Magnetic and Acoustic Techniques in Membrane or Membrane Process Characterization

The methods using electrodes to register phenomena on membrane surfaces may be the ones that are already available online or could very easily be adapted to an online situation (see Chapter 9). One of the already available techniques measures *streaming potential* (SP) online through the pores of the membrane. From the streaming potential the zeta potential can be calculated [10–12]. The measurement can be done both during pure water flux experiments and during fouling or cleaning. The result shows changes on the surfaces of the pores during the process. Nanofiltration membranes are better characterized for charge with SP measurements along the surface. Today these measurements are not yet well adapted for online measurements, but commercial devices are available for SP measurements along the membrane surface and it is most probable that it would not be very difficult to build them into a running process. It is important to notice that the charge on the pores within the membrane is not always equal to the charge on the membrane surface. A very good example of that is the track-edged membrane [13].

Another electrical measurement available is the *membrane potential* (MP) measurement [12, 14], which tells about the charge inside the membrane. MP measurements are still very tedious and will probably not be available online yet. *Impedance* measurements can also be made showing fouling, or different membrane layers (see Chapter 9) [15].

*X-ray photon spectroscopy* (XPS) is a technique which has been available for a long time. It needs dry samples, but is especially good for characterization of the top layers of membranes. Today, a new technique, *X-ray tomography* can be applied for 3D characterization of membranes (see Chapter 10). The tomography technique is very challenging for visualizing pore structures and fouling because, in the 3D mode, one can feel like a molecule passing from feed to permeate and realize the obstacles in the way.

New ways to build microelectrodes have made it possible to "implant" electrodes inside the membrane or on both sides of it. This makes measurements using *acoustics* possible (see Chapter 11) [16]. The acoustic waves can give a wave pattern that differentiates a clean membrane from a fouled membrane because the fouling spots can be identified as a new resistance. In a similar way cleaning can be studied as the removal of these resistive spots. The electrodes can be placed in such a way or be part of the membrane so that also membrane pore size could be measured or (at least with the techniques available today) pinholes could be found. Measurements made with acoustics seem to be very promising online measurements.

*Ultrasonic treatments* also belong to the acoustic methods. Already, for more than ten years ultrasound has been used for the cleaning of membranes. In most cases the transducers have been situated outside the membrane, but the membrane has sometimes also been used as the transducer. Of crucial importance are the frequency of the waves and the way the waves have been applied (see Chapter 11). The method has mostly been used to prevent the build-up of a concentration polarization layer. Polarization of dense particles has been easier to prevent than fouling or polarization of small particles [17, 18].

The ultrasound technique today can also be put up in 3D so that a fouling pattern of the membrane can be visualized continually and modeled through different statistical programs. This method is called *ultrasonic time domain reflectometry* and can be used for fouling and cleaning in situ (see Chapter 15). The system can also be used to study infrasonic sound (1–20 Hz) as waves in the study of flux characteristics like pulsing.

Ultrasound can also be used to enhance chemical processes, for instance instead of UV; or transducers and sensors can be used together for different kinds of processes. The ideal future state would be implanting micro-electrodes in the membrane and visualizing the state of the membrane and its expected life-time. Problems could arise to keep the transducers/sensors clean.

## 1.4 Process Oriented Monitoring Techniques

The methods mentioned above for characterization all involve some kind of process aspect as the membranes, virgin or fouled, are to be used in some process.

How well the process has been running can thus be distinguished by the use of methods characterizing the membrane itself. Another way to monitor the process is to use markers and sensors to measure (besides fouling) the permeate flux of compounds of interest.

The first markers to be used were radioactive markers, especially tritium or isotopic iodide-labeled proteins. The problems with this kind of markers were that radioactivity was not generally allowed to be used in experiments. Thus in order for a marker to be accepted it would have to be neither toxic nor harmful to nature. Naturally, also a marker should be easily detected, non-reactive and easily applied in the process.

Markers are very much needed for monitoring the cut-off of a membrane. For MF some monodisperse substances are available that can be monitored by microscopical methods because of their density. They are mainly used for the measurement of pore sizes, but they are very expensive. Good particles of this kind do not exist for nanofiltration today.

Online fluorescence monitoring is used in membrane bioreactors to see how well the biofilm is working (see Chapters 12 and 14). In this case the biofilm is studied through the module glass wall, using an optical well. The incoming excitation light makes the biofilm fluoresce and the emission light tells the intensity of the response. From a 2D image a 3D time-image is built up and the "fingerprint" of the biofilm can be monitored. Different fingerprints can be compared and growth or growth problems can be identified. The same type of set-up can be used using fluorescent markers or naturally fluorescent materials to study the fractionation of molecules, especially proteins.

Some of the fluorescent markers (like for instance tryptophane) change their fluorescence pattern as a result of the polarity of their environment (see Chapter 12). In this case an increase in polarity gives rise to a red shift and a decrease to a blue shift. The shifts can also be produced through adsorption of the molecule and thus fouling could be studied. Also some proteins change their structure after passing through a membrane pore. Today there exist very sensitive analysis methods for these small structural changes.

## 1.5 Future Scope of Sensors in Membrane Process Characterization

The techniques to measure and to handle data are developing very rapidly today. Computer technology has developed for mathematical simulations, equation solving and statistical analysis, so that different images can be analyzed at ever higher resolution, which makes it possible to analyze data from 2D levels to 3D or 4D levels. The limits of what can be done seem to be surpassed from year to year. This situation will be used in online monitoring of processes to make them sustainable and run at optimal conditions.

For online monitoring it is already possible today to make such monitoring where you take samples from feed and permeate and calculate how well the process is going. What is not yet done is to use this data for regulating the process conditions. If this kind of online optimization of process parameters is done, the costs and the intervals of cleaning could be regulated. Also, all the methods used to monitor the fouling layer could be used in this respect. What is needed is a window to look at the membrane so that different measurements can be done. Measurement of fluorescence was mentioned above; but also the newly marketed *particle image velocimetry/laser induced fluorescence* (PIV/LIF) equipment could be used to look at streamlines, concentration polarization and fouling on the membrane. Fluid velocity mapping for investigation of fouling has been done by Delauney et al. [19, 20].

Today, microelectrodes can be prepared and used as electrical, magnetic, or acoustic probes or sensors. In a future process the membranes, spacers or modules could be equipped with a pattern of microelectrodes for different types of measurements and the collected data would be used for optimizing the process conditions within the membrane process. It could also be possible that, in an industrial process, there would for instance be one test module equipped in this way that could be placed somewhere in the process design.

In the text above some fluorescent protein molecules have been used as markers. If the molecules to be analyzed are not fluorescent, another way to mark the molecule in the feed would be to use an implanted molecule that reacts with a target molecule and gives a complex which can be seen or which emits some measurable electrical light. These nanosensors could also be contained in controlled release packages and thus be delivered over time to find the targeted molecules in the feed or on the membrane. If the sensors are implanted molecules they could function like functionalized affinity membranes are doing today.

New types of markers are the self-assembling particles, e.g. dendrimers [21]. These molecules can be made in different sizes, even nano-sizes, and they can be spherical or have other 3D structures. They can be made with layers on the outside which have a different charge or hydrophilicity; and they can also be made to contain metals or fluorescent groups (dansylated dendrimers). In that way they can (by size) be markers for pore size, (by charge) be markers for membrane charge and (by hydrophilicity) be markers for membrane hydrophilicity interaction. By covering the markers with gadolinium (water-soluble gadomers) they become good in vivo markers, as they are inert and not visible to body reactions. These markers can be used in very different ways in monitoring. Today they are still not all on the market (if available, they demand high prices) but in the future they would probably be very useful in monitoring membrane processes.

## References


1 R. Chan, V. Chen, *J. Membrane Sci.* **2004**, *242*, 169–188.
2 V. Chen, H. Li, A.G. Fane, *J. Membrane Sci.* **2004**, *241*, 23–44.
3 J.C. Chen, Q. Li, M. Elimelech, *Adv. Colloid Interfaces* **2004**, *107*, 83–108.
4 P. Bacchin, P. Aimar, R.W. Field, Critical and sustainable fluxes: theory, experiments and applications (Review). *J. Membrane Sci.* **2006**, *281*, 42–69.
5 C. Causserand, S. Rouaix, A. Akbari, P. Aimar, *J. Membrane Sci.* **2004**, *238*, 177–190.
6 H. Li, A.G. Fane, H.G.L. Coster, S. Vigneswaran, *J. Membrane Sci.* **2003**, *217*, 29–41.
7 V. Freger, A. Bottino, G. Capanelli, M. Perry, V. Gitis, S. Belfer, *J. Membrane Sci.* **2005**, *256*, 134–142.
8 K. Boussu, B. Van der Bruggen, A. Volodin, J. Snauwaert, C. Van Haesendonck, C. Vandecasteele, *J. Colloid Interface Sci.* **2005**, *286*, 632–638.
9 K. Boussu, J. De Baerdemaeker, C. Dauwe, M. Weber, K. Lynn, D. Depla, S. Aldea, I. Vankelecom, C. Vandecasteele, B. Van der Bruggen, *Chemphyschem* **2007**, *8*, 370–379.
10 M. Nyström, M. Lindström, E. Matthiasson, *Colloids and Surfaces* **1989**, *36*, 297–312.
11 M. Nyström, A. Pihlajamäki, N. Ehsani, *J. Membrane Sci.* **1994**, *87*, 245–256.
12 P. Fievet, A. Szymczyk, B. Aoubiza, J. Pagetti, *J. Membrane Sci.* **2000**, *168*, 87–100.
13 K.-J. Kim, A.G. Fane, M. Nyström, A. Pihlajamäki, *J. Membrane Sci.* **1997**, *134*, 199–208.
14 R. Takagi, M. Hori, K. Gotoh, M. Tagawa, M. Nakagaki, *J. Membrane Sci.* **2000**, *170*, 19–25.
15 H.G.L. Coster, T.C. Chilcott, A.C.F. Coster, *Bioelectrochem. Bioenerg.* **1996**, *40*, 79–98.
16 J. Mendret, C. Guigui, C. Cabassud, P. Schmitz, *Desalination* **2006**, *199*, 216–218.
17 H. Kyllönen, P. Pirkonen, M. Nyström, J. Nuortila-Jokinen, A. Grönroos, *Ultrason. Sonochem.* **2006**, *13*, 295–302.
18 H. Kyllönen, P. Pirkonen, M. Nyström, *Desalination* **2005**, *181*, 319–335.
19 D. Delauney, M. Rabiller-Baudry, L. Paugam, *Desalination* **2006**, *200*, 205–207.
20 S. Chang, A. Yeao, A.G. Fane, M. Cholewa, Y. Ping, H. Moser, *J. Membrane Sci.* **2007**, *304*, 181–189.
21 X. Feng, D. Taton, R. Borsali, E.L. Chaikof, Y. Gnanou, *J. Am. Chem. Soc.* **2006**, *128*, 11551–11562.

# Part I
# Optical and Electronic Microscopic Techniques on Membrane Process Characterization

# 2
# Direct Visual Observation of Microfiltration Membrane Fouling and Cleaning

*Jeffrey S. Knutsen and Robert H. Davis*

## 2.1
## Introduction

Membrane fouling is generally undesirable because it leads to reductions in permeate flux (volume of permeate collected per membrane area per unit of time) and selectivity (passage of one species through the membrane while another species is retained). Typically, membrane fouling is characterized in a symptomatic way, by measuring changes in permeate flux and species transmission during an experiment. However, to provide a more fundamental understanding of membrane fouling, and hence be in the position to use this understanding to design membrane processes that minimize fouling, direct observation is desired; for example in situ observations or measurements of foulant deposition on the membrane surface (external fouling) or in the membrane pores (internal fouling). Similarly, direct observation of foulant removal, such as by backpulsing or other in situ cleaning techniques, is desired.

For microfiltration, external foulant deposition and removal can be monitored non-invasively by direct visual observation with the aid of a standard light microscope, provided that the foulant particles are micron-sized, or larger, and the filtration device is transparent. Smaller foulants may also be observed, if fluorescence is used. Figure 2.1 shows a simplified schematic of particle deposition and removal during forward and reverse filtration. Crossflow microfiltration (also called tangential flow filtration) is employed, in which the feed is introduced tangential to the membrane surface and the permeate flows through the membrane in the perpendicular direction. During forward filtration, particles are carried to the membrane surface and deposited to form an external cake layer. This cake layer grows until it reduces the permeate flux to the point where there is no longer a net deposition of particles, but the particles instead roll along the cake surface or experience a back-transport into the bulk suspension flow above the cake. During reverse filtration, which is used to clean the membrane by backpulsing or backflushing, the particles

*Monitoring and Visualizing Membrane-Based Processes*
Edited by Carme Güell, Montserrat Ferrando, and Francisco López

ISBN: 978-3-527-32006-6

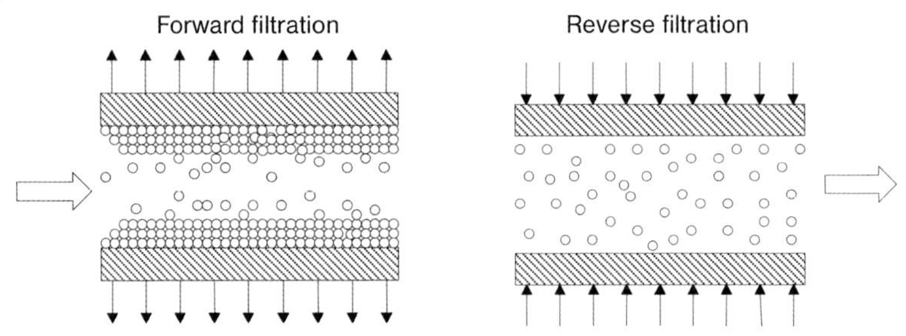


**Figure 2.1** Schematic (not to scale) of particle deposition and removal during forward and reverse filtration. (From Redkar and Davis [1]).

are lifted off the membrane surface and swept to the filter exit by the tangential feed flow.

In this chapter, direct visual observations of particle deposition and removal during crossflow microfiltration are described. We start with early work in which particle deposition and cake formation were observed from the side, and then consider more recent studies where video and still photomicrographs were taken from below or above the membrane. The chapter ends with observations of particle removal by backpulses, followed by concluding remarks. The interested reader is also referred to prior reviews of non-invasive, in situ membrane monitoring by Chen, Li and Fane [2] and Chen, Li and Elimelich [3].

## 2.2 Particle Deposition

### 2.2.1 Observations Through the Sidewalls

Early work on direct visual observation of particle motion and deposition during crossflow filtration used relatively large channels with transparent sidewalls. For example, researchers in Georges Belfort's laboratory at Renssalaer Polytechnic Institute performed single-particle trajectory experiments to observe the motion of a particle flowing above a porous wall [4]. The lateral migration of the particle toward or away from the wall due to a combination of inertial lift and the permeate crossflow was found to be in good agreement with inertial lift theory. Initial work in Robert Davis' laboratory at the University of Colorado also used devices with transparent sidewalls, though the focus was instead on measuring cake thickness as a function of distance from the channel entrance [5, 6]. To improve the ease of visual observation, they used relatively large channels of 0.5–1.0 cm in height and particles of 50–200 μm in diameter. They also used relatively low feed velocities (2.5–7.2 $cm\,s^{-1}$), resulting in wall shear rates in the range of 16–60 $s^{-1}$.

At the University of Cambridge, Mackley and Sherman [7] also used a relatively large channel (1.0 cm in height) and large particles (125–180 μm in diameter), along with a video camera at 15 × magnification to observe cake growth and particle trajectories. During the early stages of filtration, the permeate flux was relatively high and the particles advected to the filter cake remained in the position where initial impact occurred. In the later stages, however, the permeate flux was reduced by the cake resistance and the particles were observed to roll along the cake surface until they encountered an edge or crevice that caused capture.

Wakeman [8] at the University of Exeter subsequently performed systematic experiments with smaller particles (0.5–25 μm diameter) and higher crossflow velocities (1–4 m $s^{-1}$) that are more typical of practice. A high-speed video camera with a high magnification achieved by a zoom lens was used to observe cake formation through optical quality glass sidewalls. He showed how the steady cake thickness increases (and hence the permeate flux decreases) with decreasing crossflow velocity and increasing feed solids concentration. The rate of cake growth during the transient phase also increases with increasing feeds concentration. With increased particle size, the permeate flux as well as the cake thickness increases, due to the lower resistance of a cake formed with larger particles. In contrast to some theories, the cake thickness was found to be nearly uniform along the channel length, except for slight dips near the entrance and exit.

Wakeman [8] also performed some experiments with larger particles at very low concentrations, which enabled the motions of individual particles to be observed. Figure 2.2 shows a glass bead of approximately 100 μm in diameter that is rolling along a woven mesh with a weak flow of liquid through the mesh and a cross flow above the mesh. As also observed by Mackley and Sherman [7], rolling particles came to rest when they reached a stable position.

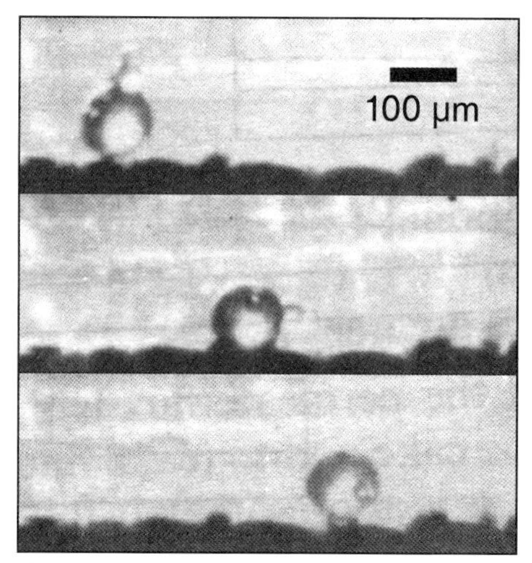


**Figure 2.2** Motion of a 100 μm glass bead, with an attached asperity, moving along a woven wire mesh. There is tangential fluid flow along the mesh, as well as a slow liquid flow through the mesh. (From Wakeman [8]).

### 2.2.2
### Observations Through the Membrane

A major advance was made by Tony Fane and co-workers at the University of New South Wales when they developed a method to perform visual observation from the permeate side of the membrane. This technique, called *direct observation through membrane* (DOTM), was first described by Hodgson, Pillay and Fane [9] and then discussed more fully by Li et al. [10]. The key idea is to use a membrane which becomes transparent when wetted, and then observe particle deposition by placing a microscope on the permeate or "back" side of the membrane. Because of the relatively short focal length, compared to looking into the center of a membrane molecule from the side, high magnification may be used to observe individual particles as small as 1–10 μm in diameter.

Figure 2.3 is a schematic of a DOTM apparatus described by Li et al. [10]. An anodized aluminum membrane with high porosity and straight pores was used, and the membrane module was made of transparent Plexiglas to allow for lighting from the feed side. The channel height is only 0.2 cm. The microscope was then positioned on the permeate side and focused through the transparent membrane on its "front" surface in contact with the feed suspension. Images were then viewed through the microscope at high power by a video camera. In the initial work, yeast cells of approximately 5 μm in diameter and latex beads of 3–12 μm in diameter were used.

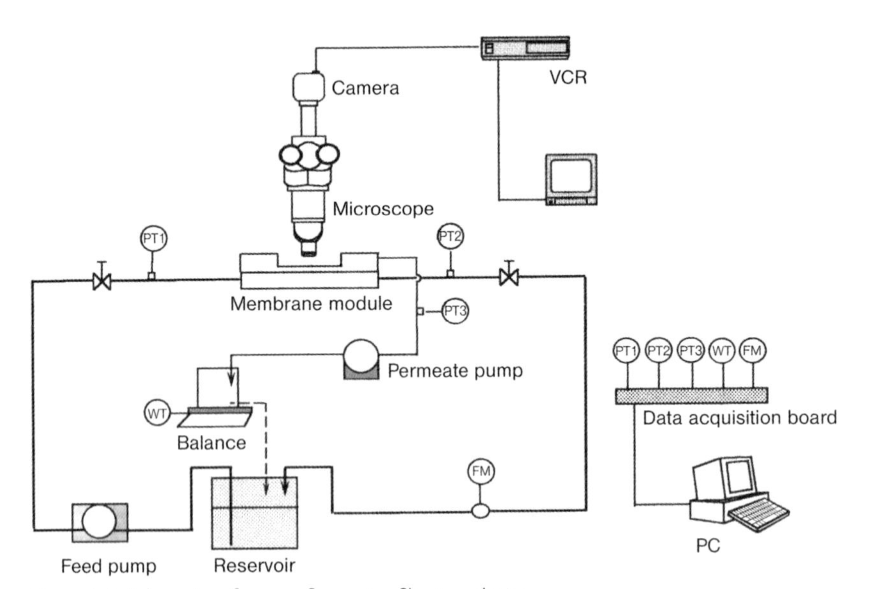


**Figure 2.3** Schematic of a crossflow microfiltration device employing direct observation through the membrane. (From Li et al. [10]).

Using DOTM, Li et al. [10] observed the membrane surface at fixed crossflow velocity and with step increases in permeate flux. Significant particle deposition occurred when the permeate flux exceeded a critical value. The underpinnings of the concept of a critical flux, below which membrane fouling is negligible, had been introduced in the 1980s but the term did not gain use until definitive work was published in 1995 by the laboratories of Pierre Aimar in Toulouse, Robert Field at Oxford and John Howell in Bath – see the recent review by Bacchin, Aimar and Field [11] and the original papers by Field et al. [12], Bacchin, Aimar and Sanchez [13] and Howell [14]. While the critical flux is often measured by a change in membrane resistance (flux decline at fixed transmembrane pressure, or pressure rise at fixed permeate flux), Li et al. [10] noted that particle deposition observed by DOTM is a more sensitive indication of incipient cake formation. The deposition of particles was found to be reversible, so that the deposited particles were swept downstream (Figure 2.4) when the permeate flux was reduced below the critical value. Additional studies by Li et al. [15, 16] used DOTM to show that the critical flux for the onset of particle deposition increases with increasing crossflow velocity and particle size. Reasonable agreement for some, but not all, particle systems was found between the observed values of the critical flux and those predicted by a shear-induced diffusion model.

When the permeate flux was controlled at a value near the critical flux, Li et al. [10] observed the motion of individual particles along the membrane surface. Individual latex beads of 6.4 μm and 11.9 μm in diameter, as well as yeast cells of 5.0 μm diameter, would roll along the membrane surface and then stop when encountering previously deposited particles or else move off the surface and back to the bulk flow. For 3.0 μm latex particles, however, the movement along the membrane surface was primarily in clumps. Newly depositing particles tended to accumulate around existing deposits, and then an entire cluster (which tended to be only one or two particle layers thick) would dislodge and begin to move along the membrane surface toward the filter exit. Figure 2.5 shows the movement of a cluster from one image to the next, as well as the appearance of a second cluster in the second image.

Additional applications of DOTM include observing deposition and critical fluxes for yeast, algae and bacterial cells [10, 15–17], and particle deposition patterns that occur when spacers are used in the membrane channels [18]. Spacers were shown to increase the critical flux by up to two times [19]. Observations of the deposition and removal of submicron bacteria required the use of fluorescence. Removal of the bacterial cake when the flux was reduced below the critical value was observed to occur in flocs [16]. In a recent application of DOTM, Zhang, Fane and Law [20] showed that the presence of larger particles increased the critical flux of smaller particles in a mixture and that only the smaller particles selectively deposited from a mixture when the imposed permeate flux was above this critical flux. This finding is consistent with the earlier observation of Li et al. [10] that the smaller particles are preferentially deposited from a feed distribution of particle sizes.

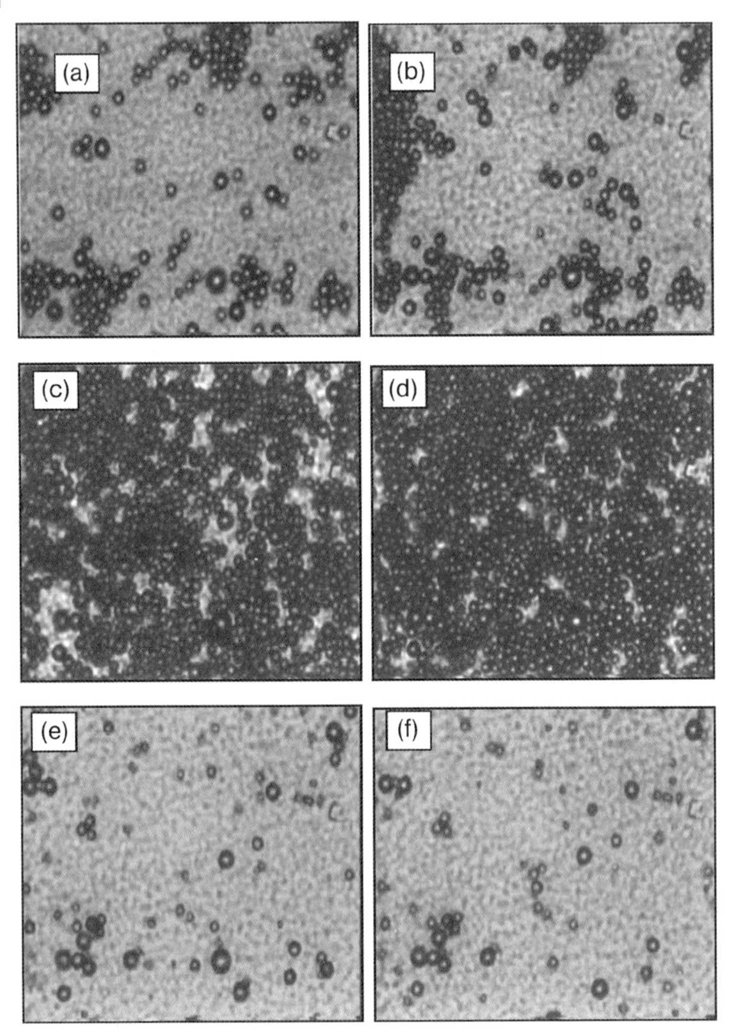


**Figure 2.4** DOTM images showing 6.4 μm latex particle deposition above the critical flux (images a–d) and removal below the critical flux (images e, f). Images (a–d) were taken after 3, 5, 17 and 30 min, respectively, of filtration with 15 $\mathrm{L\,m^{-2}\,h^{-1}}$ (1 $\mathrm{L\,m^{-2}\,h^{-1}}$ = $2.8 \times 10^{-7}\,\mathrm{m\,s^{-1}}$) permeate flux and average crossflow velocity of 0.23 $\mathrm{m\,s^{-1}}$. Images (e) and (f) were taken at 1 min and 13 min, respectively, after the permeate flux was subsequently decreased to 13 $\mathrm{L\,m^{-2}\,h^{-1}}$ and the average crossflow velocity was increased to 0.42 $\mathrm{m\,s^{-1}}$. (From Li et al. [10]).

### 2.2.3
### Observations of the Membrane Surface

As an alternative to the DOTM technique to observe membrane fouling from "below" the membrane on the permeate side, microscopic observation may be

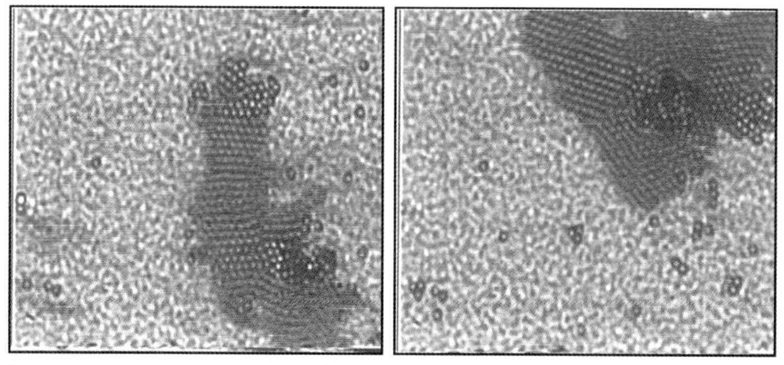

**Figure 2.5** DOTM images of 3 μm latex particles in moving clumps for filtration near the critical flux with 61 $L\,m^{-2}\,h^{-1}$ permeate flux and an average crossflow velocity of 0.85 $m\,s^{-1}$. The image on the right was taken 33 s after the image on the left. (From Li et al. [10]).

made from "above" the membrane on the feed or retentate side. Figure 2.6 shows a schematic of this technique, which Mores and Davis [21] refer to as direct visual observation (DVO), while Fane, Beatson and Li [22] call it direct observation of the surface of the membrane (DOSM). Advantages of DVO (or DOSM) include that the membrane does not need to be transparent and that tubular instead of flat membranes may be used (provided that the filtration is from outside in, so that the observed cake layer is on the outside of the hollow tube or fiber). However, the feed suspension must be dilute, so that it is not opaque, and a combination of a narrow channel and an objective lens with long working distance is required when high magnification is desired.

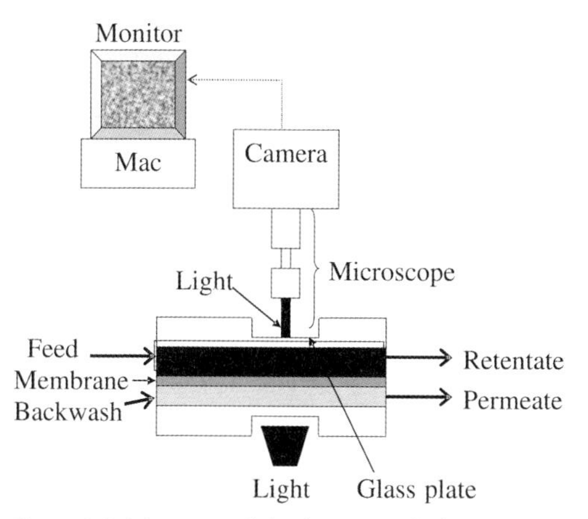


**Figure 2.6** Schematic of the direct visual observation (DVO) apparatus. (From Mores and Davis [21]).

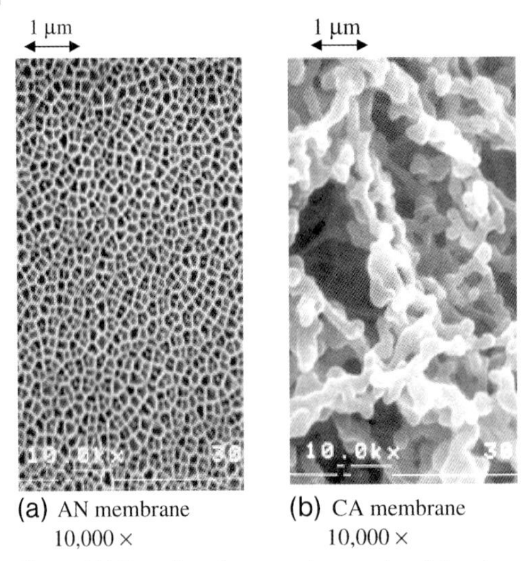


**Figure 2.7** Scanning electron micrographs of the clean membranes made of (a) anodized aluminum and (b) cellulose acetate. (From Mores and Davis [21]).

Mores and Davis [21] studied the deposition of yeast cells from dilute feed suspensions (0.05–0.1 g $L^{-1}$) during crossflow microfiltration at conditions well above the critical flux. The two types of membranes used include Anopore anodized aluminum (AN) and Micron Separations cellulose acetate (CA), both with a nominal pore size of 0.2 μm. As seen in Figure 2.7a, the AN membrane (which is the same type as used in DOTM studies by Li et al. [10], except the pores are bigger) has a smooth surface and regular, straight pores. The CA membrane (Figure 2.7b) has an irregular surface with large pore mouths.

Figure 2.8 shows a series of photomicrographs of yeast deposition on the AN membrane. The deposition occurs uniformly and reaches a complete monolayer within 2–3 min. After 5 min of filtration, the cake has at least two layers of yeast cells, as evident from the black image.

In contrast to the AN membranes, the CA membranes exhibit nonuniform particle deposition during forward filtration. As seen in Figure 2.9, the deposition occurs in localized patches, reflective of the uneven structure of the CA membrane surface. These DVO images were taken at lower magnification than those for the AN membrane, as otherwise the rough surface of the CA membrane would be out of focus.

Kang et al. [23] constructed a DVO apparatus similar to that of Mores and Davis [21] and used image analysis to provide rapid quantification of the deposition rates of yeast, bacteria and latex beads under different conditions. Figure 2.10 shows images after 30 min of filtration of yeast at feed concentrations of 25, 50 and 100 mg $L^{-1}$. As expected, the fractional coverages increase with increasing feed concentration, corresponding to approximately 49%, 62% and 86%, respectively.

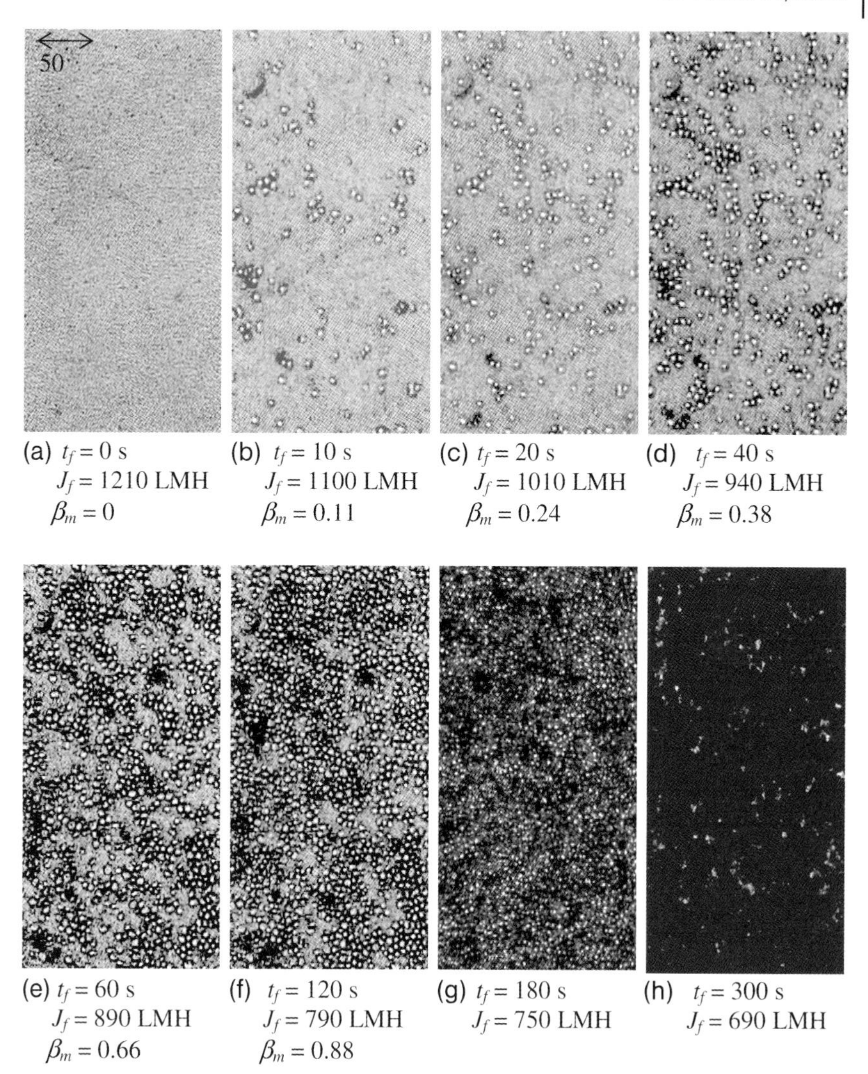


**Figure 2.8** Videomicrographs of an AN membrane taken using DVO at various times ($t_f$) of forward filtration of 0.05 g $L^{-1}$ yeast at a transmembrane pressure of 2 psi (1 psi = 6.9 kPa) and a nominal wall shear rate of 2500 $s^{-1}$. Shown are the fractional coverage ($\beta_m$) and the measured permeate flux ($J_f$), where 1 LMH = 1 L $m^{-2}$ $h^{-1}$ = $2.8 \times 10^{-7}$ m $s^{-1}$. (From Mores and Davis [21]).

The authors also showed that the initial deposition rate was nearly independent of crossflow rate, increased in proportion to the permeate velocity, and could be manipulated by changing solution or surface chemistry. Operation was above the critical flux, and the authors conclude the initial deposition was governed by permeate drag and electrostatic double-layer forces.

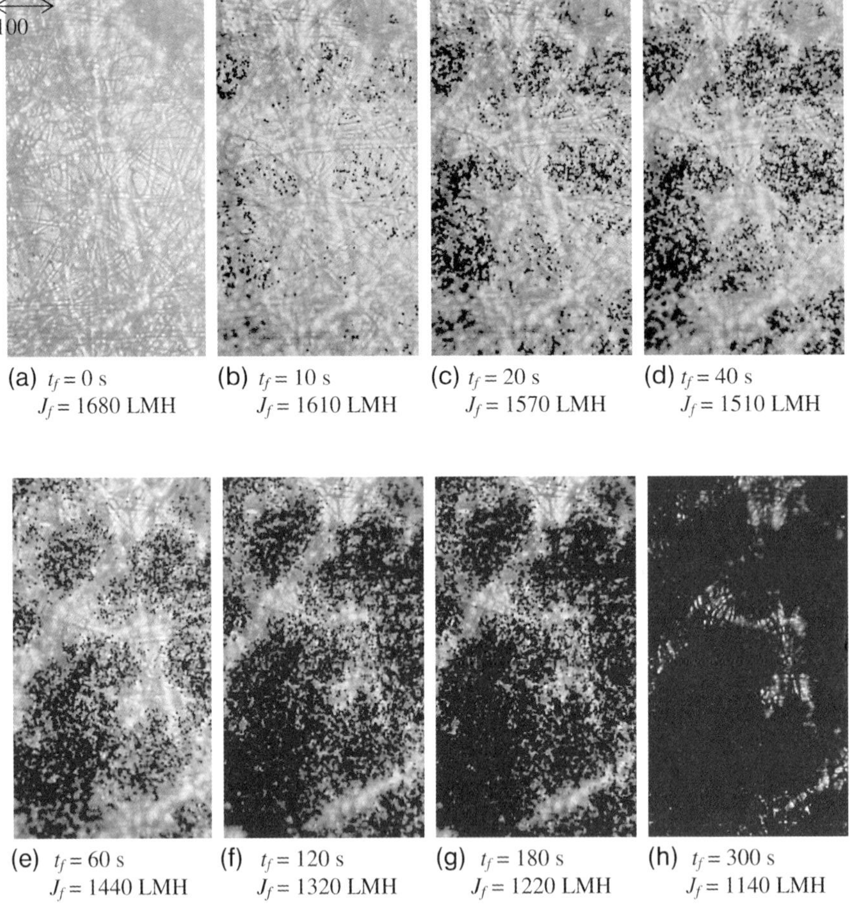


**Figure 2.9** Videomicrographs of a CA membrane taken using DVO at various times ($t_f$) of forward filtration of 0.1 g $L^{-1}$ yeast at a transmembrane pressure of 3 psi and a nominal wall shear rate of 6100 $s^{-1}$. Shown at each time is the measured permeate flux ($J_f$). (From Mores and Davis [21]).

More recently, Knutsen and Davis [24] used a similar DVO apparatus to quantity the deposition of yeast cells and latex particles during crossflow microfiltration. Rather than following the fractional coverage of the membrane, they used automated image analysis to count individual particles on the membrane as a function of time during the early stages of filtration. They then defined a fractional deposition, $\theta = C_w(t)/(J_o c_o t)$, as the fraction of those particles convected to the membrane surface by the permeate flow that deposit when reaching the membrane, where $C_w(t)$ is the number of particles deposited on the membrane surface during time $t$ per unit membrane area, $J_o$ is the permeate flux and $c_o$ is the particle concentration in the feed (number of particles, #, per volume). As

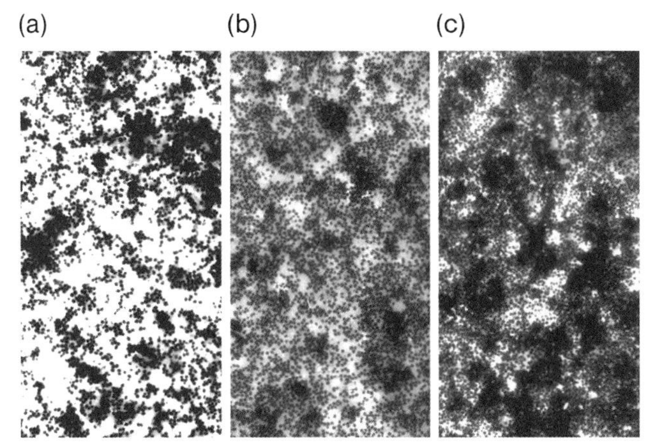


**Figure 2.10** Direct microscope images of yeast cells deposited on polyacrylonitrile ultrafiltration membranes after 30 min of filtration at a permeate flux of $72\,L\,m^{-2}\,h^{-1}$ and a crossflow velocity of $2.5\,cm\,s^{-1}$ ($150\,s^{-1}$ nominal wall shear rate), using feed solutions of (a) $25\,mg\,L^{-1}$, (b) $50\,mg\,L^{-1}$ and (c) $100\,mg\,L^{-1}$ yeast cells. (From Kang et al. [23]).

apparent from Figure 2.8, counting individual cells or particles is feasible during the early stages of filtration, but then becomes impossible when more than one layer forms.

The method of quantifying the fractional deposition allowed Knutsen and Davis [24] to make a somewhat surprising observation about the critical flux. At high fluxes, the fractional deposition was unity (100% deposited) and at low fluxes the fractional deposition was zero (0% deposited), as expected. However, at intermediate values of the permeate flux, the fractional deposition increased from 0 to 1 over an approximately 10-fold increase in permeate flux, rather than as a sharp transition at a uniquely defined critical flux, even though the particles were nearly monodisperse in size. Figure 2.11 demonstrates this phenomenon, where the surface concentration of deposited particles is shown as a function of time for several values of the permeate flux. At $J_o = 500\,L\,m^{-2}\,h^{-1}$ (LMH), there is 100% deposition ($\theta = 1$), whereas at $J_o = 5$ LMH and $J_o = 10$ LMH, there is no deposition ($\theta = 0$). At the intermediate flux of $J_o = 100$ LMH, a fraction of about 20–30% of the particles convected to the membrane surface is deposited. The lack of a unique critical flux (below which no particles deposit, above which all particles deposit, for a given shear rate) is not due to variations in particle size, as nearly monodisperse particles were employed. Instead, Knutsen and Davis [24] developed a theory based on force and torque balances on a particle at the membrane surface and showed that the fractional deposition can be explained by a distribution of roughness heights describing the membrane surface morphology. A single-particle theory, rather than a shear-induced diffusion model, was used because of the very low particle concentrations employed.

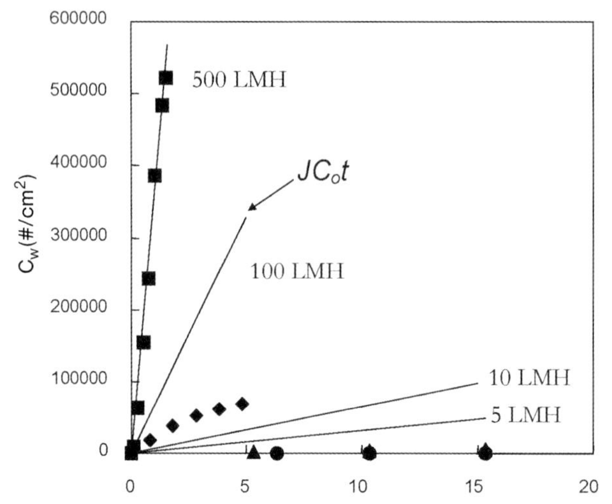


**Figure 2.11** Concentration of 10 μm latex beads deposited on a 0.2 μm anodized aluminum membrane at a nominal wall shear rate of $1000\,s^{-1}$ and various values of the permeate flux. The symbols represent experimental measurements and the solid lines give the corresponding predictions for $\theta = 1$.

Partial deposition may also occur when the crossflow velocity is varied, as shown in Figure 2.12. At a permeate flux of $J_o = 50$ LMH and a wall shear rate of $100\,s^{-1}$, 100% deposition occurs, whereas at a wall shear rate of $1000\,s^{-1}$ and the same permeate flux, the fraction of cells that adhere to the surface is decreased to about 20% as more particles are swept along the surface by the high cross flow.

Knutsen and Davis [24] were also able to use their DVO apparatus to track individual particles near the membrane surface and determine their velocities from the locations in successive video frames. As shown in Figure 2.13a, they observed particles rolling along the membrane surface, with velocity in the direction parallel to surface that remained nearly constant until the particle encountered a large asperity or another particle. They also observed (Figure 2.13b) particles slightly above the membrane surface, with velocity parallel to the surface that decreased with time as the particle approached the surface; upon contact with the surface, the particle would either stop immediately or roll along the surface. Rolling motion was most commonly observed when the permeate flux was near or below critical flux.

Similar to the DOTM observations of Zhang, Fane and Law [20] and Li et al. [10], Knutsen and Davis [24] showed that selective deposition of the smaller particles in a mixture could be obtained by proper choice of permeate flux below the critical flux for large particles and above that for small particles. The previously unpublished photos in Figure 2.14 show that all three particle sizes in a mixture (5, 10, 20 μm latex beads) deposit at a permeate flux of $200\,L\,m^{-2}\,h^{-1}$, whereas only the smallest particles deposit at $50\,L\,m^{-2}\,h^{-1}$. Of additional interest is that the

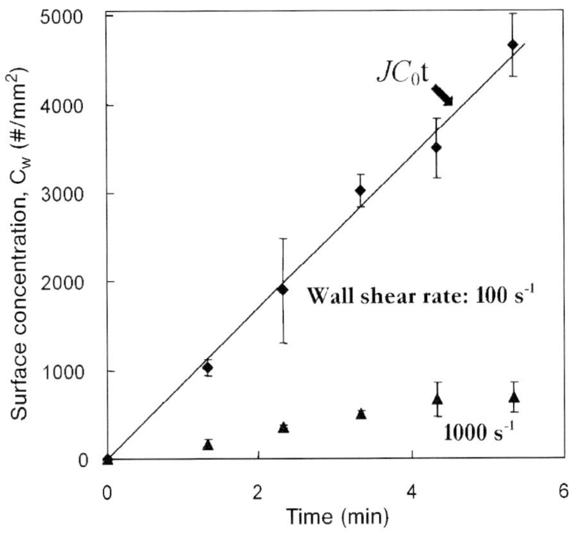


**Figure 2.12** Concentration of 5 μm yeast cells deposited on a 0.02 μm anodized aluminum membrane at a permeate flux of 50 L m$^{-2}$ h$^{-1}$ and two wall shear rates. The symbols represent experimental measurements for wall shear rates of 100 s$^{-1}$ (♦) and 1000 s$^{-1}$ (▲), and the solid line gives the prediction for $\theta = 1$.

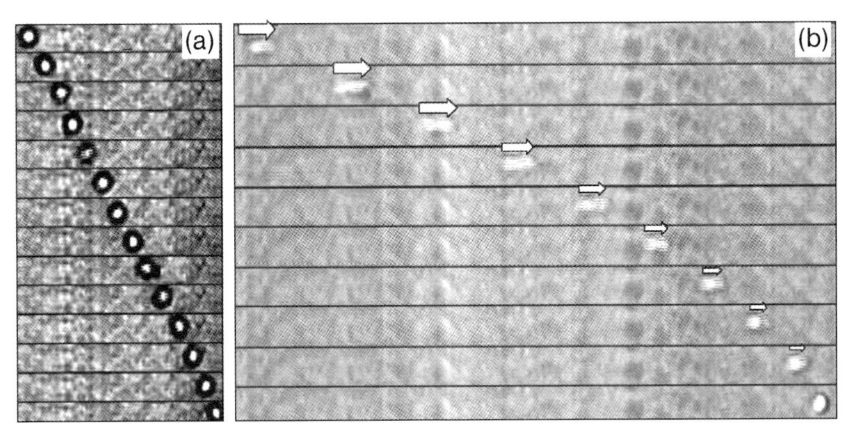


**Figure 2.13** DVO micrographs showing the motion of a yeast cell (a) rolling steadily over a 0.02 μm Anopore membrane, with the distance traveled increasing linearly with time and (b) being convected toward a 0.02 μm Anopore membrane by the flow of permeate, with decreasing velocity (highlighted by the arrows above the blurred images of the cell) as the cell approaches the membrane surface. The wall shear rate is 100 s$^{-1}$, the permeate flux is 10 LMH and the time interval between frames is 0.033 s in (a) and 0.1 s in (b). (From Knutsen and Davis [24]).

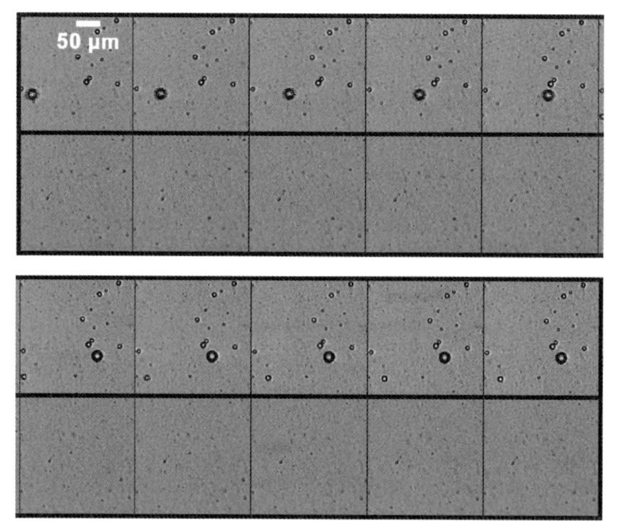


**Figure 2.14** Montage of DVO images collected during tangential flow filtration of 5 μm, 10 μm and 20 μm latex particles on an Anopore membrane with 0.02 μm pores. The two rows of images on the top and bottom were collected with a permeate flux (flowing into the page) of 200 $L\,m^{-2}\,h^{-1}$ and 50 $L\,m^{-2}\,h^{-1}$, respectively. The time interval between frames is 0.1 s and the wall shear rate is 100 $s^{-1}$. In both experiments, approximately 0.06 mL (permeate) $cm^{-2}$ (membrane) was collected prior to the time these images were taken.

rolling motion of individual particles can be observed. For example, the largest (20 μm) particle rolls along the membrane surface on the first five frames with a flux of 200 $L\,m^{-2}\,h^{-1}$, as can be assessed by the change in its location from frame to frame and by the slight blurring of the images, while it is stopped in the final five frames. There is also a 10 μm particle that enters near the bottom of the sixth frame and then is brought to rest by the eighth frame. In contrast, all of the smallest (5 μm) particles are essentially at rest on the membrane surface for the higher flux but some undergo motion at the lower flux.

## 2.3 Particle Removal

Direct visual observation can also be an effective tool for non-invasive monitoring of particle or cake removal by using methods such as backpulsing or backflushing for remediation of membrane fouling. Direct observations through the membrane (DOTM) were made by Li et al. [10, 16] of the removal of deposited particles or cells when the permeate flux was reduced below the critical flux. The removal was caused by shear from the tangential flow along the membrane surface. Here, we report on recent work in the Davis laboratory at the University of Colorado, in which backpulses were used to lift the deposited particles off the membrane

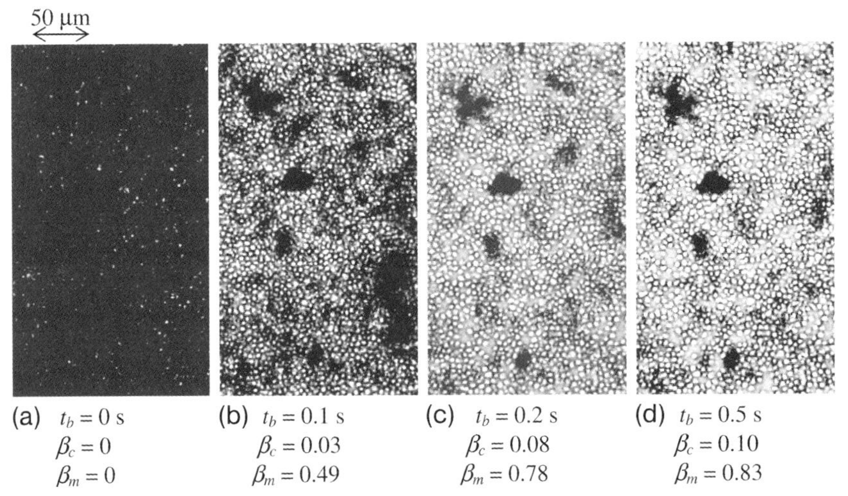


**Figure 2.15** Photomicrographs of AN membranes after backpulses at $\Delta P_b = 2$ psi and various cumulative durations. Membranes were fouled by forward filtration with 0.05 g $L^{-1}$ yeast at $\Delta P_f = 2$ psi for 5400 s prior to backpulsing. $\beta_c$ is the fraction of membrane completely cleaned with each backpulse, while $\beta_m$ is the fraction of membrane which remains covered in a monolayer of yeast cells after each backpulse. (From Mores and Davis [21]).

surface and then visual observations of the membrane surface were made from the feed side rather than the permeate side.

Mores and Davis [21] used the direct visual observation (DVO) apparatus of Figure 2.6 to observe removal of a deposited cake of yeast cells from an anodized aluminum (AN) membrane by a series of individual backpulses. The results are shown in Figure 2.15. The membrane is initially fouled by forward filtration for 1.5 h, so that it appears completely dark (representing a multi-layered yeast cake). After a single backpulse of duration $t_b = 0.1$ s, much of the cake is removed, leaving a monolayer covering 49% ($\beta_m = 0.49$) of the membrane surface, with 3% ($\beta_c = 0.03$) of the surface cleaned of all cells and the rest still covered by more than one layer of cells. After two backpulses having a cumulative duration $t_b = 0.2$ s, most of the multilayer is removed, leaving $\beta_c = 0.08$ and $\beta_m = 0.78$. After three backpulses having a cumulative duration $t_b = 0.5$ s, there is little further change and most of the membrane remains covered by a monolayer. The adherence of a monolayer of yeast cells to the membrane surface may be due to electrostatic effects (Knutsen and Davis [24]) or the presence of extracellular material.

Figure 2.16 shows the recovered flux after the series of individual backpulses depicted in Figure 2.15. Prior to any backpulsing, the permeate flux through the cake layer is less than 25% of the clean membrane flux. After three backpulses with a combined duration of $t_b = 0.5$ s, the flux is recovered to about 65% of the clean membrane flux. Additional backpulsing provides no further flux recovery, due to the adherence of a cell monolayer on the membrane surface. Thus, we

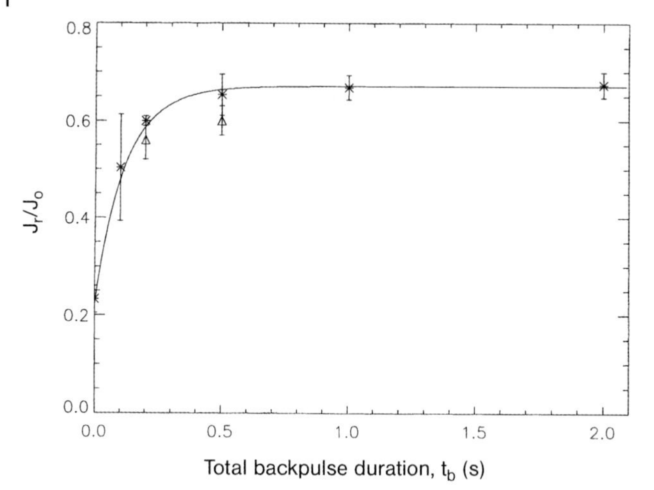


**Figure 2.16** Dimensionless recovered flux versus total backpulse duration at $\Delta P_b = 2$ psi for AN membranes. The solid line is the model fit and the symbols are the experimental data. The error bars are plus and minus one standard deviation for three repeats. Membranes were fouled by forward filtration for 5400 s with 0.05 g $L^{-1}$ yeast at $\Delta P_f = 2$ psi prior to backpulsing. (From Mores and Davis [21]).

might expect $t_b = 0.5$ s to be optimal for in situ backpulsing under these conditions, as shorter backpulses give less cleaning while longer backpulses cause additional permeate loss in the reverse direction without increased cleaning.

Mores and Davis [25] subsequently performed in situ backpulsing experiments in which cycles of forward filtration with a duration $t_f = 10$ s followed by a backpulse of duration $t_b = 0-2$ s were performed repeatedly for 1.5 h, using a feed of 0.05 g $L^{-1}$ washed yeast. Figure 2.17 shows the DVO results after these cycles were repeated for 300 s and 1800 s using AN membranes. Without backpulsing, a multilayer yeast cake accumulates on the membrane surface. When periodic backpulsing is employed, the coverage is reduced to a monolayer, or less. Longer backpulses result in more cleaning, though the difference between $t_b = 0.5$ s and $t_b = 2.0$ s is small at the longer total filtration time. By comparing the results after 5 min and 30 min, it is apparent that the effectiveness of backpulsing declines with time, due to the irreversible adhesion of a monolayer of cells to the membrane surface.

The DVO observations above are supported by net permeate flux measurements in Figure 2.18. Net flux is defined as the permeate volume collected during forward filtration minus the permeate loss during reverse filtration, divided by the membrane area and cycle time $(t_f + t_b)$. The net flux is maximized at $t_b = 0.5$ s, since longer backpulses contribute little improved cleaning and yet cause more permeate loss. The net flux declines during the course of the experiment due to gradual irreversible fouling, as will be discussed in more detail later. Nevertheless, the net permeate flux at the optimal backpulse duration is at least twice the permeate flux without backpulsing, even at the end of the experiment.

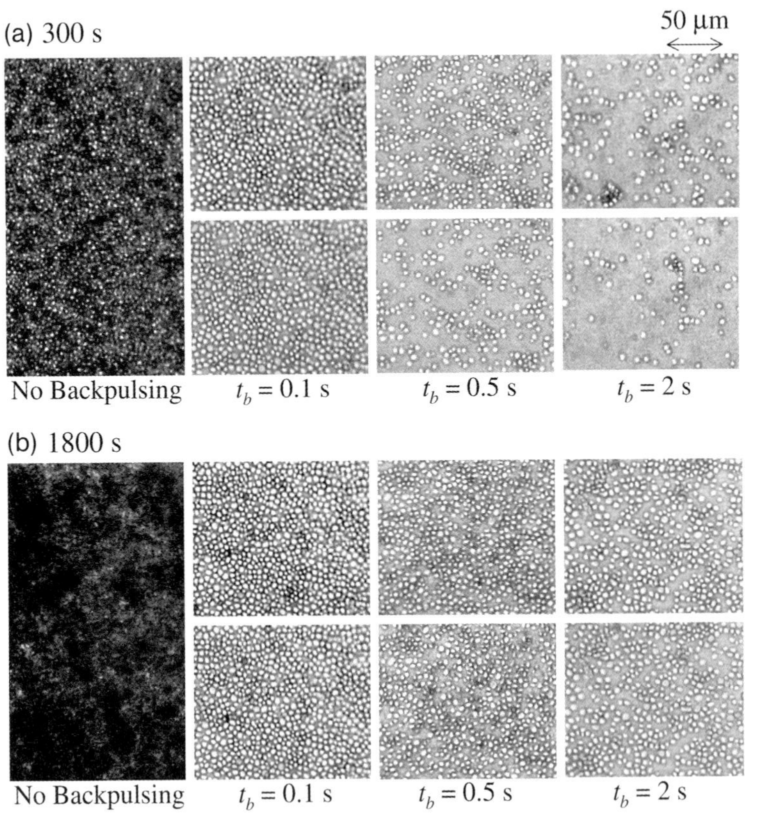


**Figure 2.17** Photomicrographs of AN membranes after (a) 300 s and (b) 1800 s of filtration with rapid backpulsing with $t_f = 10$ s, $\Delta P_f = \Delta P_b = 2$ psi and $\gamma = 2500\ s^{-1}$. The top photos depict membrane surfaces at the end of a period of forward filtration, while those on the bottom depict surfaces after the subsequent backpulse. The photos on the left show the fouled membrane surface for forward filtration without backpulsing. (From Mores and Davis [25]).

The DVO micrographs for removal of a yeast cake from cellulose acetate (CA) membranes by backpulses provide a different picture. Figure 2.19 shows the fouled membrane and then the cleaned membrane after a series of individual backpulses. The cells are removed nonuniformly in patches and there is no monolayer remaining on the membrane surface. Indeed, after four individual backpulses with a cumulative duration $t_b = 1$ s, the membrane is 94% clean. Mores and Davis [21] also found that almost complete flux recovery was obtained by backpulses of 1 s or more.

Although nearly complete cleaning of CA membranes by backpulsing was observed in the early stages of long-term experiments with cycles of forward filtration and backpulses, the performance was again observed to decline over time. Figure 2.20 shows photomicrographs of CA membranes after 10 min and 1 h of

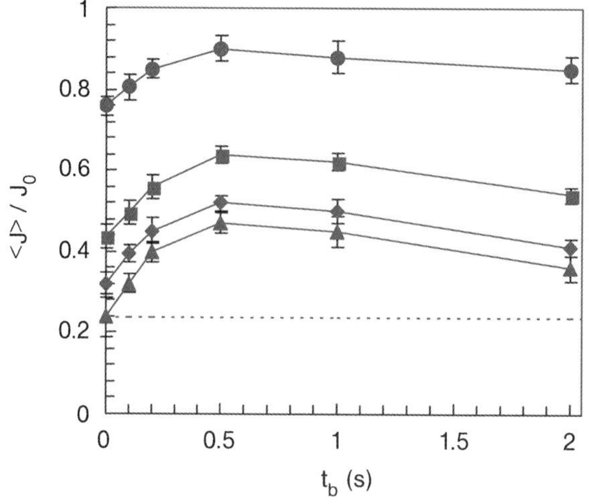


**Figure 2.18** Net flux (made dimensionless with the clean membrane flux) versus backpulse duration for rapid backpulsing experiments with $\Delta P_f = \Delta P_b = 2$ psi, $t_f = 10$ s and $\gamma = 2500\,\text{s}^{-1}$ using 0.2 μm AN membranes fouled by $0.05\,\text{g L}^{-1}$ washed yeast. Points are connected by lines for clarity. Error bars are ±1 standard deviation for 2–5 repeats. Flux data are given after 300, 1800, 3600 and 5400 s (top to bottom) of filtration. The dashed line is the membrane flux after 5400 s of forward filtration without backpulsing. (From Mores and. Davis [25]).

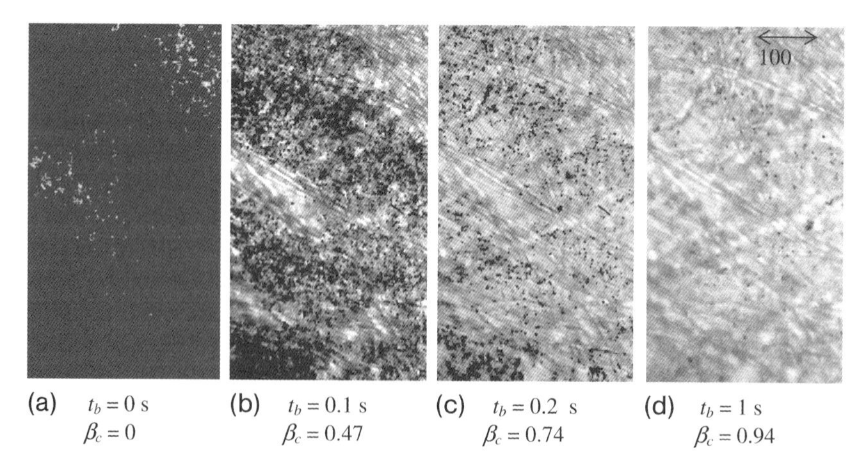


**Figure 2.19** Photomicrographs of CA membranes after backpulses of $\Delta P_b = 3$ psi and various cumulative durations. Membranes were fouled by forward filtration with $0.1\,\text{g L}^{-1}$ yeast at $\Delta P_f = 3$ psi for 9000 s prior to backpulsing. $\beta_c$ is the fraction of membrane cleaned with each backpulse. (From Mores and Davis [21]).

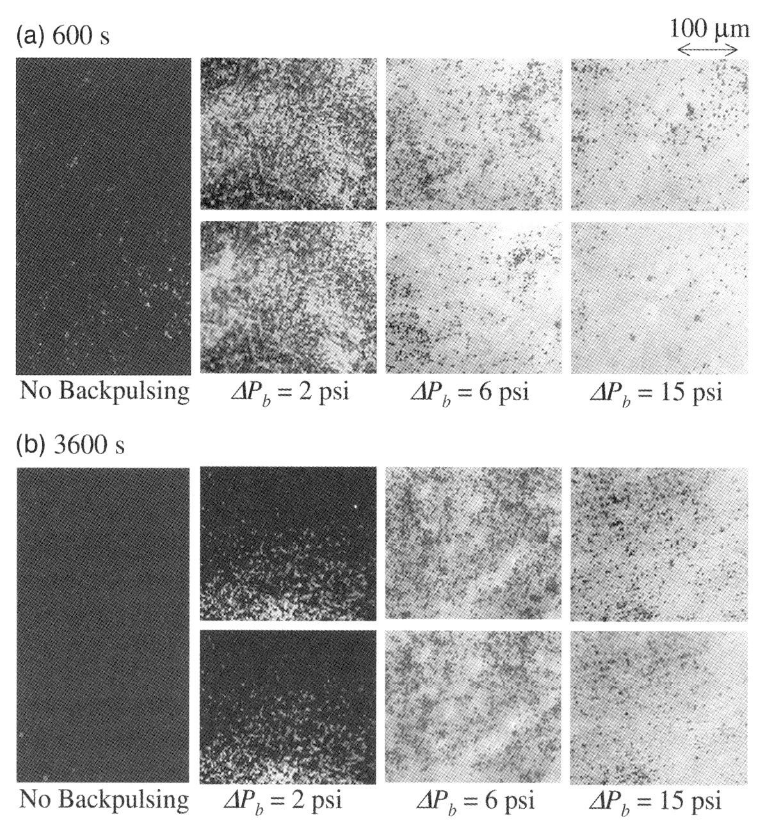


**Figure 2.20** Photomicrographs of CA membranes after (a) 600 s and (b) 3600 s of filtration with rapid backpulsing with $t_f = 10$ s, $t_b = 0.4$ s, $\Delta P_f = 3$ psi and $\gamma = 2400\ \mathrm{s}^{-1}$. The top photos depict membrane surfaces at the end of a period of forward filtration, while those on the bottom depict surfaces after the subsequent backpulse. The photos on the left show the fouled membrane surface for forward filtration without backpulsing.

crossflow microfiltration with periodic backpulsing of different strengths. As expected, a larger backpressure causes more cleaning, by forcing liquid through the membrane in the reverse direction at higher velocity and lifting the yeast cells off the membrane. Mores and Davis [26] further showed that higher wall shear rate, as well as higher backpressure, improves the cleaning efficiency of backpulses.

Figure 2.21 shows the net permeate flux as a function of the backpressure, $\Delta P_b$. At fixed backpulse duration $t_b = 0.4$ s, there is an optimum backpressure of about 3 psi (1 psi = 6.9 kPa), as smaller values of $\Delta P_b$ give too little cleaning while larger values of $\Delta P_b$ cause too much permeate loss in the reverse direction. In contrast, if the product $t_b \Delta P_b$ is fixed, representing (approximately) a fixed permeate loss per backpulse, the net flux increases with increasing backpressure. This result implies that shorter, stronger backpulses are more effective than weaker, longer backpulses in cleaning the membrane.

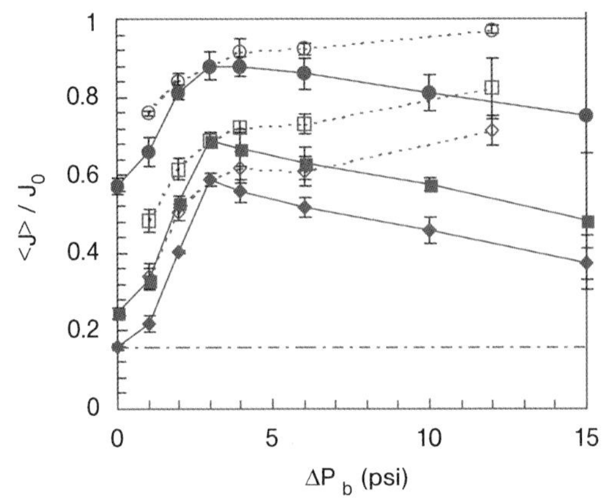


**Figure 2.21** Net flux (made dimensionless with the clean membrane flux) versus backpulse pressure for rapid backpulsing experiments with $\Delta P_f = 3$ psi, $t_f = 10$ s and $\gamma = 2400\ \mathrm{s}^{-1}$ using 0.2 μm CA membranes fouled by 0.05 g $\mathrm{L}^{-1}$ washed yeast. Points are connected by lines for clarity. Error bars are $\pm 1$ standard deviation for 2–5 repeats. Experiments were conducted using a constant backpulse duration of $t_b = 0.4$ s (solid lines) and a constant product of $t_b \Delta P_b = 1.2\ \mathrm{s} \times \mathrm{psi}$ (dashed lines). Flux data are given after 600 s, 3600 s and 7200 s (top to bottom) of runtime. The dashed dotted line is the fouled membrane flux after 7200 s of forward filtration without backpulsing. (From Mores and Davis [25]).

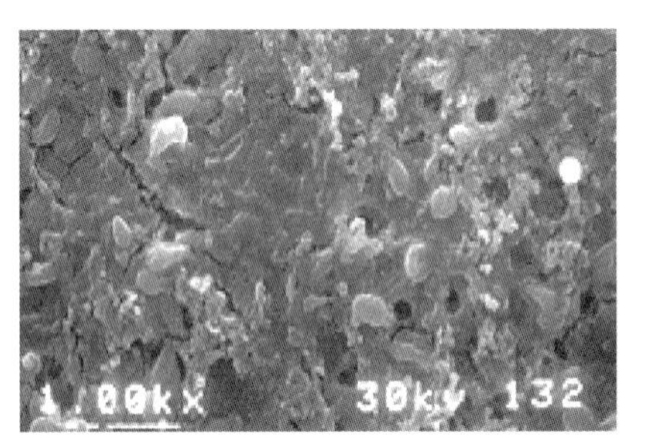


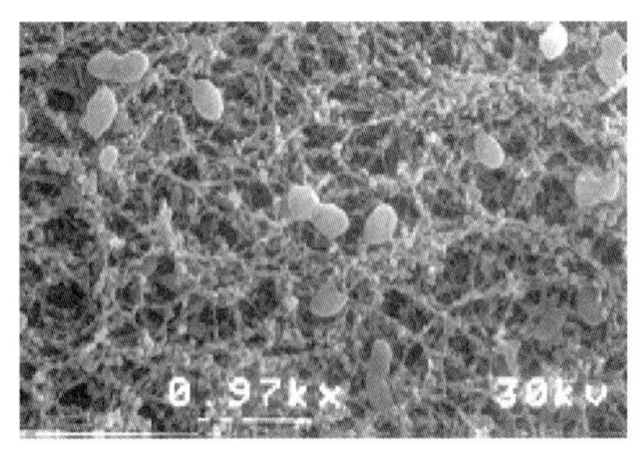


**Figure 2.22** Scanning electron micrographs of CA membranes after 2 h of filtration of 0.1 g $\mathrm{L}^{-1}$ yeast suspension with periodic backflushing using $\Delta P_f = \Delta P_b = 3$ psi, $t_f = 60$ s, $t_b = 10$ s and wall shear rate $\gamma = 640\ \mathrm{s}^{-1}$, both with (left) and without (right) recycle of the retentate to the feed reservoir.

The observed decline in cleaning and net permeate flux with backpulsing over time was studied by Mores and Davis [27]. They showed that prolonged filtration with recycle of retentate to the feed reservoir (such as done in batch concentration) leads to cell rupture. The ruptured cell debris and contents are adhesive and cause irreversible membrane fouling. Scanning electron micrographs (Figure 2.22) show that a slime layer forms on the membrane surface, but is absent when the retentate is not recycled to the feed reservoir. Mores and Davis [27] used DVO to show that nearly complete cleaning (for example, $\beta_c = 0.87$ after 2 h of operation) of the membrane occurred with periodic backflushing when the retentate was not recycled and the net flux declined only very slowly with time.

Nemade and Davis [28] also showed that gradual, irreversible fouling of the membrane occurred during filtration of mixtures of yeast and protein. Using a combination of DVO and flux measurements, they showed that the irreversible fouling could be reduced by predepositing a layer of yeast on the membrane surface with periodic removal by backpulsing. The yeast layer served as a secondary membrane that trapped protein aggregates and prevented them from fouling the primary membrane.

## 2.4 Concluding Remarks

Direct visual observation with the aid of a microscope, whether from the side, top or bottom of the channel, is a powerful tool for understanding membrane fouling and cleaning during crossflow microfiltration of cells and particles. It provides for non-invasive, in situ observations and measurements not possible with traditional methods such as monitoring permeate flux. Highlighted examples include the critical flux or other conditions for which particles deposit on the membrane, determination of whether the deposited particles remain immobile or roll along the membrane surface, selective deposition of small particles from a mixture of small and large particles, the deposit morphology (uniform or patchy, monolayer or multilayered) and how it is affected by membrane material, fractional coverage of the membrane surface, the extent of deposit removal by backpulses and more. It is hoped the reader is inspired by these methods and examples to perform additional studies involving direct visual observation of membrane fouling and cleaning.

## References


1 S.G. Redkar, R.H. Davis, *AIChE J.* **1995**, *41*, 501–508.
2 V. Chen, H. Li, A.G. Fane, *J. Membrane Sci.* **2004**, *241*, 23–44.
3 V. Chen, H. Li, M. Elimelech, *Adv. Colloid Interface Sci.* **2004**, *107*, 83–108.
4 J.R. Otis, F.W. Altena, J.J. Mahar, G. Belfort, *Exp. Fluids* **1986**, *4*, 1–10.
5 R.H. Davis, S.A. Birdsell, *Chem. Eng. Comm.* **1987**, *49*, 217–234.
6 C.A. Romero, R.H. Davis, *J. Membrane Sci.* **1991**, *62*, 249–273.
7 M.R. Mackley, N.E. Sherman, *Chem. Eng. Sci.* **1992**, *47*, 3067–3084.
8 R.J. Wakeman, *Trans. IChemE* **1994**, *72A*, 530–540.
9 P.H. Hodgson, V.L. Pillay, A.G. Fane, *Proc. World Filtration Congr.* **1993**, *6*, 607–610.
10 H. Li, A.G. Fane, H.G.L. Coster, S. Vigneswaran, *J. Membrane Sci.* **1998**, *149*, 83–97.
11 P. Bacchin, P. Aimar, R.W. Field, *J. Membrane Sci.* **2006**, *281*, 42–69.
12 R.W. Field, D. Wu, J.A. Howell, B.B. Gupta, *J. Membrane Sci.* **1995**, *100*, 259–272.

**13** P. Bacchin, P. Aimar, V. Sanchez, *AIChE J.* **1995**, *41*, 368–377.

**14** J.A. Howell, *J. Membrane Sci.* **1995**, *107*, 165–171.

**15** H. Li, A.G. Fane, H.G.L. Coster, S. Vigneswaran, *J. Membrane Sci.* **2000**, *172*, 135–147.

**16** H. Li, A.G. Fane, H.G.L. Coster, S. Vigneswaran, *J. Membrane Sci.* **2003**, *217*, 29–41.

**17** S.R. Wickramasinghe, B.B. Han, S. Akeprathumchai, V. Chen, P. Neal, X. Qian, *J. Membrane Sci.* **2004**, *242*, 57–71.

**18** H. Li, A.G. Fane, *Membrane Technol* **2000**, *123*, 10–14.

**19** P.R. Neal, H. Li, A.G. Fane, D.E. Wiley, *J. Membrane Sci.* **2003**, *214*, 165–178.

**20** Y.P. Zhang, A.G. Fane, A.W.K. Law, *J. Membrane Sci.* **2006**, *282*, 189–197.

**21** W.D. Mores, R.H. Davis, *J. Membrane Sci.* **2001**, *189*, 217–230.

**22** A.G. Fane, P. Beatson, H. Li, *Water Sci. Technol.* **2000**, *10/11*, 303–308.

**23** S.T. Kang, A. Subramani, E.M.W. Hoek, M.A. Deshusses, M.R. Matsumoto, *J. Membrane Sci.* **2004**, *244*, 151–165.

**24** J.S. Knutsen, R.H. Davis, *J. Membrane Sci.* **2006**, *271*, 101–113.

**25** W.D. Mores, R.H. Davis, *Desalination* **2002**, *146*, 135–140.

**26** W.D. Mores, R.H. Davis, *J. Membrane Sci.* **2002**, *208*, 389–404.

**27** W.D. Mores, R.H. Davis, *Ind. Eng. Chem. Res.* **2003**, *42*, 130–139.

**28** P.R. Nemade, R.H. Davis, *Appl. Biochem. Biotech.* **2004**, *113/116*, 417–432.

# 3
# Microscopy Techniques for the Characterization of Membrane Morphology

*Carles Torras, Tània Gumí, and Ricard Garcia-Valls*

## 3.1
## Introduction

Membrane processes have been increasingly gaining importance in recent decades as separation techniques, since they offer several advantages over traditional methods (like precipitation or liquid–liquid extraction), such as set-up and handling simplicity, low time consumption, including a relatively cheap technology and they are environmentally respectful. Additionally, most membrane processes can be performed at room temperature and are carried out continuously [1]. Furthermore, it must be considered that membranes can be prepared both to have a wide range of different sizes from macro to micro or nanoscale (like in the case of bio/chemical sensors) and to adopt various different configurations, from flat films to microcapsules. This versatility has further increased their attractiveness from all points of view.

As a consequence some membrane processes have already been applied at industry level; and certain of them, for instance, water treatment (including water desalination), food purification and biomedical purposes (blood dialysis), are even industrially consolidated.

Membranes may be divided, according to their composition, into organic or inorganic ones, the former (organic polymeric membranes) being mainly used in all commercial applications. Therefore, from now on, this discussion essentially refers to polymeric membrane. It is commonly agreed that the key factor for the correct development and application of polymeric membranes is the control of its polymeric morphology. Therefore, many efforts have been made within recent decades to find the relationship between membrane preparation, membrane morphology and membrane performance [2]. In that sense, membrane characterization has become fundamental for the optimization of membrane process design as membrane features, such as nature, thickness and porosity, govern the process behaviour. Best results, that is the most unfailing outputs, are encountered when combining different membrane characterization techniques.

*Monitoring and Visualizing Membrane-Based Processes*
Edited by Carme Güell, Montserrat Ferrando, and Francisco López

ISBN: 978-3-527-32006-6

The principle membrane characterization techniques that have been used up to now involve: electron microscopy, atomic force microscopy, bubble point method, permeability method, angle contact measurement, gas adsorption–desorption, thermoporometry, permporometry, liquid displacement, solute rejection measurements, physical methods, plasma etching and surface analysis methods. More recently, novel characterization methods have also been considered. These are: electron spin resonance, raman spectroscopy, nuclear magnetic resonance, neutron scattering and confocal scanning laser microscopy. Despite the vast amount of characterization techniques available, not all of them are suitable to study the membrane morphology, as certain of them focus on chemical or surface characteristics. Among all characterization techniques allowing membrane morphology characterization, microscopy-based methods are the most widely employed, for their simplicity, rapidity and precision. There are basically four main microscopy techniques, which can be used to typify membrane morphology: scanning electron microscopy (SEM), transmission electron microscopy (TEM), atomic force microscopy (AFM) and confocal scanning laser microscopy (CSLM).

This chapter is divided in two sections. The first presents a review of the different techniques used for the morphological characterization of membranes. The second explains some cases studies developed by authors to carry out the mentioned characterization, adding emphasis on software developed by the authors to achieve these purposes, providing quantitative, fast and systematic analysis.

## 3.2
## Membrane Characterization Morphology Parameters

Before presenting the main microscopy techniques employed to determine membrane morphology, it is worth listing and briefly describing the membrane morphological parameters normally measured.

The first morphological parameter to consider when characterizing a membrane is (mean) *pore size* and *pore size distribution*, as these two membrane characteristics govern most membrane applications. It should be kept in mind that the determined pore size of a given membrane will vary depending on the characterization methodology selected, as pores may have different structures along its length. Microscopy techniques normally determine the outer pore diameter, which usually corresponds to the largest pore opening, hence leading to relatively big pore diameters. In contrast, when performing solute transport experiments (solutes of exactly known molecular weight are forced to cross the membrane through its pores), the pore size measured corresponds to the smallest pore constriction [1]. The *pore tortuosity* is the morphological parameter which provides information on how far, in shape, is an internal pore structure from an ideal cylinder. Still focused on membrane pores, two other parameters should be defined: *symmetry* and *regularity*.

The symmetry corresponds to the distribution and the homogeneity of the pores, in number and size (equivalent to pore size distribution) of a membrane cross-section, and in the direction of the fluid flow. A membrane is so-called symmetric if the number and size of pores does not change through the membrane thickness (this means that there are always the same number of pores, with the same size). If not, as much as it differs, the membrane is more asymmetric. Depending on the case, it could be desirable to have a different pore number or size, in the sense of fluid advancement. For example, it could be desirable to have a thin selective region with a huge number of very small pores to give selectivity to the membrane and, after, a region with few and larger pores in order to let the fluid cross without resistance (a zone that will act just as a support).

The regularity determines the homogeneity of pores in the membrane, in size and in number, perpendicular to the fluid flow. Normally, this parameter is a necessary condition for the membranes: wherever in the region the fluid crosses the membrane, the same number and size of pores should be encountered.

Apart form pore size and pore size distribution, total *porosity*, or *surface porosity* need also to be determined. It gives an idea of the total amount of free volume in a membrane. Again, as in the case of pore size, different porosity values may be obtained depending on the techniques carried out for membrane characterization. If membrane porosity is calculated from solute transport experiments, only those pores connecting both membrane surfaces are taken into consideration. However, if gas adsorption–desorption measurements are made, the whole membrane free volume will be detected, including dead-end pores.

Another parameter to consider is *membrane thickness* and the membrane selective layer thickness. This is a key parameter in membrane processes as it influences both membrane selectivity and membrane resistance against flux.

*Surface roughness* is also a crucial morphological parameter since it seems to be directly related to membrane fouling phenomena. Membrane surface roughness is a numerical measure of the surface morphology variation of a membrane. It corresponds to the mean value of surface relative to the center plane, this is the plane for which the volume enclosed by the image above and below this plane is equal.

## 3.3 Microscopy Techniques for Membrane Morphology Characterization

### 3.3.1 Electron Microscopy

Electron microscopy allows imaging of the membrane surface and cross-section by means of hitting the membrane sample with a narrow beam of electrons of high energy (so-called primary electrons). These incident electrons lead to a high quantity of interaction between sample and the electron beam. The resulting interaction permits the identification of and discrimination between the different materials contained in the sample. Two basic electron microscopy modes are distinguished: transmission electron microscopy (TEM) and scanning electron microscopy (SEM). The former considers and compares the electron beam dispersion nature when the electron beam crosses different sample zones corresponding to different chemical compositions or physical configuration. In contrast, SEM mode uses low-energy electrons (so-called secondary electrons), which are released by the surface atoms of the sample. Therefore, in TEM techniques a very thin layer containing the sample should be prepared, while this is not required in SEM techniques. Usually, to proceed with membrane characterization a sample treatment is needed to provide the sample with a conductive surface. The pre-treatment consist normally in coating the sample surface with a thin gold layer under vacuum conditions [3]. Both modes may be used nowadays both at high- (traditional modes) and at low-vacuum conditions (refered to as environmental scanning electron microscopy, ESEM). However higher resolutions are obtained when employing high-vacuum conditions. In the case of low-vacuum conditions sample pre-treatment is not required, but at high-vacuum conditions the sample metallization process becomes compulsory.

The images obtained from electron microscopy may be used to determine both the surface and the internal membrane structure, to examine the porous morphology and to estimate membrane porosity and porous size distribution, which will determine the selectivity of the membrane as well as its permeability.

Electron microscopy (SEM, TEM) have been vastly employed in membrane characterization field [4]. The main membrane morphology applications of electron microscopy have been: the description of microporous membranes, the study of membrane formation mechanism, the characterization of the physical properties of ultrafiltration membranes, the determination of surface porosity and pore size, the investigation of surface morphology and the study of fouling mechanism. SEM has been used to characterize the morphology of polysulfone, polysulfone–lignosulfonate membranes possessing ion conductivity [5, 6], poly(ether ether ketone) hollow fibers [7] and various blend membranes, such as poly(bis phenol-A-ether sulfone) and poly(*n*-vinyl pyrrolidone) membranes [8], poly (dimethyldimethylenepiperidinium) chloride poly(vinyl alcohol) anion-exchange membranes [9], and both sulfonated poly(ether ether ketone) and cellulose acetate ultrafiltration membranes [10]. In certain cases, SEM has been applied to confirm changes in membrane morphology induced by the formation process [11], or by chemical treatment [12]. In the latest case both SEM and TEM analysis were performed, as in the case of characterizing hollow fiber polyamide membranes [13]. Surface morphology changes due to membrane usage have also been monitored by SEM [14, 15]. In that sense, polymer degradation on drug release [16] or resistance to microbial degradation were evaluated [17]. SEM has also been considered for investigating inorganic (poly-aluminum ceramic) membranes [18] and the membranes of sensors [19].

### 3.3.2
### Atomic Force Microscopy

Atomic force microscopy (AFM) allows material characterization (conductive or not) down to the nanoscale, in environmental conditions [20, 21]. It was developed by Binnig et al. [22] and its principal advantage is that no sample pre-treatment is required. Although quite new, this technique has been rapidly extended [23, 24]. AFM is a non-invasive technique that provides topographical information and allows determination of the porous size distribution. This methodology is based on a small tip, which acts as a sensor of the sample surface geometry. The tip (also called stylus) is attached to a flexible cantilever that deflects according to the tip oscillations (when interacting with the surface atoms of the sample under study), while scanning the sample in $x$, $y$ and $z$ directions. As the cantilever induces forces lower than the interatomic ones, the sample topography may be determined without atom displacement [2]. Therefore surface topography images are obtained [2]. The cantilever deflections are detected by a laser beam reflection, conveniently focalized and determined by a photo detector. The tip works usually at distances of 0.1 nm to 100 nm from the sample surface. Within this range, various force types are present, Van der Waals force being the most common. The sum of all of them defines the resulting interaction as an attractive or a repulsive force. There are three different AFM operation modes: contact mode (the sample and the tip are in contact, i.e. the distance between them is just a few Angstroms), noncontact mode (the tip is not in contact with the sample surface, thus only attractive forces are detected) and semi-contact mode (also called tapping, the tip contacts the sample only during oscillations). From AFM images, surface porous density, porosity, membrane pore and nodules size, pore size distribution and surface roughness can be obtained. AFM noncontact mode is particularly appropriate for the study of polymeric membranes. However, the process to obtain a clear micrograph with AFM is difficult (and thus, obtaining reliable numerical values is also highly complicated); and the technique is very sensitive to the zone of the sample which is being investigated and also to the condition of the tip [25]. The interested reader is referred to Chapters 5 and 6, which thoroughly review AFM applications to membranes and membrane processes.

### 3.3.3 Confocal Scanning Laser Microscopy

Confocal scanning laser microscopy (CSLM) permits non-invasive and proper optical sectioning of a specimen, perpendicular or parallel to the microscope optical axis. Imaging may be performed in two different modes: reflective or fluorescent [26]. In the second case a fluorescent agent needs to be added into the sample, unless the sample itself shows fluorescence [27]. The principle of CSLM technique is the following. A light source, the laser, is reflected by means of a dichroic mirror and focused on a minute part of the sample by the lens of the objective. Due to its longer wavelength, the fluorescent light emitted (from the corresponding sample spot) crosses the dichroic mirror, passes through the pinhole detector and is gathered by the photomultiplier. The backscattered or reflected light from structures which are not in the focalization spot are not able to pass through the pinhole detector. Therefore, conveniently moving the incident light over the sample provides a complete series of images at different positions and different depths. If computer-treated, these images lead to a three-dimensional construction. Further CSLM image analysis and use of segmentation and statistics tools may be applied to reach a more quantitative characterization of samples [28].

The main drawbacks of CSLM are related to the fluorescent labeling of the sample and to its limited magnification capability as it is an optical microscope. Chapter 4 reviews in detail the fundamentals of CSLM and its applications to membrane processes.

To sum up, AFM and CSLM allow a good number of membrane topographic parameters to be obtained, they may be operated under usual working conditions and minimal or no sample pre-treatment is needed. Additionally, CSLM is a nondestructive technique and permits three-dimensional morphology information to be obtained (not only from a thin surface). However, the (AFM, CSLM) image quality and resolution (lateral resolution approximately of 0.4 μm in the case of CSLM) [29] still needs to be further enhanced for membrane visualization applications. Like CSLM, ESEM is a nondestructive technique, but it lacks the image resolving power. In contrast, the traditional electron microscopy techniques, TEM and SEM, show a high visualizing power but typical morphological membrane parameters may not be directly obtained from images, nor quantified nor used for sample comparison. Therefore, it is not possible to correlate morphological parameters such as the mean pore size, symmetry, etc. with the membrane synthesis variables in order to optimize the process. Alternatively, a few membrane morphological parameters (such as number of pores or pore size) might be calculated manually, which implies an important time expense as well as a non-systematic procedure [1]. Furthermore, very thin sample layers are required for TEM analysis. Nevertheless, when two or more of these techniques are used, results obtained from all of them are usually comparable. Thus, they are all valid and suitable techniques for membrane morphology studies and the choice of one or another method has to be done in each particular case, depending on the membrane type to be studied and according to the exact objective of the work.

Nevertheless, when performing membrane morphological (i.e. topographical, structural) characterization it is always encouraged to use various and complementary characterization techniques.

### 3.3.4
### Analysis of SEM Images

Several approaches have been done to quantify membrane morphology parameters from SEM images. The first attempts were made in the early 1990s. They permitted more quantitative information to be obtained from membrane SEM micrographs and they showed the potential power of computer-aided SEM analysis [30, 31]. More recently, other approaches based on different model equations have been reported [32, 33]. The IFME software [32], was developed by our research group within the past years and permits numerical results of the main morphological parameters to be obtained from SEM micrographs, such as pore size distribution, porosity, symmetry, regularity and tortuosity. IFME has already been successfully employed for the characterization of various membranes [34–36].

Within the past years, our research group has focused in the preparation and characterization (mainly morphological) of polymeric membranes. Two different membrane configurations have been considered: flat membranes and microcapsules. The most common, in the research field, are flat membranes, used in static or dynamic membrane modules [37]. Commercially, they are often used in tubular modules, which include several sheets in a spiral format, thus increasing the effective area. Besides, other emerging structures are microcapsules. These are hollow microspheres with sizes usually below 800 μm. Microcapsules are well appreciated and often used in the pharmaceutical or the consumer goods industry to encapsulate drugs, fragrances, etc. Both structures may be synthesized by similar methods.

Membrane morphology characterization ias carried out using several microscopy techniques (SEM, CSLM, AFM, etc.). Two case studies are presented and discussed next.

## 3.4 Case Studies

### 3.4.1 Flat Membranes

#### 3.4.1.1 Synthesis Considerations

Membranes are selective barriers which allow component separation between two phases. Membranes can be obtained from different materials and by different methods. Depending on the raw material employed, organic (polymeric) or inorganic membranes are found, the former being far more important. The choice of polymer to work with is a key factor, as it provides specific properties and structural parameters to the membrane. In this case, polysulfone was selected as polymeric material, due to its high chemical and mechanical resistance, low cost and industrial applicability. Various methods to prepare polymeric membranes have been described. One of the most used is the Loeb–Sourirajan phase inversion process. This method, which is well described in [1], is based on the interaction of at least three compounds: a polymer, a solvent and a nonsolvent. A polymeric mixture is prepared mixing the polymer with the solvent and then it is immersed in a coagulation bath containing the nonsolvent (which, additionally, can contain also solvent or other additives), leading to the precipitation of polymer. A mass transfer occurs (the solvent migrates to the nonsolvent and vice versa), which results in the formation of the membrane after a phase inversion process due to reaching a two-phase immiscible region. This process can be represented in a ternary diagram. Figure 3.1 shows the diagram for two systems: (a) polysulfone (PSf, polymer), dimethylformamide (DMF, solvent) and water (nonsolvent) and

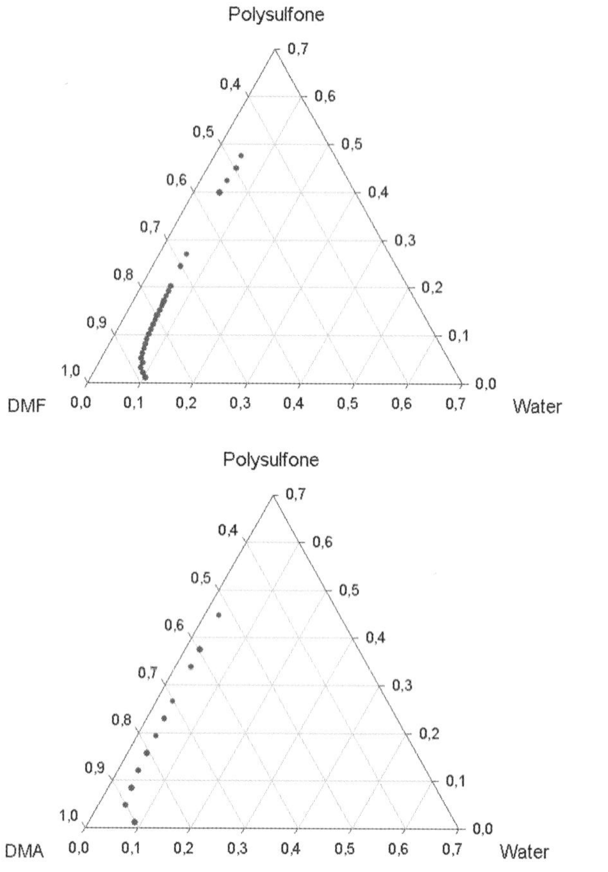


**Figure 3.1** Ternary phase diagram for PSf-DMF-water and PSf-DMA-water systems.

(b) PSf, dimethylacetamide (DMA, solvent) and water. The line of points (binodal) separates the one-phase region from the two-phase region [34].

The final membrane morphology obtained depends on the interaction of the mentioned compounds (related to the Flory–Huggins theory [38]), the coagulation bath temperature, the polymer concentration, etc. [1]. For a given polymer, the best way to obtain the widest range of membrane morphologies is to modify the composition of the coagulation bath (which includes the nonsolvent) [34]. A high interaction parameter between the nonsolvent and the polymer (e.g. water, PSf) produces a fast mass transfer, which leads to an asymmetric morphology structure with macrovoids; and a low interaction between the nonsolvent and the polymer (e.g. isopropanol, PSf) produces a delayed mass transfer, which leads to symmetric morphology structure without macrovoids. Also, the polymer concentration in the initial polymeric mixture shows a certain influence on the final membrane morphology, as can be seen in Figure 3.1.

#### 3.4.1.2 Morphological Characterization

There are several kinds of membrane characterization, depending on the information and knowledge required (morphological structure, chemical properties, permeability, selectivity, etc.). In this case, we focused on the morphological characterization of polysulfone membranes by using the SEM technique.

The morphological characterization of a flat membrane implies, basically, acquiring three micrographs: two corresponding to surface and one to cross-section. The most important is the last one. To obtain it, the sample needs to be previously treated. The membrane has to be broken, but not cut in order to preserve the porous structure. To do this, the most used technique is to dip the sample in an ethanol bath (to ensure that all the pores of the membranes are filled with the alcohol) and afterwards to immerse it in a liquid nitrogen bath to freeze the ethanol. Then, the membrane can be broken. Once the micrograph is obtained, it has to be interpreted. It is desirable to obtain numerical results of the main morphological parameters, such as the pore size distribution, porosity, symmetry, regularity and tortuosity. To obtain these parameters in a systematic and fast way the IFME software can be used [32].

The input of the software is the micrograph and the output are the mentioned properties, obtained in a numerical and a graphical way. From all the parameters determined by the software, the most interesting ones are the regularity and the symmetry, because these parameters are commonly used and defined normally in a qualitative but not a quantitative way and their relation with membrane performance is important. Additionally, while the number of pores and the pore size can be determined manually by wasting a lot of time and assuming a large error, these parameters cannot be calculated in this way.

The symmetry of the membrane is calculated from the number of symmetry groups identified in a membrane. In order to define the meaning of symmetry group, the difference between a symmetric and an asymmetric membrane can be observed in the examples shown in Figure 3.2a, b (the two images correspond to a

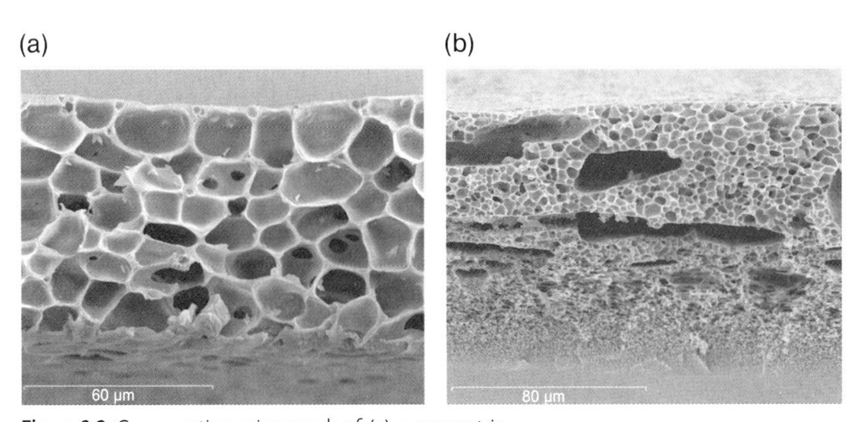


**Figure 3.2** Cross-section micrograph of (a) a symmetric membrane and (b) an asymmetric one, obtained both by SEM.

cross-section image of produced membranes). Zones with differences in pore density are identified. This is done by counting the number of pixels that conforms to a pore per file or column of the matrix (depending on the orientation of the image). The result is a vector containing inflection points that would determine what is defined as a symmetry group. In other words, the vector can be branched in several others by using the inflection points for unit separation. For example, from Figure 3.2b, the topside of the image has lots of small pores, whereas the bottom has fewer, bigger pores, so the pore density on this side is lower that on the topside.

This means that the asymmetry of the membrane increases as differences in the density of each symmetry group increase. In order to calculate the degree of asymmetry (DA) of the membrane micrograph, a factor ($f$) of each symmetry group is calculated. This factor is obtained by multiplying the length of the symmetry group and the area of it that is occupied by pores. The differences between these factors determine the DA, as calculated in Equation (3.1):

$$DA\,(\%) = \sum_{i=1}^{g} \left| \frac{\frac{100}{g} - \frac{f_i}{F} \times 100\%}{g} \right| \tag{3.1}$$

where $g$ is the number of symmetry groups and $F$ is the sum of all $f$. If the length and the number of pores of each group were the same, $f$ would be the same for each group and DA would be zero. The number of groups divides the result of the difference between the factor of each group and the one corresponding to a 100% symmetric membrane. This is supported by the fact that those membranes which contain a high number of groups are the ones that contain a large number of pores; and inversely, the membranes that have only a few symmetry groups are those which contain few pores and the differences between them are greater or more perceptible. In the case of having lots of groups, the number of subtractions is big and it yields an unjustifiably high DA because is not caused by the differences of the pores but by the number of symmetry groups.

However, in the case of few groups, the difference between the factor of each group and the one corresponding to a symmetric membrane is caused only by the differences of the pores and not by the number of sums.

The regularity analysis of the membrane is done in the perpendicular direction of the flux. For example, in case of a fluid flowing vertically (from top to bottom of the image), the analysis would be done horizontally, scanning the differences of pore distribution along the horizontal. The membrane should be regular in this sense because (independent of flux direction) the pore distribution should be the same. The results are expressed in terms of covariance and, in order to make the results comparative, the covariance is always calculated as normalized values (0–1). The covariance is calculated using Equation (3.2) [39]:

$$\mathrm{cov}(x_1, y_1) = E \cdot [(x_1 - \mu_1) \cdot (x_2 - \mu_2)] \tag{3.2}$$

where $\mu_i = E \cdot x_i$, $x_i$ represents the samples and $E$ is the mathematical expectation.

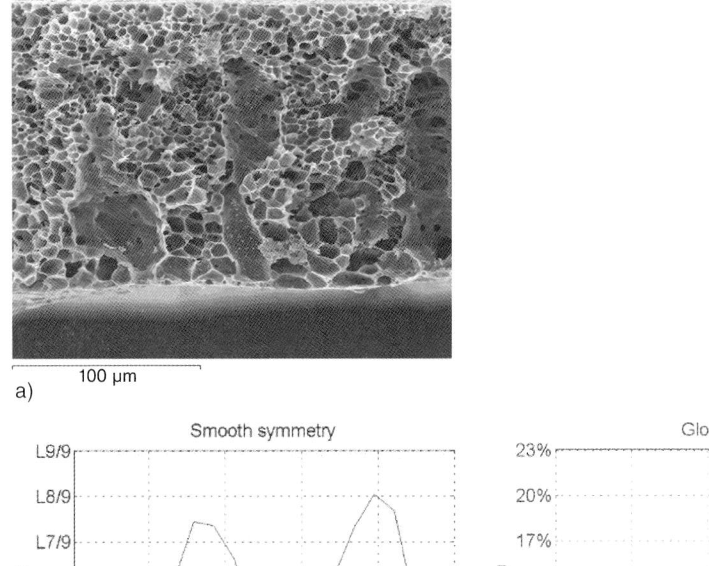


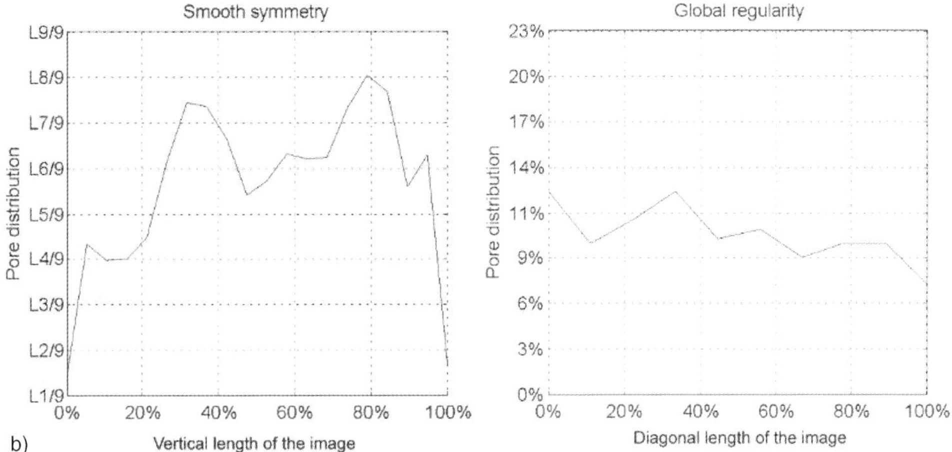


**Figure 3.3** (a) Cross-section SEM micrograph of a membrane obtained with 15% PSf in DMF and precipitated in water as a nonsolvent. Scale bar=100 μm. (b) Graphical results obtained by IFME software after interpreting a cross-section micrograph from a porous polymeric membrane.

The calculation algorithm works in a similar way to symmetry analysis. In this case, the vector is divided into several subvectors and, if the number of pores is the same in each, the membrane would be regular. The irregularity of the membrane increases with the increase in this difference.

Figure 3.3a shows a typical cross-section micrograph obtained by SEM from a polysulfone membrane. If the micrograph is being interpreted by using the IFME software, graphical results are obtained as Figure 3.3b shows and numerical results are obtained as Table 3.1 shows.

A high degree of asymmetry is obtained. This is mainly due to the presence of the macrovoids. The symmetry figure shows two clear peaks and a central zone

**Table 3.1** Morphology parameters obtained from IFME.

| Item | Result |
|---|---|
| Image area | 35629 μm$^2$ |
| Total number of pores | 340 |
| Mean size of the pores | 6.8547 μm |
| Biggest pore size | 22.594 μm |
| Smallest pore size | 3.3855 μm |
| Standard deviation | 0.1802 |
| Asymmetry | 23% |
| Global irregularity | 0.00031 |

with high pore density that corresponds to the overlapping of the three macrovoids. The regularity is relatively low, which indicates that the pores, in number and size, are homogeneous in the perpendicular direction to the flow.

The IFME software can also be applied on AFM micrographs. Nevertheless, in AFM technique, surface micrographs are usually obtained, rather than a cross-section. Thus, IFME software can determine the same morphological properties in AFM that are calculated on surface SEM micrographs. Strictly, three-dimensional data is obtained with this technique, the third dimension being the vertical one (the fluid direction, when it crosses the membrane), which has a much lower length compared with the other two. Common analysis (using AFM software or IFME, in this case not carrying out symmetry analysis but regularity analysis in a bi-dimensional plane) is performed in a two-dimensional plane and thus selects a vertical value. Or also, an analysis is carried out in each vertical plane. Results show that the values obtained for the number of pores, size, etc. change significantly depending on the vertical plane because, in the nanoscale analysis, what are called nanopores are in fact nanovoids between the polymer particles. These are close to spherical shapes and thus the nanopores have often conical shapes, the cross-section area of which depends highly on the vertical direction, as Figure 3.4c shows. Thus, considering this fact, there is an important conceptual difference between the membrane pores (obtained in the microscale) that are self-created structures and the nanopores, which correspond to voids between particles.

Additionally, from AFM technique, two types of images are obtained: phase (Figure 3.4a) and topographical (Figure 3.4b). They come from two types of data that are obtained: height data and phase data. The height data corresponds to the change in length in the vertical axis and generates the topographical map. The phase data is obtained from the changes in the phase angle due to elasticity differences within the sample (thus, from the nature of the material). Topographical micrographs, although they are more appropriate for morphological analysis, often give slightly unclear images and thus it is difficult to identify the pores and their perimeter (their limits).

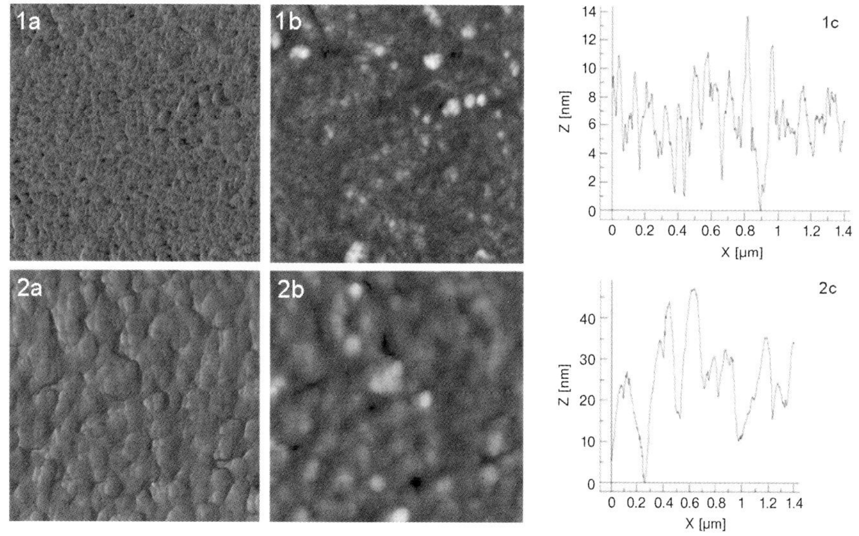


**Figure 3.4** AFM (a) phase images, (b) topographical images (both 1 × 1 μm) and (c) plots of the diagonal line from topographical images of membranes obtained with 20% PSf in a coagulation bath of DMF with either (1) pure water or (2) pure iso-propanol.

#### 3.4.1.3 Morphological Characterization Versus Membrane Performance

Often, customers who are interested in using membranes in their processes have serious problems on choosing the right membrane. When they examine the manufacturer's catalog, they often find two parameters which are supposed to describe the membrane performance: permeability and cut-off. Permeability corresponds to the normalized water flow, pressure and membrane area (usually litres per square meter per bar per hour; LMHB) and the cut-off (often in kilo-Daltons or Daltons) which, by definition, corresponds to the size of the particles that are 90% retained by the membrane.

The permeability of the test fluid that has to be filtered is very difficult to predict, since it depends on other factors. On one hand, there are fouling and polarization problems, which can reduce dramatically the flow rate. These depend on the type of sample, the concentration, etc. On the other hand but related, the design of the membrane module also plays an important role. A correct design of the module can prevent stagnant flows inside the module (which reduces the effective membrane area) and can significantly increase the shear stress on the membrane, which can prevent the mentioned phenomena.

More important is the cut-off value of the membrane given by the manufacturer. The fact that this value is widely used implies that it is almost an intrinsic property of the membrane, but reality shows a completely different view. The cut-off not only depends highly on the type of solute that has to be filtered, but also on process parameters like pressure or temperature. Considering also that membrane technology is still quite an empirical science, real intrinsic membrane parameters have to be used, although they do not provide direct data with regard to its performance. This intrinsic data is basically obtained from the morphological analysis.

An experimental way to measure the molecular weight cut-off (MWCO) of a membrane is to perform filtration experiments with a fluid containing known amounts of model substances of different molecular weight and for which the molecule radii are known. It is desirable to use substances with a narrow size dispersion to avoid overlapping. It is interesting to repeat experiments at different transmembrane pressures in order to know the correlation with the cut-off results. From the analysis of the permeate sample as well as the retentate one and after checking the mass balance, a scatter plot between the size of each component and its retention can be developed and a curve (often exponential) can be obtained. From the curve, the MWCO can be calculated (corresponding size at a retention of 90%).

Figure 3.5 shows the curve for an experiment carried out with a membrane obtained with 20% polysulfone in dimethyl formamide (DMF) and precipitated in a coagulation bath composed of 50% water and 50% DMF. The test fluid was obtained by mixing six dextrans of different molecular weights (17, 40, 70, 100, 200, 500 kDa) from Leuconostoc (Fluka). The concentration of each one in solution was 1.0 g/L. From the curve obtained, a MWCO of 28 kDa could be determined. Transmembrane pressure (driving force) was set to 9 bar.

From the experimental results obtained, the mean membrane pore diameter can be determined by using correlations found in the literature. Among them, the Ferry–Faxen equation [40] correlates the membrane retention ($R$) with the radius

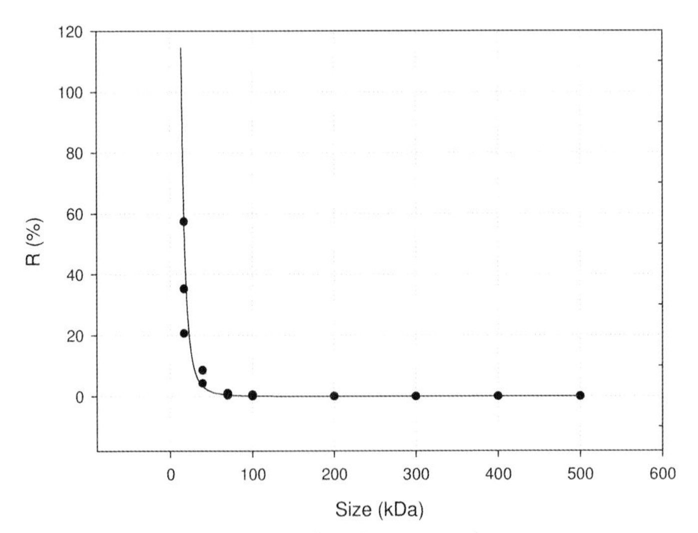


**Figure 3.5** Separation capability of a polymeric membrane obtained with 20% polysulfone in DMF and precipitated with a composition bath containing 50% water and 50% DMF.

of the solute ($r_s$) and the mean membrane pore radius ($r_p$), Equation (3.3):

$$1 - R = \left(1 - \frac{r_s}{r_p}\right)^2 \cdot \left[1 - 0.104 \cdot \frac{r_s}{r_p} - 5.21 \cdot \left(\frac{r_s}{r_p}\right)^2 + 4.19 \cdot \left(\frac{r_s}{r_p}\right)^3 + 4.18 \cdot \left(\frac{r_s}{r_p}\right)^4 - 3.04 \cdot \left(\frac{r_s}{r_p}\right)^5\right] \quad (3.3)$$

Additionally, the Aimar et al. correlation [41] proposes a relation between the solute radii and its molecular weight, Equation (3.4):

$$r_s = 0.33 \cdot M_w^{0.46} \quad (3.4)$$

Thus, if the Aimar et al. equation is combined with the Ferry–Faxen one and a retention of 90% is considered (MWCO conditions), a simple equation can be obtained, which correlates the MWCO and the pore radii, Equation (3.5):

$$\text{MWCO} = 2.2 \cdot r_p^{2.1739} \quad (3.5)$$

Thus, Equation (3.5) can be used to compare results obtained from experimental tests and from those obtained directly by morphological analysis. Considering the case of dextran separation in polymeric membranes synthesized with 20% polysulfone, there is a factor of five between the pore size obtained from experimental results and the one obtained by AFM measurements of the top membrane layer (the selective one). This factor depends on the type of solute, experimental conditions, etc. and it shows the noncorrespondence between the experimentally measured MWCO and the one calculated from morphological data. Some experiments carried out by Torras et al. [5] showed this fact. In these experiments, three polymeric membranes obtained from a polymeric solution of 20% PSf in DMF and precipitated with three different coagulation baths were used in separation experiments with dextrans. Cut-off was determined and AFM measurements were performed, showing a factor between five and eight (Table 3.2). Lindau et al. [42] performed similar experiments with commercial polymeric membranes and obtain similar results: in this case a factor of 6 was achieved.

**Table 3.2** Comparison between pore radii determined by AFM and by dextran retention experiments.

| Type of coagulation bath | Minimum pore radii (from AFM, nm) | Minimum pore radii (experimentally, nm) | Factor |
|---|---|---|---|
| 50% water – 50% DMF | 16.0 | 77.3 | 5 |
| 25% water – 75% DMF | 12.5 | 73.4 | 6 |
| 50% water – 50% IPA | 11.5 | 94.2 | 8 |

Additionally, these results point in the same direction than those obtained by Bowen et al. [43]. In all cases it has been found that experimental results give larger values for mean pore diameter than those measured by AFM, which also supports the mentioned effect. There are differences in the ratio but it has to be considered that the studies are made on membranes with different MWCO ranges (as well as other conditions, such as the solute used, which has an important influence). In our studies, we considered membranes with larger MWCO than the ones used by Bowen et al. and it is well known there is an increasing divergence between pore radii and MWCO as the MWCO increases. Thus, only morphological data can be used, up to now, to give the intrinsic properties of porous membranes.

## 3.4.2 Microcapsules

### 3.4.2.1 Synthesis Considerations

Microcapsules are non-dense spherical shaped volumes. They are formed by a shell (polymeric wall in this case) with an empty volume inside that can be used to encapsulate compounds. The structure of the shell (a thin, dense layer, or a thick, porous layer, etc.), together with its nature (chemistry, material), determines the release rate of the encapsulated compound. There are several ways to obtain microcapsules. Some of these methods are based only on physical phenomena, certain are based on polymerization reactions and others combine both physical and chemical procedures. Most authors agree in classifying them all in two different groups: chemical processes (like in-situ polymerization or desolvation in liquid media) and mechanical processes (e.g. spray drying or electrostatic deposition) [44]. A novel technology for microcapsule production, based on the employment of microdevices in continuous mode, has been presented [45]. Immersion precipitation is used in this case, in a way similar to that explained for flat membranes.

Two steps are required to obtain the microcapsules. In the first step, a micromixer is used to produce an emulsion. Two inlets are used in the microdevice: one containing the polymeric solution (polymer, solvent, compound to be encapsulated) and another containing the continuous phase, which has to be a solvent (in order to prevent polymer precipitation during this step) that is immiscible with the solvent used to dissolve the polymer in the polymeric solution (in order to produce the emulsion). When the emulsion is formed, it is pumped to a second micromixer (second step), in which it is mixed with the nonsolvent in order to precipitate the droplets and obtain the microcapsules. The phase inversion process is similar to that carried out for flat membranes. The main difference in this case is that a continuous phase is needed.

Figure 3.6 shows the experimental set-up used in order to carry out the process. The micromixers were constructed and supplied by the Institut für Mikrotechnik Mainz, Germany [46].

The main advantage of this process compared with others (and especially with batch processes) is that a narrow particle size distribution is obtained, a high

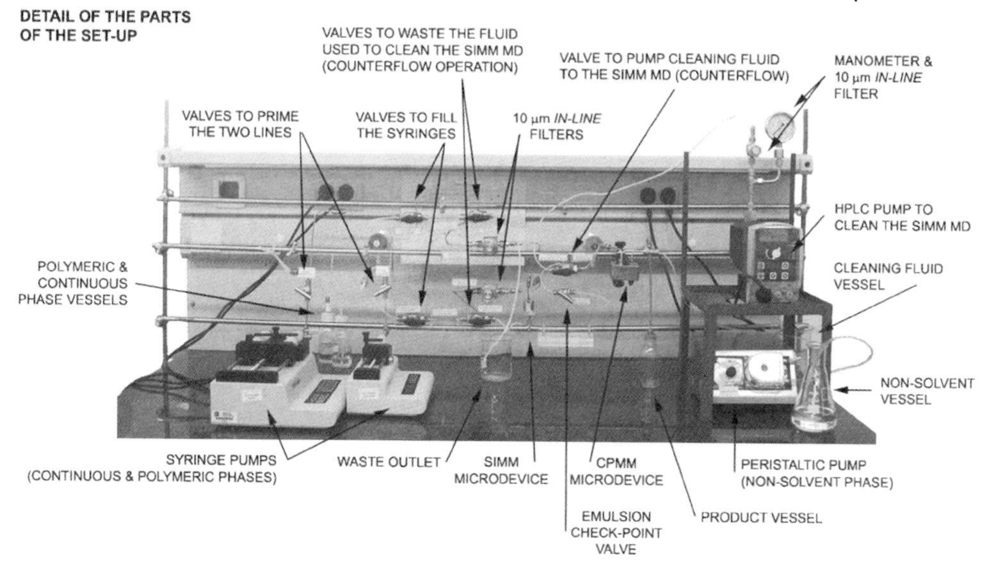


**Figure 3.6** Image of the set-up used to produce microcapsules in continuous mode by using microtechnology.

polymer efficiency is achieved (almost all the polymer is transformed into microcapsules, not in nuggets) and parameters can be controlled easier, allowing a tailored final product (in terms of size and structure).

#### 3.4.2.2 **Morphological Characterization**

Morphological characterization can be carried out in a similar way to that for flat membranes. The main difference is that the sample cannot be manipulated due to its size and shape. This is a special critical limitation when internal structures have to be characterized.

External characterization from the slurry can be easily carried out by two of the microscopic techniques described before. One technique is filtering the liquid solution and examining the sample retained in the filter by traditional SEM techniques at high- or low-vacuum: If high-vacuum is used, which is suitable for microcapsules $\leq 1\,\mu m$ that need high resolution, the sample has to be previously metalized. Another technique consists of using an environmental scanning electron microscopy (ESEM), which allows using liquids inside the microscope, with temperature control (below $5\,^{\circ}C$) and using low vacuum (4 torr). The main advantage of this method is that no sample treatment is required and it ensures that no sample is lost (by previous filtration). External characterization is used to determine the microcapsule size distribution as well as to predict the concentration of microcapsules produced. Figure 3.7 shows a typical micrograph of polymeric microcapsules. As for flat membranes, micrographs obtained from the microscopy

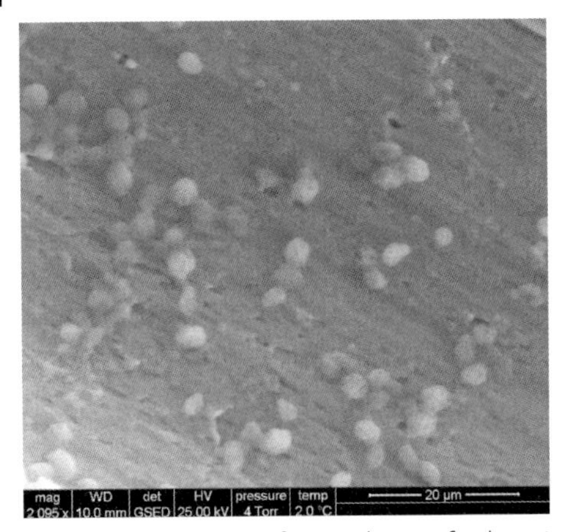


**Figure 3.7** Micrograph of external view of polymeric microcapsules obtained from a polymeric solution containing 3% PSf in DMF and precipitated in water as a nonsolvent. Cyclohexane was used as continuous phase.

techniques were interpreted using IFME software in order to obtain numerical results for the number of microcapsules and size distribution.

Internal characterization requires treatment of the sample in order to examine the cross-section of the wall. The main difference between the microcapsules and the flat membrane is that the former cannot be easily manipulated while ensuring that the porous structure is preserved; and therefore an alternative pre-treatment methodology should be applied. There are two techniques that may be used [47]. The first consists of using a cryotome. The sample is fixed over a support by using special glue, which can be easily frozen and does not damage samples. The sample is then cut inside the cryogenic chamber and deposited onto a glass support, which is metallized and examined with the SEM (cut microcapsules can be found). The second technique consists of producing a resin block containing the sample. The resin, which initially is liquid, penetrates inside the microcapsules (thus, inside the pores) and is then solidified. The sample can be cut into slices under normal conditionsfor examination by TEM, since the porous shell structure will not be damaged in the cut because the pores are filled by another solid. Figure 3.8 corresponds to a cross-section micrograph of a microcapsule, obtained with polysulfone as polymer and dichloromethane as solvent, treated with the cryogenic technique. The wall structure, which corresponds to a dense and thin layer, can be seen.

Other techniques have also been tested with microcapsules. Among them, CSLM was tested because, as a main advantage, it allows internal topographical images to be obtained without cutting or breaking the sample. To do this, a fluorescent agent was used: Rhodamine B, which is normally used as a dye. Microcapsules were prepared by using the same methodology explained before, but adding the fluorescent agent in the polymeric solution. Afterwards, the sample was analyzed by CSLM. The main problem in this case is the size of the samples. CLSM is based on optical microscopy and, although using an objective with a

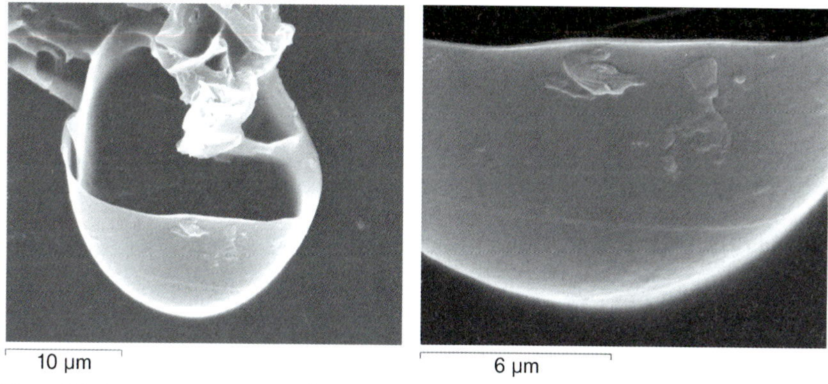


**Figure 3.8** Cross-section micrographs of a microcapsule obtained from 3% PSf in dichloromethane solution and precipitated in water as a nonsolvent.

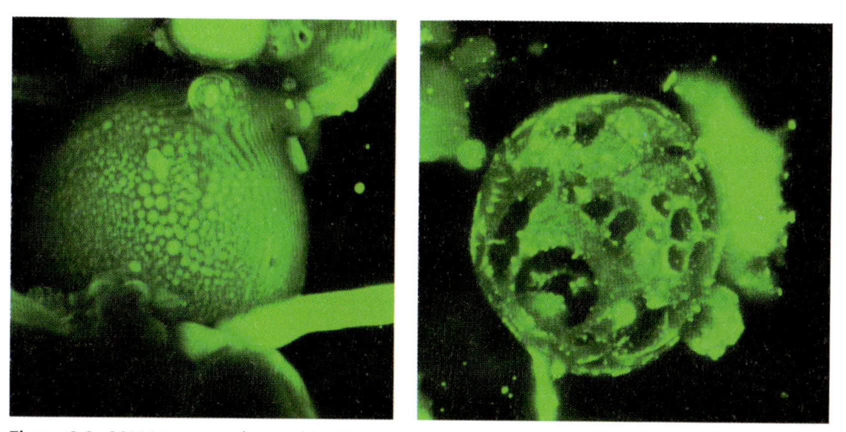

**Figure 3.9** CSLM images obtained in fluorescent mode using rhodamine B as fluorescent agent of 1 μm microcapsules. Obtained from a polymeric solution of 3% PSf in DMF and precipitated in water as nonsolvent. Cyclohexane was used as continuous phase.

magnification of 100 × with immersion oil, insufficient detail was obtained to describe the internal porous structure of microcapsules of about 1 μm. Figure 3.9 shows images obtained by CSLM of several microcapsules (without any sample preparation: they were directly observed by CSLM).

The CSLM technique was also tested with capsules of 2 mm diameter. These capsules were obtained from a solution of 15% PSf in DMF and precipitated directly in water (without using the continuous system set-up described before). Due to its size, some description of the inside structure of the capsule (Figure 3.10) was obtained without any sample treatment (also, without cutting or breaking it). Figure 3.11 shows a cross-section micrograph of a capsule obtained in the same

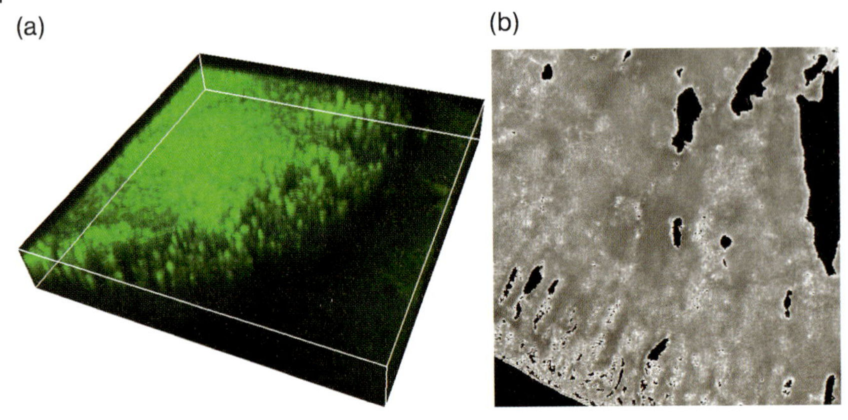


**Figure 3.10** CSLM images of 2 mm diameter capsule. Obtained from a polymeric solution of 15% PSf in DMF and precipitated in water as nonsolvent. Image (a) was obtained in fluorescent mode using rhodamine B as fluorescent agent, while image (b) was obtained in transmission mode (white light).

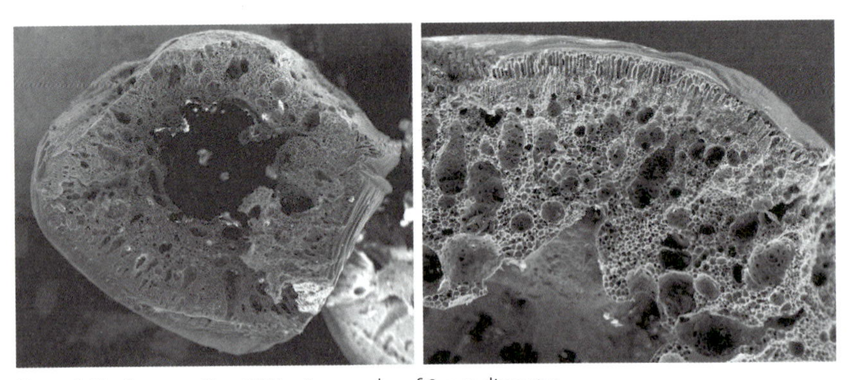

**Figure 3.11** Cross-section SEM micrographs of 2 mm diameter capsule. Obtained from a polymeric solution of 15% PSf in DMF and precipitated in water as nonsolvent.

conditions but broken in liquid nitrogen and analyzed by SEM. As can be seen, the information provided by Figure 3.11 is much more detailed than that in Figure 3.10, but in this case and as stated before, the sample was broken and vacuum was applied to get them sputtered with gold (pre-treatment required).

## 3.5 Final Remarks

Morphology characterization in membrane manufacturing is a key area in order to optimize the production process and to obtain the desired membranes for a particular process. Scanning electron microscopy (SEM) and transmission electron microscopy (TEM) are the most appropriate techniques. To achieve so, not only have micrographs to be obtained, but further processing is required to obtain numerical results from the images. The IFME software provides the tool to make the desired data processing. With this data, numerical correlations between synthesis conditions and morphological parameter can be calculated and a specific permeability and selectivity behavior of the membrane can be achieved.

Before obtaining results from the different microscopy techniques, samples should be treated. Pre-treatments are usually needed to obtain a conductive sample in order to use a high vacuum and to obtain good resolution and information related to the cross-section of the sample. The techniques used to cut the sample and observe the inside structure should guarantee that the porous structure is not damaged, although the difficulty increases with decreasing sample size. Nevertheless, several traditional techniques are being used for flat membranes and for capsules with diameters in the millimetric order of magnitude and novel techniques are being developed for nano- and microcapsules.

## Acknowledgments

The authors thank the Universitat Rovira i Virgili for research facilities (especially the Research Resources Service, Mercè Moncusí, Dr. Mariana Stefanova) and also the Basic Medical Science Department of URV (especially Mrs. Amparo Aguilar). Finally, the authors thank the Spanish Ministry of Science, Catalan Government (Departament d'Educació i Universitats) and the European Commission (project FP6-2003-NMP-IN-3) for funding.

## References


1 M. Mulder, *Basic Principles of Membrane Technology*, Kluwer Academic Publishers, Dordrecht, **1997**.

2 K.C. Khulbe, T. Matsuura, *Polymer* **2000**, *41*, 1917–1935.

3 J.I. Golstein, A.D. Jr. Romig, D.E. Newbury, C.E. Lyman, P. Echlin, C. Fiori, D.C. Joy, E. Lifshin (Eds.), *Scanning Electron Microscopy and X-ray Microanalysis*, Plenum Press, New York (USA), **1992**.

4 K.J. Kim, M.R. Dickson, A.G. Fane, C.J.D. Fell, *J. Microsc.* **1991**, *162*, 403–413.

5 C. Torras, F. Ferrando, J. Paltakari, R. Garcia-Valls, *J. Membrane Sci.* **2006**, *282*, 149–161.

6 X. Zhang, J. Benavente, R. Garcia-Valls, *J. Power Sour.* **2005**, *145*, 292–297.

7 F. Tasseli, J.C. Jansen, F. Sidari, E. Drioli, *J. Membrane Sci.* **2005**, *255*, 13–22.

8 R. Bhattacharya, T.N. Phaniraj, D. Shailaja, *J. Membrane Sci.* **2003**, *227*, 23–37.

9 C.O. M'Bareck, M. Metayer, Q.T. Nguyen, S. Alexandre, J.J. Malandain, *J. Membrane Sci.* **2003**, *221*, 53–68.

10 G. Arthanareeswarasn, K. Srinivasan, R. Mahendran, D. Mohan, M. Rajendran, V. Mohan, *Eur. Polym. J.* **2004**, *40*, 751–762.

11 J. Kurdi, A. Kumar, *J. Membrane Sci.* **2006**, *280*, 234–244.

12 X. Zhang, S. Liu, J. Yin, *J. Membrane Sci.* **2006**, *275*, 119–126.
13 S. Kazama, M. Sakashita, *J. Membrane Sci.* **2004**, *243*, 59–68.
14 A.M. Barbe, P.A. Hogan, R.A. Johnson, *J. Membrane Sci.* **2000**, *172*, 149–156.
15 T. Gumi, J. Fernandez-Delgado Albacete, D. Paolucci-Jeanjean, M.-P. Belleville, G.M. Rios, *J. Membrane Sci.* **2008**, in press.
16 L.K. Chiu, W.J. Chiu, Y.-L. Cheng, *Int. J. Pharm.* **1995**, *126*, 169–178.
17 G. Qunhui, H. Ohya, C. Liankai, H. Jicai, *J. Membrane Sci.* **1995**, *100*, 217–228.
18 H. Weiguang, L. Shaokun, Z. Kancheng, L. Xiaogang, L. Senshu, *J. Membrane Sci.* **1999**, *155*, 185–191.
19 Y. Liu, J. Qian, X. Fu, H. Liu, J. Deng, T. Yu, *Enzyme Microb. Tech.* **1997**, *21*, 154–159.
20 S. Singh, K.C. Khulbe, T. Matsuura, P. Ramamurthy, *J. Membrane Sci.* **1998**, *142*, 111–127.
21 J.I. Calvo, P. Prádanos, A. Hernández, W.R. Bowen, N. Hilal, R.W. Lovitt, P.M Williams, *J. Membrane Sci.* **1997**, *128*, 7–21.
22 G. Binnig, C.F. Quate, Ch. Gerber, *Phys. Rev. Lett.* **1986**, *12*, 930.
23 W.R. Bowen, N. Hilal, R.W. Lovitt, P.M. Williams, *J. Membrane Sci.* **1996**, *110*, 229–232.
24 N. Hilal, H. Al-Zoubi, N.A. Darwish, A.W. Mohammad, M. Abu Arabi, *Desalination* **2004**, *170*, 281–308.
25 W. Yoshida, Y. Cohen, *J. Membrane Sci.* **2003**, *215*, 249–264.
26 C. Charcosset, A. Cherfi, J.-C. Bernengo, *Chem. Eng. Sci.* **2000**, *55*, 5351–5358.
27 M. Ferrando, A. Rożek, M. Zator, F. López, C. Güell, *J. Membrane Sci.* **2005**, *250*, 283–293.
28 M.A. Snyder, D.G. Vlachos, V. Nikolakis, *J. Membrane Sci.* **2007**, *290*, 1–18.
29 C. Charcosset, F. Yousefian, J.-F. Thovert, P.M. Adler, *Desalination* **2002**, *145*, 133–138.
30 Q. Wu, B. Wu, *J. Membrane Sci.* **1995**, *105*, 113–120.
31 J.I. Calvo, A. Hernández, G. Caruana, L. Martínez, *J. Colloid Interface Sci.* **1995**, *175*, 138–150.
32 C. Torras, R. Garcia-Valls, *J. Membrane Sci.* **2004**, *233*, 119–127.
33 L. Wang, X. Wang, *J. Membrane Sci.* **2006**, *283*, 109–115.
34 C. Torras, *Obtenció de membranes polimèriques selectives*. PhD dissertation, Universitat Rovira i Virgili, Tarragona, **2005**. Available online at: www.carlestorras.info
35 T. Gumí, C. Torras, R. Garcia-Valls, C. Palet, *Ind. Eng. Chem. Res.* **2005**, *44*, 7696–7700.
36 A. Conesa, T. Gumi, C. Palet, *J. Membrane Sci.* **2007**, *287*, 29–40.
37 R. Bouzerar, L.H. Ding, M.Y. Jaffrin, *J. Membrane Sci.* **2000**, *170*, 17–141.
38 P.J. Flory, *Principles of Polymer Chemistry*, Cornell University Press, Ithaca, NY, **1953**.
39 J. Domingo, *Estadística tècnica: una introducció constructivista*, Universitat Rovira i Virgili, Tarragona, **1997**.
40 J.D. Ferry, *J. Gen. Physiol.* **1936**, *20*, 95–104.
41 P. Aimar, M. Meireles, V. Sanchez, *J. Membrane Sci.* **1990**, *54*, 321–338.
42 J. Lindau, A.-S. Jönsson, A. Botino, *J. Membrane Sci.* **1998**, *149*, 11–20.
43 W.R. Bowen, T.A. Doneva, *Surf. Interface Anal.* **2000**, *29*, 544–547.
44 C. Thies, *Microencapsulation* **1996**, *73*, 1–19.
45 IMPULSE Project. European integrated project from the sixth European framework programme: FP6-2003-NMP-IN-3, **2003**, http://www.impulse-project.net
46 Institut für Mikrotechnik Mainz, Mainz, Germany: http://www.imm-mainz.de
47 C. Torras, L. Pitol-Filho, R. Garcia-Valls, *J. Membrane Sci.* **2007**, *305*, 1–4.

# 4
# Confocal Scanning Laser Microscopy: Fundamentals and Uses on Membrane Fouling Characterization and Opportunities for Online Monitoring

*Montserrat Ferrando, Maria Zator, Francisco López, and Carme Güell*

## 4.1
## Introduction

The industrial application of pressure-driven membrane processes has gained significance in recent decades. Several technological innovations addressed to develop highly effective and permeable membranes and to improve efficiency in module design have spread the use of membrane separation processes to (among others) chemical, food, pharmaceutical and biomedical industries. Some important drawbacks of pressure-driven membrane processes, however, are still limiting their wide industrial acceptance.

Membrane fouling and the subsequent permeate flux decline is one of the major drawbacks. A number of theories about concentration polarization, cake formation and flux decline have been developed to predict membrane fouling. Generally those models are based on an assumed behavior and validated by experimental determinations. Nevertheless, typical experimental measurements are focused on macroscopic parameters, such as permeate flux, pressure drop and solute rejection, which provide little information about microscopic phenomena and, thus, are not able to validate mechanistic models.

Recently, several novel non-invasive techniques have been applied to characterize membrane fouling [1, 2]. Among them, optical and electron microscopic techniques have been recently used to visualize particle deposition, cake formation and membrane fouling. Conventional optical microscopy enables direct visualization of particle deposition during filtration, but its low resolution limits its applications. Additionally, conventional optical microscopy cannot be used to visualize cake structure and morphology since it has a very low axial resolution and requires ultrathin specimens to obtain a good visualization.

Electron microscopy, in particular scanning electron microscopy (SEM) and transmission electron microscopy (TEM), has been extensively applied to describe the structure and morphology of membranes. More recently, environmental

*Monitoring and Visualizing Membrane-Based Processes*
Edited by Carme Güell, Montserrat Ferrando, and Francisco López

ISBN: 978-3-527-32006-6

scanning electron microscopy (ESEM) has been used to characterize membrane fouling, mainly because it allows the observation of hydrated samples and does not require the sample preparation (drying, coating) necessary for conventional SEM. With ESEM, lateral resolution (10–20 nm) is high, suitable to visualize big colloids, but lower than that of SEM (1 nm). ESEM presents, however, an important limitation: it does not differentiate among several foulants when a multicomponent suspension is filtered.

Confocal scanning laser microscopy (CSLM), commercially developed in the early 1990s, is an optical microscopic technique that has some major advantages over the aforementioned microscopic techniques. CSLM has better axial resolution than conventional optical microscopy and, additionally, it provides high resolution images obtained at different depths within a three-dimensional object. This optical sectioning of the sample means that invasive procedures are not needed in sample preparation, since physical slicing is no longer necessary. In addition, by combining fluorescent and reflection modes, CSLM enables separate visualization of not only the membrane but also several species involved in membrane fouling.

This chapter covers aspects related to CSLM and its uses in pressure-driven membrane processes. The first part deals with the fundamentals of the technique, the conditions required to obtain a proper visualization either of the membrane or the foulants and some examples of image analysis that lead to quantitative information on membrane structure and the extent of fouling. The uses of CSLM for membrane characterization, adsorption of target compounds to membranes and, in particular, fouling characterization are thoroughly reviewed in the second part of this chapter.

## 4.2
## Fundamentals of Confocal Scanning Laser Microscopy

### 4.2.1
### General Principle

In conventional optical microscopy, the whole sample is illuminated and the image can be observed directly by eye or projected directly onto an image capture devise or photographic film. In contrast, the method of image formation in a confocal microscope is essentially different. In CSLM, an excitation beam from the laser is reflected by a dichroic mirror and focused by the objective lens onto a microvolume of the sample (Figure 4.1). Light is emitted from the irradiated microvolume of the sample, passes through the dichroic mirror (a chromatic beam splitter) and an adjustable imaging aperture or pinhole – placed confocally with respect the illumination aperture – and is collected by a photomultiplier. In order to build an image, the focused microvolume must be scanned across the sample in some way. The most common alternative is to scan the illumination beam across a stationary specimen. CSLM uses single-beam scanning, that is, the scanning of

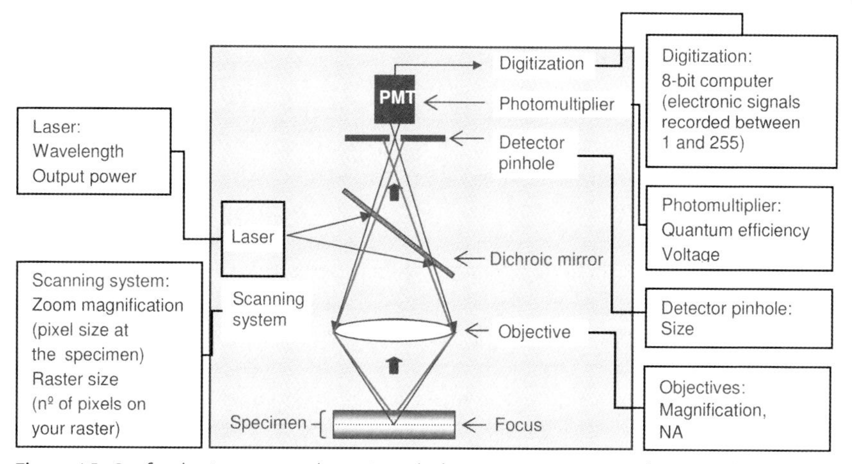


**Figure 4.1** Confocal microscope schematic with the main parameters to be controlled.

the beam is achieved by computer-controlled mirrors driven by galvanometers at a rate of one frame per second.

CSLM can be used in transmission, reflection and fluorescence modes, but fluorescence imaging is the most extensively used because it enables selective labeling of individual structures/macromolecules in a multicomponent complex sample. In reflection mode, the reflected or backscattered light from the sample allows it to be viewed in an unstained state. In some cases, the sample can be labeled with a probe that is highly reflective, such as gold or silver grains.

In order to obtain well resolved images in fluorescence mode with CSLM, some parameters have to be observed. Although a thorough revision of all factors that have to be considered is not possible in this chapter, a short list of those having a major effect on spatial resolution is included. Covering the identified parts of the microscope showed in Figure 4.1, we will try to encompass the main parameters. For an extended and complete revision of CSLM fundamentals we recommend the works published by Pauley [3] and Wilson [4].

The laser unit is the illumination source; and the wavelength of emission, power at each wavelength and stability are important parameters to take into consideration. The wavelength of emission affects optical performance since, together with the absorption spectrum of the fluorescent dye, which is incorporated into the sample, it determines the amount of fluorescence produced. In addition, the fluorescence measured is proportional to the laser output power.

Zoom magnification and raster size are two important parameters of the scanning system. Zoom magnification determines the area scanned on the sample and, together with the number of pixels along the edges of the raster (raster size), it determines the pixel size. The pixel size has to be adjusted considering the smallest structure that we expect to image in the sample and the photobleaching effect (the vanishing of the fluorescence emission as a consequence of exposure to the emission light). The pixel size must be small enough to avoid undersampling, but big enough to reduce photobleaching.

Magnification and numerical aperture (NA) of the objectives affect the resolution. As the resolution is inversely proportional to the numerical aperture of the imaging lens, objectives of high NA (>1) are advantageous.

In order to obtain an image of good quality without subjecting the sample to extremely high levels of exciting light, not only the zoom magnification but also the pinhole size (through which the signal passes) must be set up properly. The optimal size of the pinhole to obtain both a high signal-to-noise and a low blur of additional signals from other planes (high axial resolution) is a sensitive function of the light wavelength and the NA and magnification of the objective lens. According to these factors, the pinhole size is usually set to be equal to the diameter of the Airy disk at the plane of the pinhole: the usual values of the pinhole diameter can be between 10 μm and 1000 μm. Current CSLM equipment sets the optimum pinhole diameter automatically.

A photomultiplier tube is the detector unit of CSLM equipment. The detected signal is directly proportional to the quantum efficiency and is amplified by the photomultiplier voltage. Quantum efficiency, which describes the chance that energy absorbed from an exciting photon will be re-radiated as a fluorescent photon, strongly depends on wavelength, the pH and ionic strength of the sample. Most commercial microscopes are equipped with several photomultipliers (commonly three) and a set of filters that enable the emission signal to be split, based on its wavelength. In this way, fluorescent signals emitted from different fluorophores can be collected simultaneously.

The typical 8-bit digitization used in most CSLM equipments leads to electronic signals of a size to be recorded between 1 and 255 per pixel. Because of statistical noise, values between 10 and 220 per pixel are recommended.

By setting properly all the aforementioned parameters, a proper imaging of the sample will be obtained. At those conditions, lateral or $x/y$ resolution (minimum spacing that can just be resolved) is about 1.4 times better than in widefield fluorescence microscopy (180–200 nm are typical values), while the improvement in viewing axis or $z$ resolution at optimal conditions is about 3.0 times poorer than the lateral resolution (500–800 nm). This represents a marked improvement compared with conventional microscopy, which arises from the rejection of fluorescence light from out-of-focus regions of the specimen.

The imaging depth or the distance below the surface of the sample which can be accurately imaged is restricted, first of all, by the working distance of the objective, which is the distance between the focal plane and the objective. Its value can be hundreds of microns or some millimeters, depending on the NA and magnification of the lens and also the immersion medium. However the real limitation of the imaging depth is imposed by the opacity and inhomogeneity of the sample. These cause a reduction in the signal intensity detected as the focal plane penetrates deeper into the sample. The illuminating beam is scattered or absorbed as it passes through the upper layers of the sample, limiting the ability to reach a further focus plane within the sample. Likewise, fluorescent emission which is dispersed or absorbed between the focal plane and the detector cannot be detected. In addition, any optical inhomogeneity in the sample above the focal plane will tend to defocus the beam of illumination and the returning signal, thus decreasing the signal which can pass through the pinhole. The penetration depth can be several hundreds of microns, but it is reduced in samples with an imbibing medium having an index of refraction very different to the solid components of the sample.

### 4.2.2
### Sample Preparation and Fluorescent Labeling

Fluorescent emission is required to visualize a macromolecule or structure by CSLM in fluorescent mode. In membrane applications, the fluorescent signal usually comes from the components of the filtrated solution; and their fluorescent signal enables that they are visualized and located on the top or inside the membrane.

Some macromolecules present autofluorescence or have an inherent fluorescence capacity but, mostly, fluorescence has to be induced by chemical treatment. This is also known as secondary fluorescence and can be achieved in two ways: (a) by treating the sample with specific stains (fluorophores) or producing a specific fluorescent reaction products in the sections of interest, (b) by applying fluorescent labeled biological molecules with a binding affinity for specific constituents. The number of fluorophores with an absorption wavelength close to the emission wavelengths of the available lasers has been increased during the past years. Up to 28 fluorophores could be listed as of current practical importance [5] and they cover the overall range of commercial laser sources available in CSLM.

The general requirements that a fluorescent probe has to fulfill to be used in CSLM are the following:

1. The maximum absorption of the fluorescent probe has to be close to the emission wavelength of the laser.
2. The fluorescent emission must be intense (fluorescence intensity is directly proportional to the product of the molar extinction coefficient for absorption and the quantum yield for the fluorescence emission of the dye).
3. Multilabeling experiments (in which more than one macromolecule or structure is stained) require fluorescent compounds whose excitation/emission wavelengths overlap minimally.

Some of these fluorescent probes have extreme sensitivity to their environment and are able to change their fluorescence amplitude and wavelength shifts of excitation or emission as a response to changes in the ion concentration or pH. This may suggest a more extended application of those probes to characterize molecular reactions and interactions occurring during membrane processes. However, it is important to note that the size of the fluorophores does not alter the adsorption/interaction behavior of the molecules.

### 4.2.3
### Image Analysis

Quantitative and qualitative information can be obtained from CSLM images. The results of CSLM analysis are $z$ series or stacks of 2D images taken at a known $z$ axis distance. Commercial software can be used to generate 3D reconstructions. There are several ways to visualize these 3D reconstructions; two of the most common ones are the following:

- Orthogonal planes, which make it possible to examine cross-sectional slices through the image volume in the $xz$, $yz$ and $xy$ planes. This helps to view 3D structural information about the specimen. Sliders are used to select the $x$, $y$ and $z$ coordinates and the views in the image windows give the cross-sections at that position.
- Volumetric 3D reconstruction, consisting of rotated views from a stack of through-focus images.

In addition, an analysis of CSLM digitized images may be used to describe them in quantitative terms. For measurements of feature-specific parameters related to object/structure dimensions (such as area, length, width, perimeter, surface area, volume, gray level) and shapes (such as form factor, convexity, number of holes, etc.), the object of interest is usually segmented from the image. The simplest form of segmentation relies on a single-intensity threshold to define the boundary of the object. Drawbacks related to threshold selection are mentioned in Section 5.2.3. Segmentation and automated analysis of different features can be carried out with commercial software from either 2D or 3D images. Computational requirements vary, depending (among others) on the parameters to be measured, the raster size, the number of images in a z series and whether calculations are based on 2D or 3D images.

Considering the applications reported later (see Section 4.3), two image analysis methodologies were followed to characterize membrane fouling.

#### 4.2.3.1 Surface Porosity Measurements

To describe fouling in membrane pores, $P_s$ was the parameter quantified from 2D images [6]. $P_s$ is the fraction of the pore surface in which a fluorescent foulant has been detected:

$$P_s = \frac{A_f}{A_P} \tag{4.1}$$

where $A_f$ is the surface of the pore in which a fluorescent foulant has been detected and $A_P$ is the surface occupied by pores. By calculating $P_s$ of 2D CSLM images at different depths, the $P_s$ profile for selected compounds could be obtained from the membrane surface to a depth of some microns inside.

The processing of the CSLM images included identifying the features to be analyzed and then segmenting and extracting the measurements of interest. The digital images, with an integer value ranging from 0 to 255, were segmented/binarized into foreground and background by setting a threshold, which separated the pixels of interest from the rest of the image. The threshold was the same for samples from the same filtration conditions. $A_p$ values were obtained from binarized reflection images while $A_f$ values were obtained by arithmetic operations of binarized reflection and fluorescent images.

Within the context of membrane fouling, different fluorescent foulants may fill the same pore, while in the context of digital imaging the colors emitted by the fluorescent molecules occupy the same pixel in the image. By analyzing separately the images obtained from the same sample but collected in different photomultipliers (each collecting the signal of a different fluorescent foulant), $P_s$ of different foulants in the same voxel could be measured.

#### 4.2.3.2 Cake Characterization

The combination of a multiple staining protocol with CSLM enabled the visualization of the 3D structure of the main components of a biofouling layer (details are reviewed in Section 4.3.3). By applying a complex image analysis, several morphological parameters were determined, such as the porosity and fractal dimension of the fouling layer.

In the work by Yang et al. [7] and Chen et al. [8], $z$ series of CSLM images were converted into binarized images and the boundaries in the binarized images were classified according to its connectivity and whether they lay in the object or its complement. The resulting surfaces were triangulated and each region defined as a pore was filled with tetrahedral grids. These grids were allocated in a mesh model to describe the interior architecture of the fouling layer. On this basis, the porosity of a selected volume of the fouling layer was calculated as (void space)/(selected volume); and from that the vertical distribution of the porosity in the fouling layer was obtained.

Other authors [9–11] have calculated 3D parameters of the cake structure by Image structure analyzer (ISA 2). This software was particularly developed to quantify structural parameters from CSLM images [12] and has been used to measure (among others) porosity, biovolume, cake volume and average run length of the cake structure.

## 4.3
## Applications of CSLM to Membranes and Membrane Processes

The application of CSLM to membrane and membrane process characterization has been mainly developed during the past decade. The earliest applications of CSLM found in the literature were addressed to membrane characterization; and the adsorption of proteins onto ion exchange membranes was the first membrane process visualized by CSLM. More recent uses of CSLM still focus on membrane characterization, but some other applications have been reported on membrane fouling characterization; and some attempts to apply CSLM for online visualization have been also published. The following sections review the main results obtained by the use of CSLM on each of the aforementioned applications; and special attention is paid to research on membrane fouling characterization in microfiltration processes performed by our research group during the past years.

### 4.3.1
### Membrane Characterization

Membrane characterization by CSLM has been rather limited when compared with other microscopic techniques such as SEM and atomic force microscopy (AFM). The earliest work found in the literature [13] records how van den Berg et al. used a combination of AFM and CSLM to study qualitative differences in the pore geometry of different brands of polypropylene membranes. The first reported applications that used only CSLM for membrane characterization [14, 15] were by Charcosset et al. who used CSLM to characterize microporous membrane morphologies and to obtain values of surface porosity and pore size. The conclusions of those studies were that CSLM gave some characteristics on membrane morphology that SEM, which views only surfaces, cannot provide. However, as also mentioned previously in this chapter, they pointed out low resolution for membrane characterization as the main drawback of CSLM. This restricts the use of CSLM to the characterization of microfiltration membranes if measurements on pore size and surface porosity have to be performed.

Thomas et al. [16] used CSLM to observe in situ the formation of polyamide membranes; and the measurements were used to study polymer precipitation kinetics. Turner and Cheng [17] applied CSLM and hydrophilic fluorescent probes of varying molecular weights to image the size distribution of poly(methacrylic acid) (PMAA) hydrogel domains in polydimethylsiloxane (PDMS)-PMAA interpenetrating polymer networks. The combination of CSLM with AFM, SEM and X-ray spectroscopy allowed characterization of the structure of stimuli-responsive polymeric composite membranes [18].

A more recent application of CSLM for membrane characterization has been presented by Green et al. [19]. In this work, CSLM and 3D image reconstructions were used to visualize the interconnecting pores that result from the bi-axial stretching of $CaCO_3$/polyolefin composite membranes. CSLM has also been used in the characterization of organic–inorganic microporous membranes [20]. The membranes were formed by dispersing small quantities of fine alumina particles in a solution of polyvinylidene fluoride and dimethylacetamide. Membrane characteristics such as surface morphology and structures, surface roughness, porous dispersion and the distribution of alumina particles were obtained from CSLM analysis using different image analysis software packages. Quantitative data on polycrystallinity of zeolite membranes have been obtained by Snyder et al. [21], based on CSLM images. They developed and applied new protocols for image analysis basically to rationally segment the CSLM images and used image calibration based on fluorescent intensity of the fluorophores to estimate the size of sub-resolution features. Finally they linked the polycrystallinity characteristics of the zeolite membranes with their separation performance.

Most of the studies reviewed in this section apply CSLM to morphological membrane characterization and relate the results with the membrane formation parameters and transport properties. Considering that the type, nature, preparation method and final use of the membrane can differ significantly, the CSLM analysis protocol has to be particularized for each application.

### 4.3.2 Membrane Fouling Characterization by CSLM

The increasing interest of using CSLM to characterize membranes has also coincided with a growing number of works that apply CSLM to characterize membrane processes. One of the first applications was the study of protein binding on ion exchangers and protein adsorption on ion exchange membranes. Ahmed and Pyle [22] used CSLM to study protein binding at the ion exchanger surface. These authors obtained pictures and intensity profiles that show that the two proteins studied (bovine serum albumin (BSA)–Bodipy FL and ovalbumin–Texas red) bind preferentially to the surface and not to the core of the spherical ion exchangers. They concluded that CSLM was an excellent tool for selecting between different ion exchangers for the optimization of protein purification processes. Reichert et al. [23] established a method for locating and identifying proteins bound to an ion exchange membrane. They observed the adsorption of fluorescent BSA and lysozyme onto a cation exchange membrane and the adsorption of formate dehydrogenase onto an anion exchange membrane. Even though the resolution available only allowed them to carry out a qualitative analysis, they pointed out that advances in confocal microscopy would increase the level of detail. Regarding the depth penetration of the analyses, they obtained images of about 25% of the thickness of commercially available adsorptive membranes. Reichert et al. [23], together with Ljunglöf and Thömmes [24] and Linden et al. [25], are pioneers in the use of CSLM to visualize and study protein/membrane interactions.

More recently, other applications of CSLM in the characterization of membrane processes have been reported, such as the studies of Hayama et al. [26, 27] who visualized the distribution of an endotoxin trapped in an endotoxin-blocking filtration dialysis membrane. In their first study [26], they found that endotoxins were practically rejected by the outer skin layer of a polyester–polymer alloy (PEPA) membrane, hence playing a vital role in endotoxin removal from the dialysate fluid. In a second work [27], CSLM contributed successfully to visualize the distribution of endotoxin fluorescently labeled in six kinds of dialysis membranes. They concluded that a dialysis membrane favorable for endotoxin blocking should have a double skin layer structure, composed of a hydrophilic inner skin layer and a completely hydrophobic void and outer skin layers.

The aforementioned studies using CSLM to characterize protein/membrane interactions have been the basis of the research carried out by our research group to characterize membrane fouling during the microfiltration of protein mixtures

(a)

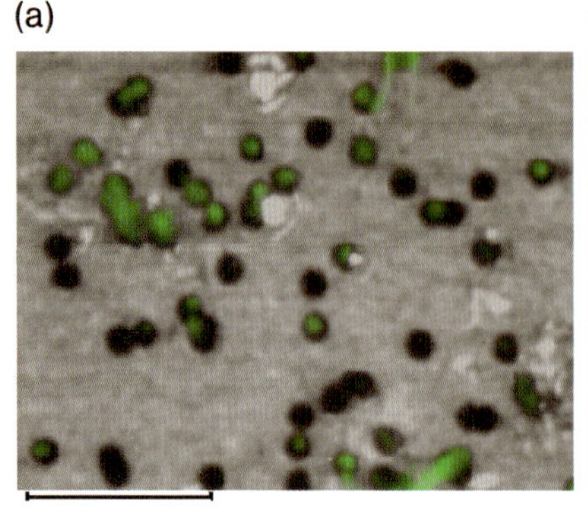

(b)

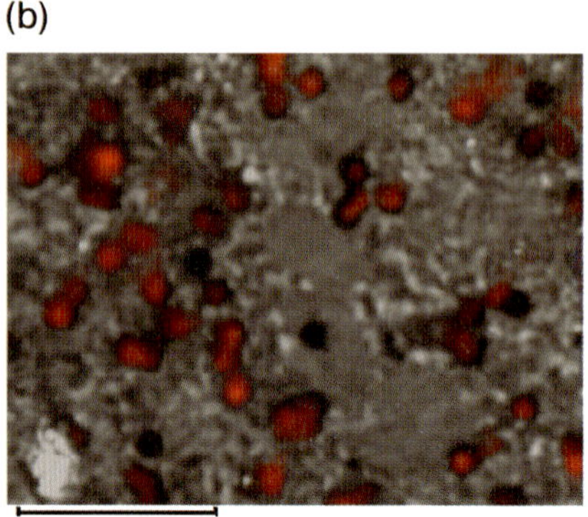

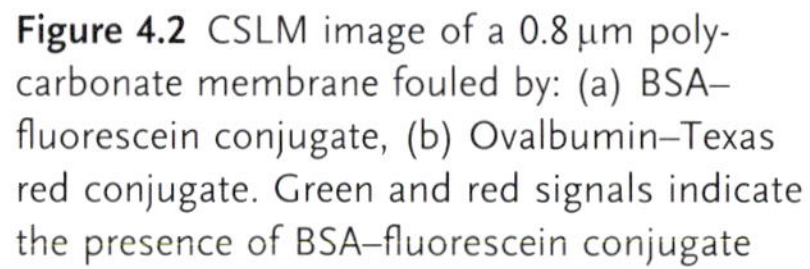

**Figure 4.2** CSLM image of a 0.8 μm polycarbonate membrane fouled by: (a) BSA–fluorescein conjugate, (b) Ovalbumin–Texas red conjugate. Green and red signals indicate the presence of BSA–fluorescein conjugate and ovalbumin–Texas red conjugate, respectively. The images correspond to a $z$ projection of a stack of 20 images along the axis perpendicular to image plane, inside the membrane. Scale bar = 5 μm.

and protein–polysaccharide solutions by CSLM [6, 28–29]. From the first studies on the use of CSLM to characterize membrane fouling, Ferrando et al. [6] concluded that CSLM allows the locating and identifying of fouling agents (proteins) on the membrane surface and inside the pores. Figure 4.2 presents a $z$ projection of a stack of 20 images, combining reflection images (membrane visualization) and fluorescence images (green signal corresponds to BSA–fluorescein, red signal corresponds to ovalbumin–Texas red). Quantitative information about the presence of proteins on the surface and inside the pores could be obtained by applying image analysis which allowed the calculating of the fraction of pore surface in which protein was detected ($P_s$), defined previously in Equation (4.1).

The simultaneous analysis of the results obtained through CSLM imaging combined with image analysis and macroscopic filtration data (permeate flux evolution) enabled to identify the prevailing fouling mechanism and also to identify the major fouling agent. The results obtained during cross-flow microfiltration, using a polycarbonate membrane, of single and binary 0.5 g $L^{-1}$ protein solutions of BSA and/or ovalbumin showed internal fouling as the prevailing mechanism. The ovalbumin–Texas red $P_s$ values were always higher than those of BSA–fluorescein isothiocyanate (FITC). The maximum $P_s$ due to ovalbumin–Texas red was about 100% at $z$ positions close to the membrane surface, whereas the maximum $P_s$ due to the BSA–fluorescein conjugate was between 34% and 70%. It was found, in agreement with previous results [30], that ovalbumin caused considerably more fouling than BSA, in both single and binary protein solutions, even though the molecular weight of ovalbumin is lower than the molecular weight of BSA.

Membrane fouling characterization of protein/dextran mixtures has also been studied [29]. The first step of this work was addressed to examine the influence of fluorescent labeling on membrane fouling. Permeate fluxes of BSA/dextran

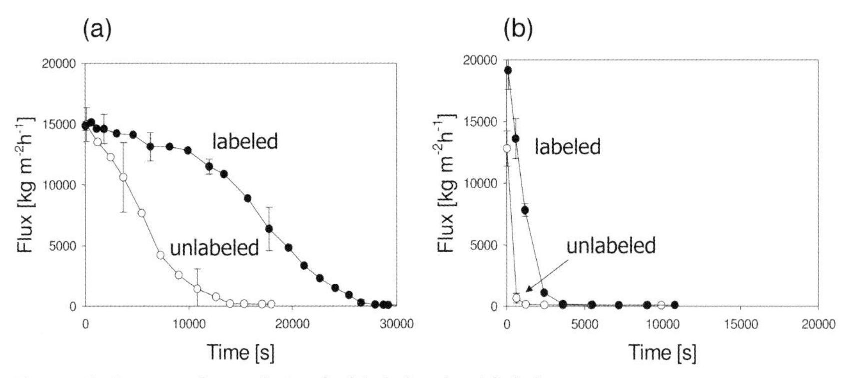


**Figure 4.3** Permeate flux evolution for labeled and unlabeled BSA/150 kDa dextran (0.5 g L$^{-1}$) solutions filtered using: (a) 0.45 μm mixed esters and (b) 0.40 μm polycarbonate membranes.

labeled and unlabeled solutions were obtained for 0.45 μm mixed esters and 0.40 μm polycarbonate membranes. The results (Figure 4.3) prove that labeled solutions had a reduced tendency to foul the membranes. However, for the polycarbonate membrane (which has a much lower surface porosity than the mixed esters membrane) both the labeled and unlabeled solutions presented a very rapid permeate flux decrease. Since the fluorescent tags used to label BSA and dextrans have aromatic rings, they add a slight negative charge to the molecules. This is thought to cause some repulsion among the molecules, which would eventually result in a decrease in fouling. The effect of fluorescent labeling can also be seen through the $P_S$ profiles, which show the amount of foulant detected in the surface and at different depths within the membrane. Figure 4.4 shows the $P_S$ values obtained for labeled BSA (BSA-FITC) after filtration of solutions containing the labeled protein together with labeled (dextran–tetramethyl rhodamine isothiocyanate; TRITC) or unlabeled 150 kDa and 70 kDa dextrans. As can be seen by the $P_S$ profiles, it is clear that, when the solution contains both labeled protein and labeled dextran, the $P_S$ values for BSA-FITC are smaller, indicating a lower coverage of the pores by the protein. However, when the solution contains unlabeled dextran, the $P_S$ profiles obtained for the protein show up to 95% coverage of the pores until a depth of 1 μm inside the membrane for the mixed esters membrane. For the polycarbonate membrane, the $P_S$ values are smaller in both situations, but the tendency is the same, with the highest $P_S$ values corresponding to the solution of labeled protein and unlabeled dextran. The main conclusion of this research was that the membrane fouling characterization by CSLM can be performed for membranes with low surface porosity, since in this situation fouling occurs very rapidly.

As the second step of this work, CSLM was used to determine how the dextran molecular weight affected the fouling pattern of protein/dextran solutions. It was observed that, independently of the fluorescent labeling, the solution with BSA

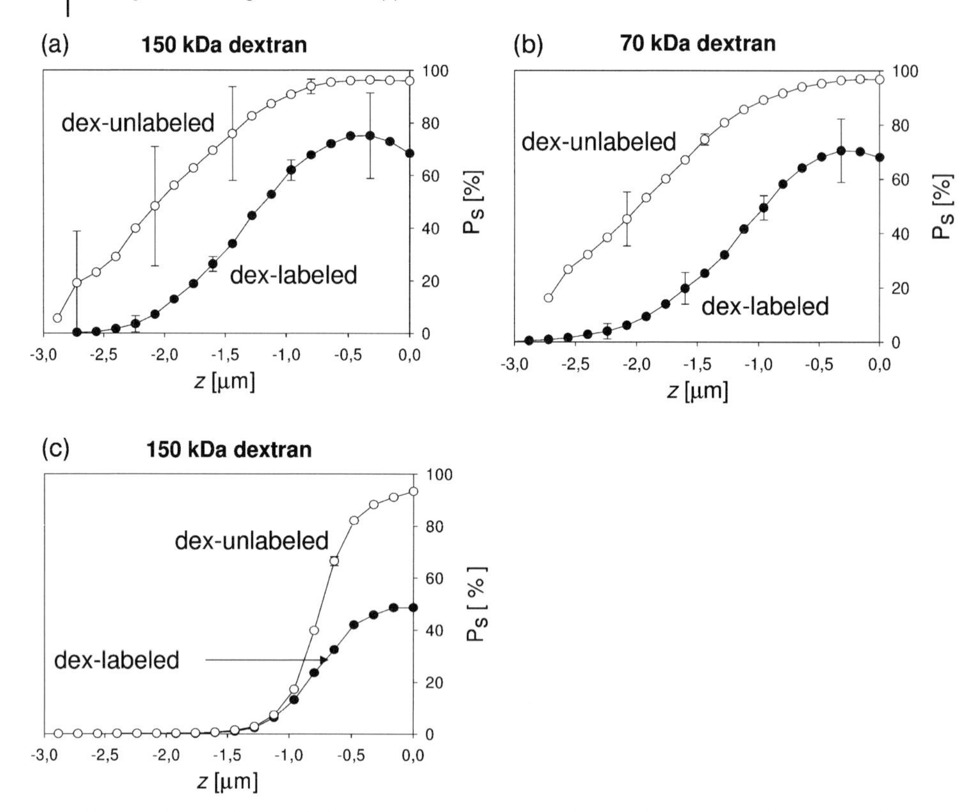


**Figure 4.4** $P_s$ profiles inside the membrane obtained for labeled BSA (BSA-FITC) after the filtration of solutions containing the labeled protein together with labeled (dextran-TRITC) and unlabeled dextran with a final concentration of 0.5 g $L^{-1}$: (a) 0.45 μm mixed esters membrane, BSA-FITC/150 kDa dextran labeled or unlabeled; (b) 0.45 μm mixed esters membrane, BSA-FITC/70 kDa dextran labeled or unlabeled; (c) 40 μm polycarbonate membrane BSA-FITC/150 kDa dextran labeled or unlabeled.

and the biggest dextran (150 kDa) needed longer times to reach the steady state and fouling was reduced.

CSLM together with macroscopic results has also been used to study membrane cleaning protocols. The studies have been carried out using 0.80 μm polycarbonate membranes and solutions of BSA-FITC with 70 kDa or 150 kDa dextran-TRITC. First of all, the effect of a rinsing step with water on the recovery of pure water permeate flux ($J_0$) was studied. The results show that independently of the molecular weight of the dextran, the rinsing step did not contribute to the recovery of $J_0$. Pure water fluxes obtained after the filtration of the BSA-FITC/dextran-TRITC solutions were always higher than the ones obtained after a 30 min rinsing step. This reduction in the permeate water flux after the rinsing step can be explained by considering the $P_s$ profiles for BSA-FITC. From Figure 4.5a, it is possible to observe that $P_s$ values for BSA-FITC slightly increased after the rinsing step,

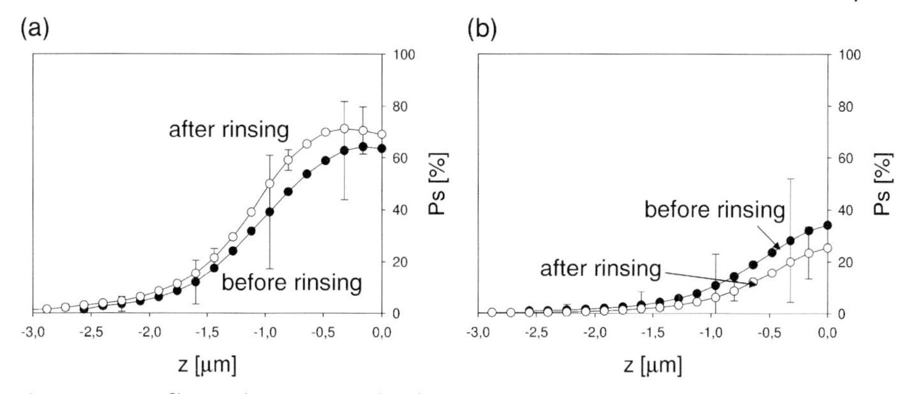


**Figure 4.5** $P_s$ profiles inside a 0.80 μm polycarbonate membrane after filtration of 0.5 g L$^{-1}$ solutions of BSA-FITC/dextran-RITC before and after a rinsing step: (a) BSA-FITC; (b) dextran-RITC.

meaning that more protein was detected on the surface and inside the pores after washing the membrane with water for 30 min. Therefore, as a result of the rinsing step, more protein is driven inside the membrane, which does not improve its water permeability. The $P_s$ profiles plotted in Figure 4.5 confirm that the protein has the main responsibility for the observed membrane fouling, since the rinsing step removed part of the dextran both from the surface and from inside the pores (Figure 4.5b).

Chemical cleaning protocols have also been studied using CSLM. An enzymatic cleaning agent (P3-Ultrasil 53) was used to clean polycarbonate membranes after the filtration of BSA-FITC/dextran-RITC solutions. The main parameters studied were the cleaning time and the cleaning agent concentration. Permeate fluxes and flux recovery (FREC) calculated according to Equation (4.2) are presented in Table 4.1:

$$\mathrm{FREC} = \frac{100^{*} J_{\mathrm{filtration}}}{J_0} \tag{4.2}$$

where $J_{\mathrm{filtration}}$ is the pure water flux after filtration/cleaning and $J_0$ is the pure water flux of the original non-used membrane.

As can be seen by the flux recovery values in Table 4.1, the optimum cleaning time (which results in maximum water flux recovery) for the smallest concentration of the chemical cleaning agent (0.1%) is 15 min for the 0.5 g L$^{-1}$ BSA-FITC/dextran-RITC solution. When the concentration of the cleaning agent is increased to 0.5%, the optimum cleaning time is reduced to 5 min. The existence of an optimum combination of cleaning agent concentration and time could be explained by how the cleaning agent interacts with the foulants. It is assumed that Ultrasil breaks the protein and protein aggregates into smaller parts and these parts are driven inside the membrane, leading eventually to some internal blocking of the pores. The permeate fluxes and the recovery values correlate with

**Table 4.1** Pure water flux after filtration of BSA/dextran 70 kDa 0.5 g $L^{-1}$ solutions and after chemical cleaning with P3-Ultrasil 53. WF: Water flux; F REC: flux recovery, calculated according to Equation (4.2).

| | WF after filtration ($L h^{-1} m^{-2}$) | WF after 5 min P3-US 53 ($L h^{-1} m^{-2}$) | F REC (%) | WF after 15 min P3-US 53 ($L h^{-1} m^{-2}$) | F REC (%) | WF after 30 min P3-US 53 ($L h^{-1} m^{-2}$) | F REC [%] |
|---|---|---|---|---|---|---|---|
| BSA-FITC/dex-RITC 70, US 53 0.1% | 1229 ± 346 | 5726 ± 1157 | 28 | 8019 ± 311 | 39 | 6333 ± 1004 | 31 |
| BSA-FITC/dex-RITC 70, US 53 0.5% | 1229 ± 346 | 6099 ± 277 | 30 | 5746 ± 225 | 28 | N/A | N/A |

N/A: Not available.

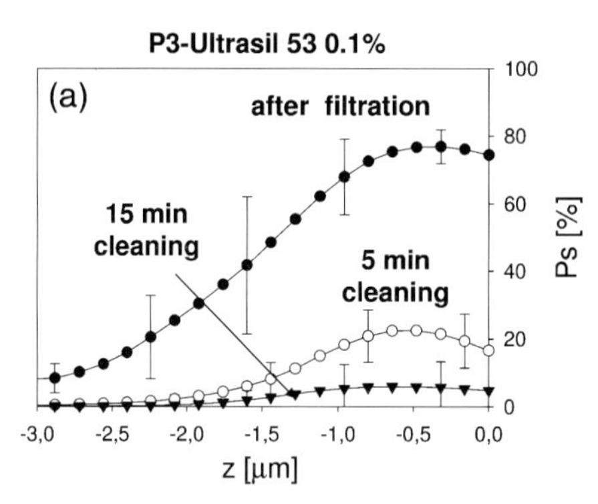


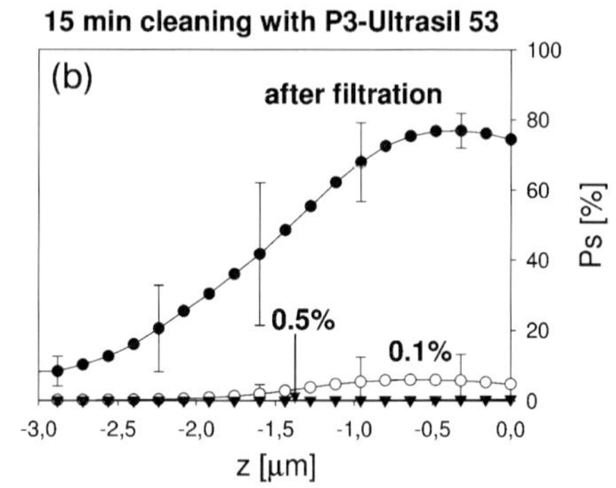


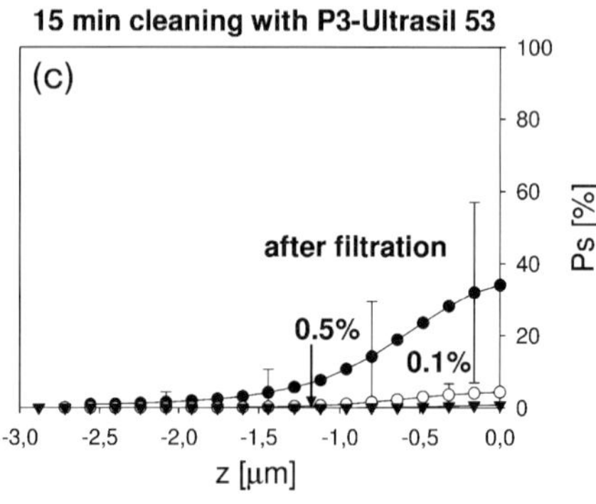


**Figure 4.6** $P_s$ profiles inside a 0.80 μm polycarbonate membrane after filtration of 0.5 g $L^{-1}$ solutions of BSA-FITC/dextran-RITC 70 kDa before and after chemical cleaning: (a) BSA-FITC profiles with P3-Ultrasil 53 0.1% at different cleaning times; (b) BSA-FITC profiles for 15 min and two P3-Ultrasil 53 concentrations; (c) dextran-RITC profiles for 15 min and two P3-Ultrasil 53 concentrations.

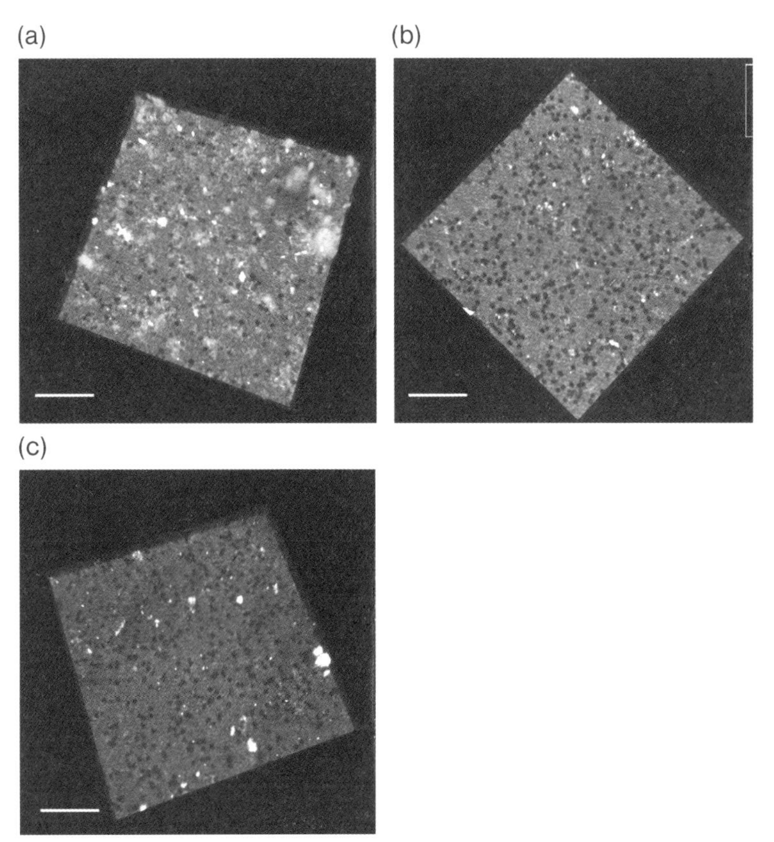


**Figure 4.7** 3D volumetric reconstructions of polycarbonate membranes after filtration of a 0.5 g $L^{-1}$ BSA-FITC/70 kDa dextran-RITC (a); after 5 min cleaning with 0.1% P3-Ultrasil 53 (b); after 15 min cleaning with 0.1% P3-Ultrasil 53 (c).

the $P_s$ profiles obtained for BSA-FITC and dextran-RITC (Figure 4.6). As a result of the chemical cleaning applied, the $P_s$ values obtained for both protein and dextran were reduced to values close to zero in the surface and inside the membrane pores, which shows the efficiency of the cleaning protocol. CSLM analysis can also provide more qualitative information using 3D reconstructions of the image stacks obtained. Figure 4.7 compares a 3D volumetric reconstruction of image stacks obtained after the filtration of a 0.5 g $L^{-1}$ solution of BSA-FITC/70 kDa dextran-RITC with the image stacks obtained after 5 min and 15 min cleaning with 0.1% P3-Ultrasil 53. As can be observed by the images, the membrane surface (gray signal) after filtration is partially covered by BSA-FITC (green signal) and dextran-RITC (red signal). It is also possible to observe that in some areas both the protein and the dextran are found together, which is shown by a yellow signal. After 5 min of chemical cleaning (Figure 4.7b) a noticeable reduction in the green and red signals can be observed on the membrane surface, indicating the removal of protein and dextran due to the cleaning applied. After 15 min of cleaning, the 3D volumetric reconstruction (Figure 4.7c) shows practically only gray signal, corresponding to the membrane, therefore meaning that most of the protein and dextran was washed away from the top of the membrane. These results agree with the $P_s$ profiles presented previously, but it should be noted that images correspond to a single field in one membrane, while the $P_s$ profiles presented throughout this chapter correspond to the averages of three membranes and three fields per membrane.

From the results presented in this section it is possible to conclude that CSLM can provide useful information for membrane fouling characterization, through quantitative and qualitative data obtained by image analysis, combined with macroscopic results such as permeate fluxes or resistances. It is also clear that any evaluation of the effect of fluorescent labeling on fouling has to be taken into account (as is also mentioned in Section 8.5.2). It has been shown that membranes with high surface porosity foul less rapidly if the solutions have previously been fluorescently labeled. However, for membranes with a lower surface porosity, which tend to foul more rapidly, the effect of fluorescent labeling is less noticeable.

### 4.3.3 Characterization of Bio-fouling Layers in Membrane Bioreactors by CSLM

There are some recent studies that used CSLM to characterize membrane fouling in MBRs [8–11]. Chen et al. [8] used a quadruple staining protocol combined with CSLM to study the contribution to membrane fouling made by cells (nucleic acids), proteins, and α- and β-polysaccharides. From CSLM images of the fouling layer located at 1 μm above the membrane they found a first deposit with a mixture of the four foulants. The β-D-glucopyranose polysaccharides formed a continuous layer, proteins were found aggregated into clusters, cells were distributed in dispersed form and α-D-glucopyranose polysaccharides appeared only in some spots. Above this initial deposit layer (about 10 μm above the membrane surface), they found a more developed layer, where proteins still appeared in clusters, with a few α-D-glucopyranose polysaccharides. In this upper layer, cells and β-D-glucopyranose polysaccharides were largely clustered, presenting larger channels than in the initial deposit, which were believed did not significantly impact on filtration resistance. From CSLM side-view images they obtained a fouling layer thickness of 6.8–17.8 μm with a very non-uniform spatial distribution. The initial deposit was found to have a thickness of 5 μm; and it consisted of a mixture of all four foulants. Above the initial deposit layer, the distribution of the foulants was found to be highly stratified, with an increase in proteins and nucleic acids at the top of the fouling layer. Regarding porosity of the cake, they found a low porosity ($<0.28$) of the fouling layer near the membrane surface. Moving away from the membrane, the average porosity increased reaching values close to 0.5 at 6 μm above the membrane surface. The box-counting fractal dimension value indicated that most foulants were packed on the membrane surface following similar mechanisms at all scales, with no particular preference.

Hwang et al. [9] used three fluorescent dyes to stain bacterial cells (nucleic acids), polysaccharides (α-mannose, α-glucose) and proteins to study the effect of a membrane fouling reducer (a cationic polymeric material) on flux enhancement in a submerged MBR. CSLM images enabled analysis of the 3D architecture of the cakes and the spatial distribution of the cellular and polymeric constituents. The information acquired from the CSLM images provided data on cake porosity, biovolume and a parameter associated with the growth of bacterial cells and clusters in each spatial direction [10]. The results showed that the cake formed with the presence of the membrane fouling reducer exhibited larger porosity and agglomerated particles than the cake without the membrane fouling reducer, which enhanced its permeability. Based on the analysis of CSLM images and comparing data on porosity, biovolume and average run length (parameter associated with the growth of bacterial cell clusters in each direction), the study concluded that, for the MBR with the fouling reducer, the loss of permeability with time was due to the accumulation of biomass caused by cake growth as well as the deposition of microbial flocs.

Yang et al. [7] used a six-fold staining protocol combined with CSLM to study the biofouling layer formed on top of a mixed cellulose ester membrane. From CSLM images scanned at 5 μm above, it was possible to visualize all the stained compounds (proteins, total and dead cells, α- and β-D-glucopyranose polysaccharides, lipids) distributed in clusters over the fouling layer. The β-D-glucopyranose polysaccharides, lipids and proteins appeared to be mostly bound to cell membranes, while the α-D-glucopyranose polysaccharides may have contributed to the free extracellular polymeric substances, which were not tightly bound to the cell membrane. The cake thickness found in this work is within the same range as the one reported by Chen et al. [8]; and the fouling layer was also highly stratified. CSLM has also been used to obtain biofilm porosity data in a more general study on the effect of dissolved oxygen on biofilm structure [11] in submerged membrane bioreactors. The results show a higher porosity of biofilms formed under conditions of high dissolved oxygen concentrations than for biofilms formed under low dissolved oxygen concentrations. These data were verified in terms of specific cake resistance by comparing experimentally measured values with the semi-theoretically computed values.

Meng et al. [31] centered their study on the analysis of cake formation during the operation of a submerged MBR treating synthetic municipal wastewater. The main purpose was to gather information to help define the optimum range of operational variables. They used CSLM together with particle size analysis, SEM, X-ray fluorescence, energy-diffusive X-ray analysis and FTIR spectroscopy. Regarding CSLM, they used two different probes, one to stain bacteria and another to stain polysaccharides with D-glucose or D-mannose. The main information gathered by CSLM allowed reporting that both bacteria (as clusters) and polysaccharides were present on the membrane surface. From CSLM images it seemed that the polysaccharides could be adsorbed onto the membrane surface and adhered on bacterial cells. The cake thickness calculated from CSLM was found to be about 100 μm.

### 4.3.4 Online Monitoring of Membrane Processes by CSLM

Online monitoring by CSLM of membrane processes has recently been reported in studies by Kromkamp et al. [32] and Brans et al. [33]. Both studies used the same experimental set-up, which consisted of a parallel plate device with a channel height of 0.25 mm and a channel width of 26 mm, provided by a transparent upper plate. The permeate flux through the membrane was monitored with an analytical balance and the particle deposition was monitored using CSLM. In the study by Kromkamp et al. [32], the main goal was to study the behavior of bidisperse suspensions in the shear-induced diffusive back-transport regime of microfiltration. The bidisperse suspensions consisted of mixtures of dyed polystyrene spherical particles of different diameter. The suspensions were filtered using a polyethersulfone membrane with a molecular weight cut-off of 100 kDa. CSLM images scanned at different depths of the membrane were used to monitor particle deposition in the transient flux regime; and the particle deposition on the membrane was calculated as the surface load (which corresponds to the volume of particles per surface area of membrane). The results on the measured deposition rate of the small particles and the flux showed that the back-transport flux of small particles did not depend on the composition of the bidisperse suspension and was equal to the back-transport flux of the monodisperse suspension with the same total particle volume fraction. The explanation proposed by the authors is that particle size segregation occurs in the feed flow, which leads to an enrichment with small particles of the suspension in the vicinity of the membrane wall. This conclusion is relevant when applied to fractionation processes, since fractionation of a polydisperse suspension can be achieved using a membrane with a pore size larger than the largest particle present.

Brans et al. [33] studied the deposition of particles on top of the membrane and inside the pores during dead-end and cross-flow filtration with polymer membranes and polymer microsieves. CSLM was used to investigate the interaction of fluorescent polystyrene microspheres and fluorescent sulfate microspheres with the membranes, by online monitoring of the particle transmission. The results showed a different fouling behavior, depending on the type of membrane. For the polymeric membrane, small particles were adsorbed at random places in its tortuous structure or became trapped in the pores, while for the microsieves, in-pore fouling and adhesion of the particles to the membrane pore edges was observed. Regarding permeate fluxes, it was found that the flux increased linearly with transmembrane pressure for the polymeric membrane, but for the microsieve the flux decreased after each pressure step. However, it has to be mentioned that the initial flux for the microsieve was one order of magnitude higher than for the polymeric membrane. The images obtained by CSLM allowed quantifying the percentage of the membrane surface area covered by particles at different times. For both the polymeric membrane and the microsieves it was found that, during the first stages of filtration, the deposited particles showed a linear increase with the cumulative permeate volume. For the polymeric membrane, this increase was kept until the end of filtration, while for the microsieves a maximum coverage of 10% was reached qat 1.5 kPa; and it did not increase at higher pressures. The authors concluded that, for fractionation applications, a suitable combination of transmembrane pressure and cross-flow velocity should be applied, ensuring full particle transmission and keeping the larger particles away from the membrane. They also reported that, for high demanding separations, microsieves seem to be more suitable.

## 4.4 Evaluating Limits and Prospects of CSLM in the Characterization of Membranes and Membrane Processes

As seen in previous sections of this chapter, CSLM can provide very interesting insights for the characterization of membranes and membrane processes. The applications reviewed herein show that CSLM can be a very powerful tool as a non-destructive in situ or ex situ characterization technique. Images obtained by CSLM allow the visualization of different membrane structures and foulant deposition inside the pores (internal fouling) and on the membrane surface (cake). Image analysis, using any of the software available, has provided data on pore size, membrane and cake thickness, roughness, size of defects, porosity of cake, etc. CSLM has been proven to be very useful in the characterization of membrane processes, helping to locate and identify foulants, to understand about the formation and structure of a cake and even to perform online monitoring of membrane processes. Although this chapter is mainly devoted to applications of CSLM to characterize membranes and pressure-driven membrane processes, the increasing interest in membrane technologies applied to other fields, such as membrane emulsification, offers a good opportunity for new uses of CSLM. All the research performed up to now offers a solid background for new applications and a starting point to overcome the current limits of the technique.

As mentioned previously in this chapter, the main limits of CSLM are related to its resolution, which is primarily a function of the numerical aperture of the optical system and the wavelength of the light. However, sample preparation is minimal allowing the performance of in situ and online experiments. The current most important drawbacks preventing a much wider application of in situ monitoring of membrane processes by CSLM are related to commercially available objectives. They should combine values of working distances that allow working with miniaturized filtration modules and high NA. However, considering that most confocal scanning laser microscopes take approximately 1 s to acquire a single optical section, the monitoring of very fast processes is still not possible.

Regarding sample preparation, it is extremely important to consider the interaction between the fluorophore and all the components present in the system (membrane, inlet stream, outlet stream), when using fluorescence mode. Therefore, it is advisable to find staining protocols to minimize the amount of fluorophore in the target component and to validate experimental results obtained with and without fluorophores.

In addition, it is important to consider what kind of information can be obtained from CSLM images. Visualization of the membrane/cake/foulants can be obtained through 3D reconstructions performed with the available commercial software. The 3D reconstructions offer a view of a single field in the analyzed sample and therefore, unless the sample is very homogeneous, the 3D reconstructions should be considered to provide qualitative or local information. To obtain quantitative information statistically representative of the whole analyzed specimen, a suitable sampling design has to be performed. Besides, it should be considered that the kind of information obtained is very valuable mainly if it is processed using software that enables the characterization of 3D structures (pore connectivity, tortuosity, porosity).

There is growing interest in the use of CSLM in the characterization of membranes and membrane processes, shown by the increasing number of publications in the past few years. CSLM is becoming more and more a complementary characterization method used together with the most accepted microscopic techniques that will provide very useful information to improve membrane manufacturing and process performance.

## Acknowledgments

The authors would like to acknowledge financial support from Spanish Ministry of Science and Education, project CTQ2004-01369.

## References

1 J.C. Chen, Q.L. Li, M. Elimelech, *Adv. Colloid Interfaces* **2004**, *107*, 83–108.
2 P. Le-Clech, Y. Marselina, Y. Ye, R.M. Stuetz, *J. Membrane Sci.* **2007**, *290*, 36–45.
3 J.B. Pawley, *Handbook of Biological Confocal Microscopy*, 3rd edition. Springer, New York, **2006**.
4 T. Wilson, *Confocal Microscopy*, Academic, San Diego, **1990**.
5 R.P. Haugland, *Handbook of Fluorescent Probes and Research Products*, 9th edition. Molecular Molecular Probes, Eugene, **2002**.
6 M. Ferrando, A. Rŏzek, M. Zator, F. López, C. Güell, *J. Membrane Sci.* **2005**, *250*, 283–293.
7 Z. Yang, X.F. Peng, M.Y. Chen, D.J. Lee, J.Y. Lai, *J. Membrane Sci.* **2007**, *287*, 280–286.
8 M.Y. Chen, D.J. Lee, Z. Yang, X.F. Peng, J.Y. Lai, *Environ. Sci. Technol.* **2006**, *40*, 6642–6646.

9 B.K. Hwang, W.N. Lee, P.K. Park, C.H. Lee, I.S. Chang, *J. Membrane Sci.* **2007**, *288*, 149–156.

10 W.N. Lee, I.S. Chang, B.K. Hwang, P.K. Park, C.H. Lee, *X. Process Biochem.* **2007**, *42*, 655–661.

11 Y.L. Jin, W.N. Lee, C.H. Lee, I.S. Chang, X. Huang, T. Swaminathan, *Water Res.* **2006**, *40*, 2829–2836.

12 H. Beyenal, C. Donovan, Z. Lewandowski, G. Harkin, *J. Microbiol. Methods* **2004**, *59*, 395–413.

13 R. van den Berg, D. Schulze, J.A. Bolt-Westerhoff, F. de Jong, D.N. Reinhoult, D. Velinova, L. Buitenhuis, *J. Phys. Chem.* **1995**, *99*, 7760–7765.

14 C. Charcosset, J.C. Bernengo, *J. Membrane Sci.* **2000**, *168*, 53–62.

15 C. Charcosset, A. Cherfi, J.C. Bernengo, *Chem. Eng. Sci.* **2000**, *55*, 5351–5358.

16 J.L. Thomas, M. Olzog, C. Drake, C.H. Shilh, C.C. Gryte, *Polymer* **2002**, *43*, 4153–4157.

17 J.S. Turner, Y.L. Cheng, *Macromolecules* **2003**, *36*, 1962–1966.

18 K. Zhang, H. Huang, G. Yang, J. Shaw, C. Yip, X.Y. Wu, *Biomacromolecules* **2004**, *5*, 1248–1255.

19 D.L. Green, L. McAmish, A.V. McCormick, *J. Membrane Sci.* **2006**, *279*, 100–110.

20 L. Yan, L. Hui, S. Xianda, L. Jianghong, Y. Shuili, *Desalination* **2007**, *217*, 203–211.

21 M.A. Snyder, D.G. Vlachos, V. Nikolakis, *J. Membrane Sci.* **2007**, *290*, 1–18.

22 M. Ahmed, D.L. Pyle, *J. Chem. Technol. Biotechnol.* **1999**, *74*, 193–198.

23 U. Reichert, T. Linden, G. Belfort, M.R. Kula, J. Thömmes, *J. Membrane Sci.* **2002**, *199*, 161–166.

24 A. Ljunglöf, J. Thömmes, *J. Chromatogr. A* **1998**, *813*, 387–395.

25 T. Linden, A. Ljunglöf, M.-R. Kula, J. Thömmes, *Biotechnol. Bioeng.* **1999**, *65*, 622–628.

26 M. Hyama, T. Miyasaka, S. Mochizuki, H. Asahara, K. Tsujioka, F. Kohori, K. Sakai, Y. Jinbo, M. Yoshida, *J. Membrane Sci.* **2002**, *210*, 45–53.

27 M. Hayama, T. Miyasaka, S. Mochizuki, H. Asahara, K. Yamamoto, F. Kohori, K. Tsujioka, K. Sakai, *J. Membrane Sci.* **2003**, *219*, 15–25.

28 M. Zator, M. Ferrando, F. López, C. Güell, *Desalination* **2006**, *200*, 203–204.

29 M. Zator, M. Ferrando, F. López, C. Güell, *J. Membrane Sci.* **2007**, *301*, 57–66.

30 C. Güell, R.H. Davis, *J. Membrane Sci.* **1996**, *119*, 269–284.

31 F. Meng, H. Zhang, F. Yang, L. Liu, *Environ. Sci. Technol.* **2007**, *41*, 4065–4070.

32 J. Kromkamp, F. Faber, K. Schroën, R. Boom, *J. Membrane Sci.* **2006**, *268*, 189–197.

33 G. Brans, A. van Dinther, B. Odum, C.G.P.H. Schroën, R.M. Boom, *J. Membrane Sci.* **2007**, *290*, 230–240.

# 5
# Scanning Probe Microscopy Techniques in the Investigation of Homogeneous and Heterogeneous Dense Membranes: the Case for Gas Separtion Membranes

*Antonio Hernández, Petro Prádanos, Laura Palacio, Roberto Recio, Ángel Marcos-Fernández, and Ángel Emilio Lozano*

## 5.1
## Introduction

Membrane processes are being applied in many fields. It is clear that membranes developed for such a wide variety of processes must necessarily have very different properties and structure. Membranes could be classified according to different aspects [1].

In any case, morphological and structural aspects are crucial as far as they determine the barrier properties of a membrane along with their physicochemical properties. According to their structure membranes can be classified as porous membranes or as nonporous or dense ones. A porous membrane presenting equal pores should be called homoporous; actually all membranes have pores with a more or less wide range, so being heteroporous.

Really, the concept of a pore could be generalized to include all interstices in the materials of a solid film. According to this, strictly speaking all membranes should be porous. Even if we accept that pores are only supra-molecular, more or less interconnected paths, with an estimable contribution to the transport through the membrane, many membranes usually considered as dense, as for example gas separation or reverse osmosis membranes, should be reinterpreted as porous as far as microvoids in amorphous or semicristalline polymeric films play a relevant role in permeability.

In any case, what can be understood as dense membranes depends on the scale to which they are studied. In dense membranes, the performance (permeability, selectivity) is determined by the intrinsic properties of the material by solution–diffusion through the molecular interstices in the membrane material. When this is not the main mechanism, the membranes can be called porous. In this case, the selectivity is mainly determined by the dimension of the pores. Actually it can also

*Monitoring and Visualizing Membrane-Based Processes*
Edited by Carme Güell, Montserrat Ferrando, and Francisco López

ISBN: 978-3-527-32006-6

play a key role in selectivity through its charge and dielectric constant when the solute is charged. When this is not the case the material also has very important effects through phenomena such as adsorption and chemical stability under the conditions of actual application and membrane cleaning [2].

The processes where porous membranes find their main applications are pressure-driven ones, such as microfiltration, ultrafiltration and nanofiltration. These processes are also especially interesting due to their wide range of practical applications. They can be used for the processing of fine particles, colloids and biological materials such as protein precipitates and microorganisms [3]. Membranes used are commonly polymeric materials but innovative development has been made in the fields of ceramic and inorganic membranes.

However, membranes can be symmetric or asymmetric. Many porous or dense membranes are asymmetric and have one or several more porous supporting layers and a thin skin layer which, in fact, gives selectivity. If these two layers are made of different materials, the membrane is a composite one. On some occasions, dense membranes have inclusions of other materials; these are, of course, also composite membranes. In the case of gas separation membranes it has became usual to include inorganic charges in a polymeric membrane to get what is called a mixed matrix composite membrane.

The prediction of the process performances of these membranes for industrially relevant separations ultimately rests on the development and application of effective procedures for membrane characterization.

Most manufacturers of porous membranes describe their products by giving a single pore size or a molecular weight cut-off value. These data are usually obtained by measuring the rejection of macromolecules or particles of increasing hydrodynamic diameter or molecular weight. Of course, such a single value does not determine totally their structure nor their separation properties. In any case, the molecular weight cut-off can be accepted as a useful datum only for preliminary selection. However, all membranes must be assumed to contain size-distributed pores.

There are several independent methods that can be used to study pore statistics [2, 4–7]. The major ones are summarily presented below [8].

1. Electron microscopy uses several available electronic microscopy techniques to view the top or cross-sections of membranes: scanning electron microscopy (SEM), transmission electron microscopy (TEM), field effect scanning electron microscopy (FESEM), etc. The corresponding images are analyzed to obtain pore size distributions [9] (see Chapter 3).

2. Scanning probe microscopy (SPM) and specifically atomic force microscopy (AFM) are techniques allowing a study of the surfaces of nonconducting materials, down to the scale of nanometers [10]. The main advantage of such techniques is that no previous preparation of the sample is needed [9]. Although it is a relatively novel technique, application to membranes, both biological and synthetic, is growing rapidly [11, 12] (see Chapter 6).

3. Bubble pressure methods [13] are based on the measurement of the pressure necessary to force a fluid to pass through the pores previously filled with

another fluid. These include both gas–liquid and liquid–liquid displacement techniques [14–18]. In the pure bubble point method, air is pushed until its first appearance on the permeate side of a previously swollen membrane. This method has been frequently used for an estimation of a representative pore size of many commercial membranes.

4. Mercury porosimetry is a method based on the same principles as the bubble pressure method; but now mercury (a nonwetting fluid) is used to fill a dry membrane [19].

5. Adsorption–desorption methods [20] allow analysis of pore size distribution. The technique is based on the Kelvin equation, which relates the reduced vapor pressure of a liquid with a curved interface to the equilibrium vapor pressure of the same liquid in bulk with a plane liquid–vapor interface [21]. The BET adsorption theory is frequently applied to gas adsorption in order to obtain specific surface areas.

6. Permporometry is based on the controlled blocking of pores by condensation of the vapor present as a component of a gas mixture, with the simultaneous measurement of the flux of the noncondensable gas in the mixture [22]. If the Kelvin equation is used, the pore size distribution is obtained.
7. Thermoporometry, first suggested by Brun et al. [23], is based on the fact that the solidification point of the vapor condensed in the pores is a function of the interface curvature. By using a differential scanning calorimeter (DSC), the phase transition can be easily monitored and the pore size distribution calculated.

8. The solute retention test, where rejection is measured under more or less standardized conditions for various solutes of increasing molecular weights or hydrodynamic sizes, allows an evaluation of the pore size distribution [24, 25].

Other techniques can also be used for determining pores and pore sizes in filters such as, for example, NMR measurements [26], wide angle X-ray diffraction [27], small angle X-ray scattering [28] and electrical conductance [29]. These are methods that have been used to get pore sizes of different membranes.

When dealing with the free volume or interstices in dense membranes, small angle X-ray spectroscopy (SAXS) is very helpful, along with positron annihilation lifetime spectroscopy (PALS) [30], or even ellipsometry [31, 32], which measures density locally giving a first insight on the space free volume distribution in depth.

Of course there are many other experimental facilities that can give important physical and chemical parameters, other than pore size distributions, of porous or dense materials, such as electron spectroscopy for chemical analysis (ESCA), Fourier transform infrared spectroscopy (FTIR) and contact angle determinations. Also thermal or thermomechanical analysis, including differential scanning calorimetry (DSC) and dynamic mechanical analysis (DMA) customarily used to characterize polymers in materials science, can be very valuable when applied to dense or porous membranes.

A description of these techniques is not our objective here. It is clear, nevertheless, that they can be placed in two main groups: some of these methods (the ones developed for the characterization of general porous materials) directly obtain morphological properties, while others give parameters related to membrane permeation (those designed specifically to characterize membrane materials) [33].

Appropriate elucidation of structure is not only relevant to describe sieving effects but also to study the interactions of the membrane material with the feed to be treated, as far as the corresponding interfaces are placed inside the pores as well as on the membrane external surface. Thus, electrically determined membrane properties act on the solutes inside the pores and transport is affected by these properties (zeta potential, surface charges, etc.) in such a way that makes necessary a detailed knowledge of the geometry of both the inner and external surfaces of the membrane to adequately correlate interactions with their effects on flux (see Chapter 9).

Actually most of the techniques outlined above can also be useful to characterize dense membranes. In effect, solid dense membranes are usually fixed on a porous support whose structure can penetrate into the dense layer, leading to defects in the dense material and corrections to the expected flow. Even if these defects are not present, transport through the support material can play a significant role in process features.

Those techniques not referring specifically to porous morphology are especially relevant when concerning dense membranes. For example microscopy, including SEM, TEM, FESEM and especially SPM, can give information on the characteristics of nonporous surfaces of dense membranes and thus can be of great help.

Among the dense membranes, those designed to perform gas separations are very important. In effect, there is a big market for gas separation through membranes. Of course, the material structure and, thus, synthesis processes must be considerably improved to reach optimization of selectivity and permeability for target gases. Long-term research has been dedicated to this aim. The fractional free volume in the most restrictive layer, usually that just on the active surface layer, plays a key role in the selectivity and permeability of dense membranes in gas separation. As a consequence it is very important to characterise the membrane surface.

Our objective here is to describe how techniques of microscopy (namely scanning probe microscopy; SPM) can help in the investigation of homogeneous and heterogeneous dense membranes. In order to show their potentialities, we focus specifically on gas separation membranes.

## 5.2 Microscopic Techniques

Visual inspection of microscopical structure is an invaluable tool for a deep knowledge of filters. This is why the visualizing techniques were very early used to characterize them. Nevertheless, as the developed filters included relevant structures with submicron sizes, optical microscopy was no longer useful to achieve a real picture of the membrane topography, given that the resolution is limited by the light diffraction pattern. Only the development of non-optical microscopic techniques made it possible to solve this problem. Originally, electron microscopy [34], and afterwards scanning probe microscopy [10], were shown to be priceless tools for the characterization of membranes, so that we can now have information of membrane surfaces covering the full range of membranes.

### 5.2.1 Electron Microscopy: SEM and TEM

When high-energy electrons collide with a solid, the interactions between the solid material and the electron beam can be used to identify the specimen and the elements present in it, but also to characterize physically the solid surfaces and bulk, including the holes, pores or inclusions appearing through it.

Transmission electron microscopy (TEM) operates by flooding the sample with an electron beam, most commonly at 100–200 keV and detecting the image generated by both elastically and inelastically scattered electrons passing through the sample. TEM operates in the magnification range from $600\times$ to $10^6\times$.

Similarly, a fine beam of medium-energy electrons (5–50 keV) causes several interactions with the material, secondary electrons being used in the SEM technique. SEM equipment is able to achieve magnifications ranging from $20\times$ to $10^5\times$, giving images marked by a great depth of field and thus leading to considerable information about the surface texture of the target.

The main problems and difficulties of the microscopic observation by both TEM and SEM are how to prepare a membrane sample without any artefact. The first step of the preparation is a careful drying of the sample, and in order to avoid collapse of the original structure, the freeze-dry technique using liquid nitrogen or the critical-point drying method with carbon dioxide is usually employed.

In order to observe cross-sections by SEM, the dried membrane must first be fractured at the liquid nitrogen temperature and then fixed perpendicularly onto the sample holder. Usually, samples are afterwards covered by a thin metallic layer (normally a gold film of some hundreds of angstroms), increasing the production of secondary electrons and therefore improving the image contrast [35]. For TEM observation, a more complicated procedure is required. The dried sample is usually firstly embedded and then cut by using a microtome. Of course the embedding medium has to be adequate in order to avoid any influence on the membrane. The section must be thin enough for electrons to penetrate, that is, less than 50 nm. If only the surface is being analyzed, a replica technique can be used by coating the membrane with a carbon film and then removing the membrane material (by dissolving it for example) and analyzing this replica [36].

When high electron beam energy is applied, the maximum resolution of TEM is $\sim$0.3–0.5 nm, while SEM has 10 times greater resolution. In these conditions, the sample surface can be seriously damaged, especially when dealing with polymeric membranes, which makes observation difficult. In the early 1980s field emission scanning electron miscroscopy (FESEM) was developed and used to observe the surface pores of ultrafiltration membranes [37]. Nowadays FESEM achieves very high resolution (to 0.7 nm), even at low beam energy, with accelerating voltages in the range 1.5–4.0 kV.

### 5.2.2
### Scanning Probe Microscopy: STM and AFM

Atomic force microscopy (AFM) covers a range of recently developed techniques that can be used to characterize membranes. Actually it presents very high possibilities of development and application in the field of microscopic observation and characterization of various surfaces. A very small tip scans the surface and moves vertically according to its interaction with the sample, similar to the technique used in scanning tunneling microscopy (STM).

The two techniques that can be called scanning probe microscopy differ in the method they use to detect interactions. In STM the tip is so close to the sample (both being electrically conducting) that it allows a current to flow by tunnel effect and the sample or tip moves to keep this current constant. In AFM the tip is placed on a cantilever whose deflection can be detected by the reflection of a laser beam appropriately focused. This allows the analysis of nonconducting materials, which makes the method more convenient to study membrane materials [38, 39].

Several operation procedures can be used in AFM:

1. Contact mode AFM. This measures the sample topography by sliding the probe tip across the sample surface. The tip–sample distance is maintained within the repulsive range of the atomic forces.
2. Noncontact mode AFM. The topography of the sample is measured by sensing the Van der Waals attractive forces between the surface and the probe tip held above the surface. Of course, worse resolution than contact mode is achieved. Nevertheless, the risk of sample damage is avoided or minimized.
3. Tapping or intermittent contact mode AFM. This is a variation of the contact mode and operationally it is similar to noncontact mode; thus it features the best characteristics of both methods. The cantilever is oscillated at its resonant frequency with high amplitude (over 100 nm) allowing it to touch the sample during the oscillation. The topographical information is obtained from the register of the vertical displacements of the set-point needed to keep the oscillation amplitude constant. This method maintains the high resolutions achieved in contact mode, but minimizes surface damage, as far as it eliminates the lateral friction forces.

These previously commented techniques give an account of the sample topography. Other information on the surface can be acquired by AFM, such as for example heterogeneities that can be detected attending to differences in adhesion and in elasticity or stiffness. Techniques that spot these differences are:

1. Phase contrast. In the tapping mode technique the tip of the probe is oscillating near the surface of the sample at a given frequency and amplitude. When the tip approaches the surface there is a shift in the phase of the oscillating cantilever. If in this situation there is a variation in the interaction, when scanning the surface, the corresponding phase shift changes in accordance. This modification in the interaction, and thus in the phase shift, can be caused by surface topographic accidents and by the appearance of interaction forces, that could be for example magnetic or electric. They can also be caused by a change in the surface material due, in this case, to the viscoeleastic properties of the surface.
2. Force modulation. In this technique the cantilever is forced to oscillate at a very low frequency (far below the cantilever natural frequency) when it is close to the surface. In this way the amplitude of the of oscillation decreases when the constant energy supplied has to be spent in overcoming the strong interaction appearing when the surface is somehow sticky, while it gives big amplitudes when the surface is hard enough.
3. Lateral force microscopy. In this case, the frictional force between the probe tip and the sample surface is analyzed. The sample is swept in such a way that the cantilever tends to bend, with a this torsion force which is proportional to the friction force. Of course this frictional force depends on the material and reveals easily the inhomogeneities on the membrane surface.

Other properties of the surfaces can be analyzed by measuring different forces between sample and tip. Magnetic and electric force microscopy (MFM, EFM) both measure magnetic (or electric) force gradient distribution above the sample surface. Surface potential microscopy measures differences in local surface potential across the sample surface. Finally, electrochemical microscopy measures the surface structure and properties of conducting materials immersed in electrolyte solutions with or without potential control.

In many of these techniques, an appropriate treatment of the measured forces is necessary to eliminate the contribution of the topographical images. This is necessary to minimize or to eliminate the influence of the pure topographical information, to isolate the information concerning the specific interaction studied. This method consists of registering the topographical information and reproducing the scan at a constant distance on the surface, now revealing only the deflections caused by the interaction being studied.

### 5.2.3 Computerized Image Analysis

Image analysis can be carried out by means of a plethora of software packages, some of which are supplied by the main optical or electronic microscopies manufacturers (Jeol, Leica, Karl Zeiss, Nikkon, etc.) as a complement to their devices.

In all cases each photograph is first digitized by assigning to each pixel a gray level ranging from 0 (black) to 255 (white). Then, a clearfield equalization is made to each image field to eliminate parasite changes in gray levels due to uneven illumination. Obviously, a perfect clearfield equalization should require a blank image with a perfect flat sample of the same material equally treated and acquired in the same way. In fact this is impossible and even inconvenient, as far as uneven illumination can be due to the roughness of the sample itself. What can be done is to use what is called pseudoclearfield equalization by dividing the original image into a convenient number of rectangles. Then we can assign to all pixels the intensity such that 95% of the original pixels have a lower intensity. Finally these rectangles are placed together by linear interpolation from rectangle to rectangle and substracted from the original picture.

Once illumination effects are eliminated, the image gray spectrum is spanned to get the maximum contrast and definition. Then the images are redefined according to an assigned gray threshold level under which every pixel is assigned to 1 and the rest to 0. The resulting binary picture is improved by scrapping isolated pixels, in such a way that all the remaining 1 s in the matrix are assumed to belong to a pore, an inclusion or a segregated phase of the material. Finally the borders are smoothed in order to reduce the influence of the finite size of pixels and low definition.

Of course a correct selection of threshold gray level is fundamental to performing a correct analysis of accurate assigned pores. Customarily the gray spectrum is analyzed and the threshold placed centered in the peak to peak valley of the almost bimodal distributions obtained. Unfortunately sometimes the spectra are so flat that this technique is only of relative help to make a correct threshold election [40]. In any case, eye inspection facilitates the process of selection of several reasonable threshold candidates whose outcomes are conveniently averaged.

### 5.2.4 Roughness and Fractal Dimension

Average roughness ($R_q$) can be directly obtained from AFM images for different explored areas using the definition expressed by the following equation:

$$R_q = \sqrt{\frac{1}{n}\sum_{i=0}^{n}(Z_i - Z_m)^2} \tag{5.1}$$

where $Z_m$ is the mean value of the tip height in each point of the image ($Z_i$) over a reference baseline ($Z$) [41].

The roughness versus scanned area pattern is characteristic of a given material and defines a fractal dimension, $d_{fr}$, which is evaluated as: $d_{fr} = 3 - \alpha$ where $\alpha$ is the so-called roughness exponent that can be calculated as the slope of roughness versus scan size in a double log plot [42].

This fractal dimension experimentally gives a good represention of how roughness increases with scan size, with accurate fittings to experimental data [43]. Thus, a value of fractal dimension close to 2.0 means that the membrane has a reasonably flat surface; while, if $d_{fr}$ is close to 3.0, the membrane surface has a 3D-like interface.

## 5.3
## Gas Separation Membranes

In the competition with other more traditional processes of gas separation, such as cryogenic distillation or pressure swing adsorption, polymeric membranes have the advantage of their simplicity, continuous working ability, low energy consumption and low capital costs [44].

A high selectivity leads to a high purity of products and allows a reduction in the number of operation steps. A high permeability involves a high process velocity and a lower membrane area. Nevertheless, high selectivity is normally obtained with low permeability and vice versa [45]. Of course, it should be convenient to reach simultaneously high permeabilities and selectivities or at least to increase one of these parameters without decreasing the other.

Permeability of the $i$-th gas can be obtained as the product of its diffusivity ($D_i$) and its solubility ($S_i$):

$$P_i = D_i S_i \tag{5.2}$$

with selectivity for a given pair of gases:

$$\alpha_{ij} = \mathrm{P_i}/\mathrm{P_j} \tag{5.3}$$

Unfortunately, the Robeson bound [45] gives an upper limit for the selectivity versus permeability correlation that can be reached by increasing in chain stiffness and/or fractional free volume of a polymeric material. This limit is based on empirical data and seems to underestimate slightly the predictions of a relatively simple model by Freeman [46, 47].

The main physicochemical characteristics that rule gas permeability and thus selectivity of polymers are:

- The intersegmental spacing and the corresponding polymer free volume.
- The mobility of the polymer chains that can be correlated with the glass transition temperature, $T_g$.
- The interaction between the polymer and the penetrant gas or its solubility [48].

It seems clear that there is a direct correlation between gas diffusivity in the polymer and its free volume [49]. In turn, solubility should increase with the inner surface of the voids where the gas can enter. Thus it seems that selectivity should be caused by differences in the free volume accessible to each gas along with the differences of gas-surface interactions. Furthermore the restricting fraction of free volume should be very relevant in building up selectivity. Thus, the size and characteristics of the narrower necks connecting bigger voids are very important.

In summary, both selectivity and permeability characteristics can be interpreted in terms of a series of parameters that are substantially determined or basically controlled by the characteristics of the free volume of the polymer [50].

Attending to the membrane manufacture procedures, these aspects should be more restrictive at the membrane surface, where free volume should have an influence in the roughness of the resulting surface of the membrane. We give two examples below on how an analysis of the resulting fractal dimension to increase gas permeability-selectivity performances.

Another possible way to improve the performance of a membrane relies on increasing flux for a given permeability and selectivity by simply reducing the thickness of the active layer. An example on how AFM can help in this research program is worthy of comment here. This is the procedure to make Matrimid asymmetric membranes.

Finally, other possible way to produce useful membranes is to mix a certain material (usually a polymer with good mechanical and chemical properties) with another material (an activated carbon for example) with better selectivity-permeability performances but which it is not advisable to use alone in the actual gas environment for a given application or which is difficult to obtain in a resistant film layer. This constitutes what is called a mixed matrix composite membrane (MMCM).

For a system where the filler is very close to a disposition of dispersed spheres and there are no significant changes in permeability caused by the interaction of the filler with the continuous polymeric phase, the theory of Maxwell can be applied to give:

$$P_{i,\text{mem}} = P_{i,c}\left[\frac{P_{i,d} + 2P_{i,c} - 2\phi(P_{i,c} - P_{i,d})}{P_{i,d} + 2P_{i,c} + \phi(P_{i,c} - P_{i,d})}\right] \tag{5.4}$$

for the permeability of the membrane (MMCM) for the $i$-th gas and with $d$ and $c$ subindexes referring to disperse (filler) and continuous (polymer) phases [51–53]. This equation is usually expected to hold for low volume fractions of the filler [54].

The possible effects of the interaction of the filler with the continuous phase where it is inserted can produce an interfacial layer in between them that, depending on their compactness, can have different effects [55], as shown in Figure 5.1.

The good dispersion, agglomeration and the amount of filler included in the continuous phase as well as the quality of the interfaces between the two phases forming MMCM can be analyzed by using scanning probe microscopy. An example concerning two kinds of activated carbon in an acrylonitrile butadiene styrene (ABS) matrix is explained in the next section.

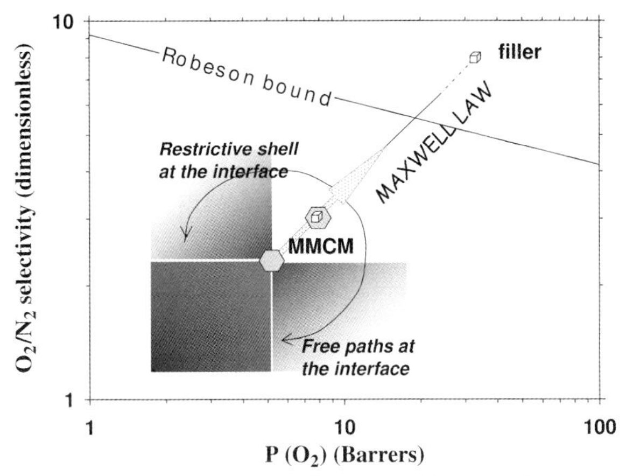


**Figure 5.1** An scheme of the different possible ways of evolution of a given polymer by introduction of a filler depending on the quality of the interface formed in the resulting mixed matrix composite membrane.
1 Ba = 1 Barrer = $7.5005 \times 10^{-18}$ $m^2/s$ Pa.

## 5.4 Case Studies

### 5.4.1 Phase Segregated Membranes

The electronic interactions of $CO_2$ with the oxygen in the ethylene groups in polyethylene glycols (PEG) increase the solubility of this gas in the polymeric matrix and thus should lead to an improved transport and better permeability and selectivity of $CO_2$. In order to explore this possibility, several polymers were tested. They were obtained by a two-steps polyimidation of 3,3′,4,4′-biphenyltetracarboxyl (BPDA) or 3,3′,4,4′-benzophenonetetracarboxyl (BKDA) dianhydride with mixtures of different diamines. The diamines in these mixtures were an aromatic one, benzidine (BNZ), oxydiamiline (ODA) or phenylendiamine (PPD), and an aliphatic one with PEG. Several proportions of aromatic to aliphatic diamines have been tested. The films were thermally dehydrated at 120 °C and the corresponding complete imidation was tested by FT-IR.

The resulting polymer was solved in dimethylacetamide (DMAc) and evaporated at 60 °C during 5 h. Finally the obtained membranes were heated from 160 °C to 265 °C under a $N_2$ atmosphere.

In all cases the resulting permeability and selectivity were very interesting [56, 57]. As an example the permeability and selectivity of two of these membranes are shown in Figure 5.2. The notable increase in permeability and selectivity, when the membrane was thermally treated at temperatures below degradation, should be explained in terms of the changes in the membrane morphology.

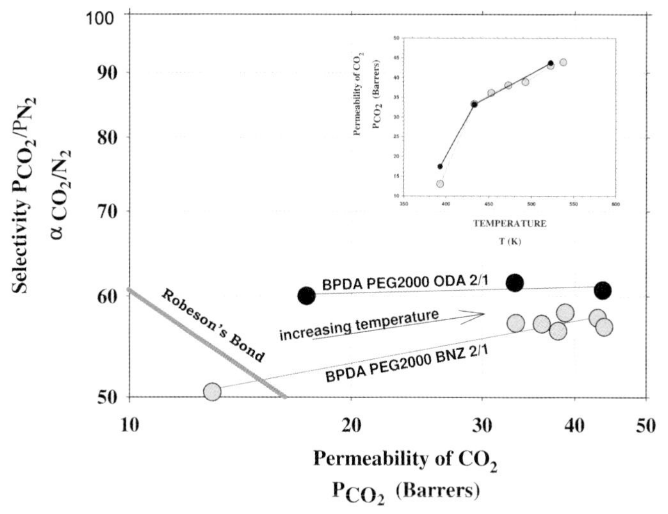


**Figure 5.2** Robeson diagram showing $CO_2/N_2$ selectivity versus $CO_2$ permeability for the membranes BPDA-PEG2000-ODA 2/1 and BPDA-PEG2000-BNZ 2/1 after different temperatures of treatment. An insert showing their $CO_2$ permeability as a function of temperature is also included.

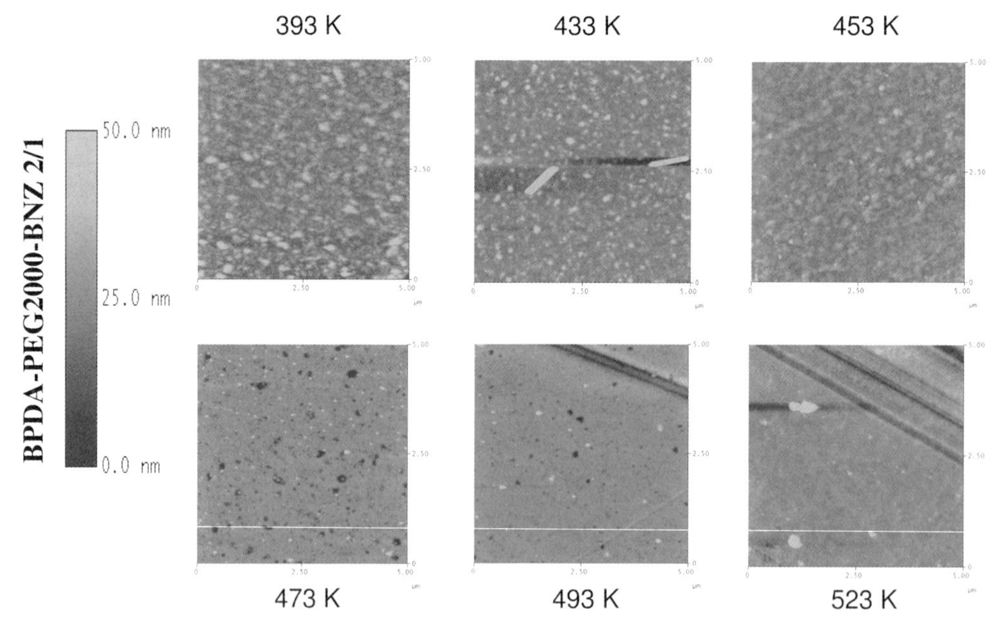


**Figure 5.3** AFM topographic tapping mode pictures of the BPDA-PEG2000-BNZ 2/1 membrane after treatment at different temperatures.

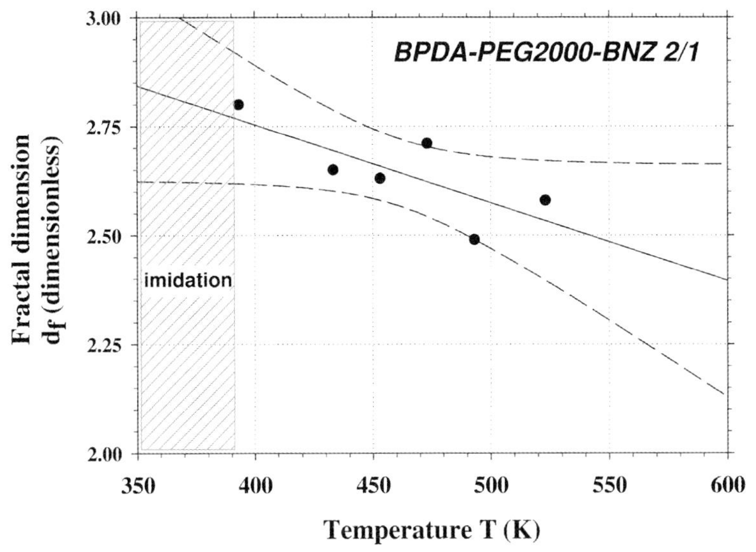


**Figure 5.4** Fractal dimension of the BPDA-PEG2000-BNZ 2/1 membrane as a function of the temperature of treatment. The temperature range of imidation is also shown.

In order to investigate the corresponding changes in the membrane material after thermal treatement of these membranes, AFM pictures were analyzed. Figure 5.3 shows an example of AFM pictures. These images were obtained for membranes treated at different temperatures and using the tapping mode. The corresponding fractal dimension is shown as a function of the treatment temperature in Figure 5.4. From these results it seems clear that the membranes are becoming more and more flat when heated.

A certain phase segregation is seen to appear by SAXS [58, 59]. An example is shown in Figure 5.5, where the mean distance from phase to phase is shown versus the treatment temperature. Note that the corresponding phase segregation should start at temperatures only slightly over the imidation one. The phase segregation is also seen, for example, in the TEM image shown in Figure 5.6.

### 5.4.2 Solvent Evaporation

It has been shown that the introduction of bulky groups in the chains of glassy polymers makes their structure stiffer and hinders an efficient packing of chains [60, 61]. This should lead to an increase in free volume. An example of such a kind of polymer is 6FDA-6FpDA, polymerized from 2,2-bis(4-aminophenyl) hexafluoropolylidone and 2,2 bis-(3,4 dicarboxyphenyl) hexafluoropropylidine dianhydride [62].

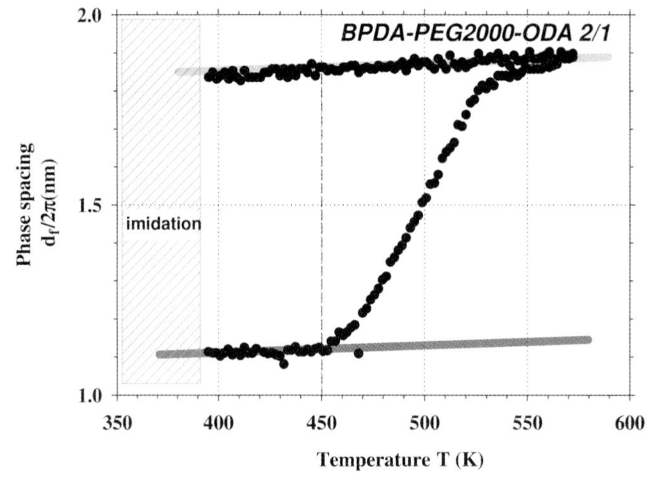


**Figure 5.5** SAXS results showing the phase spacing for the BPDA-PEG2000-ODA 2/1 membrane as a function of the temperature of treatment. The range of imidation and the starting of phase separation are also shown.

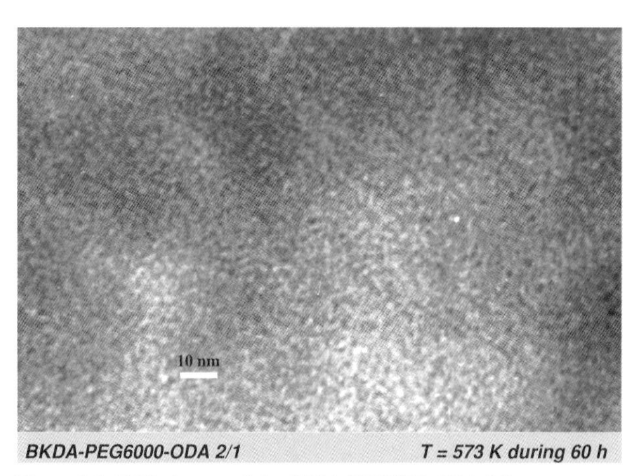


**Figure 5.6** TEM picture of the BPDA-PEG2000-ODA 2/1 membrane showing that at this temperature PEG is totally segregated.

Fluorinated polyimides are particularly interesting for gas separation because they have good mechanical, thermal and transport properties [63]. They also have a certain resistance to plasticization [64].

In most cases it has been assumed that the solvent used in the casting does not influence the resulting structure of the membrane once it has been evacuated. Nevertheless some studies have been done on this dependency [65, 66] without conclusive explanation for the differences in permeation found.

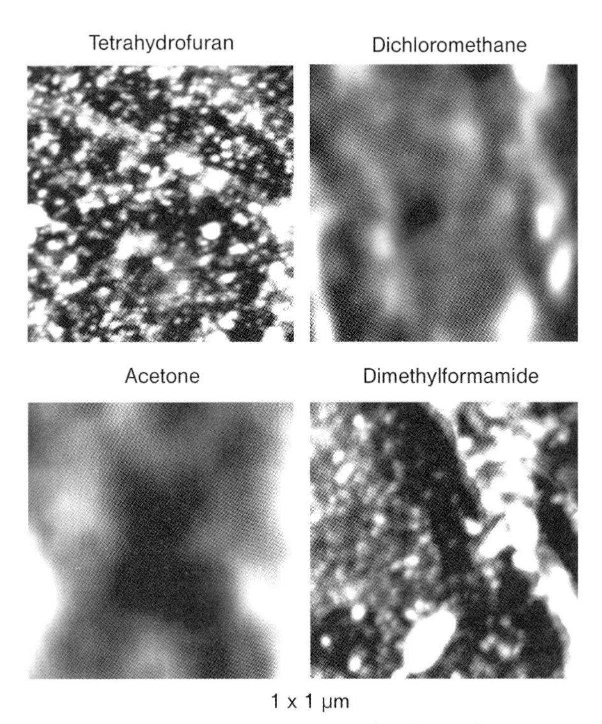


**Figure 5.7** AFM picture, in tapping mode, of several membranes made from 6FDA-6FpDA with different solvents.

The surface of the membranes made with different solvents and/or at different evaporation temperatures (for DMAc) show different conformations. Figure 5.7 shows examples of these different conformations for different solvents. Note that, when the solvent is good enough, the surface of the membrane is much flatter: while for not so good solvents the surface shows a granular structure.

Differences in selectivity and permeability have also been detected on 6FDA-6FpDA membranes, depending on the solvent used in their manufacture. Moreover, the morphology of the membrane, both on the surface and in the bulk, along with permeability and selectivity, could be correlated with the solvent characteristics.

The quality of a solvent–polymer system can be correlated with their Hildebrand solubility parameters ($\delta$) in such a way that, when these parameters are similar: $|\delta_s - \delta_p| \leq 2.5\ (\mathrm{cal/cm^3})^{1/2}$, the solubility is acceptable (where $\delta_s$ and $\delta_p$ are the Hildebrand solubility parameters for the solvent and the polymer, respectively). The above criterion comes from the approximate Flory–Huggins theory.

Figure 5.8 plots the fractal dimension versus $\delta_s$ for the solvents used, as obtained in the literature [43, 67, 68]. A parabolic plot should be obtained in this kind of plot [68], with a minimum in fractal dimension (flat membrane surface). Here, this minimum appears at 10.2 $(\mathrm{cal/cm^3})^{1/2}$ when 6FDA-6FpDA dissolves best.

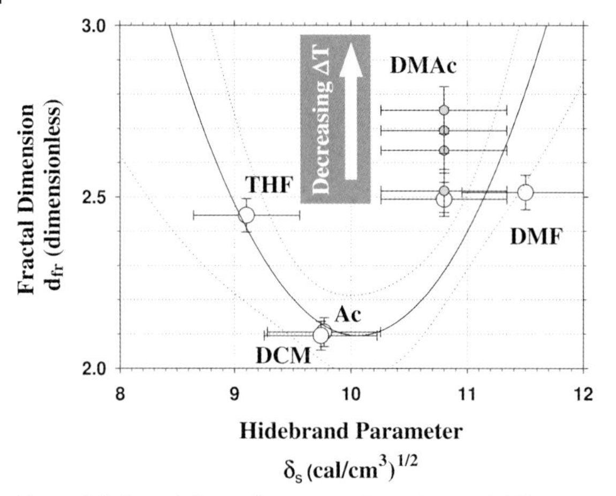


**Figure 5.8** Fractal dimension versus the solvent solubility parameter for 6FDA-6FpDA membranes manufactured with different solvents or evaporated at different temperatures (for DMAc) below the solvent boiling point, $\Delta T$.

The minimum in fractal dimension is mainly due to the polar contribution of the solubility Hildebrand parameter. This can be proved by the reproducibility of the minimum when the polar solubility parameter of Hansen is used, while no correlation between fractal dimension and the dispersive or hydrogen bond Hansen parameters can be found [69]. Moreover a similar tendency, but oppositely now showing a maximum, can hold for the glass transition temperature of the resulting membrane [67], as shown in Figure 5.9 with a very similar extremal Hildebrand parameter. The maximum glass transition temperature, or minimum fractal dimension, appears at a solvent Hildebrand parameter that very finely estimates that of the polymer.

The Flory–Huggins theory can be improved to incorporate new terms. For example, the Prigogine–Flory–Patterson theory introduces an entropic contribution due to free volume effects in order to obtain more realistic results [70]. This contribution does not modify the solubility criterion as stated above. When the Flory interaction parameter ($\chi$) is considered to have both an enthalpic component $\chi^H$ and an entropic (or residual) component $\chi^S$, such a Flory parameter [71] appears as:

$$\chi = \chi^S + \chi^H = 0.34 + \frac{V_S}{RT}\left(\delta_s - \delta_p\right)^2 \tag{5.5}$$

This dimensionless interaction parameter characterizes the interaction energy per solvent molecule normalized in terms of kT. Assuming that the molar volume of the solvent is almost independent of temperature, like the Hildebrand's parameters of solubility for both solvent and polymer, Equation (5.5) can be used to take into account the influence of the temperature of membrane fabrication.

In order to study in a more quantitative way how a change in the solvent can increase permeability and selectivity, the distance from the point in a selectivity

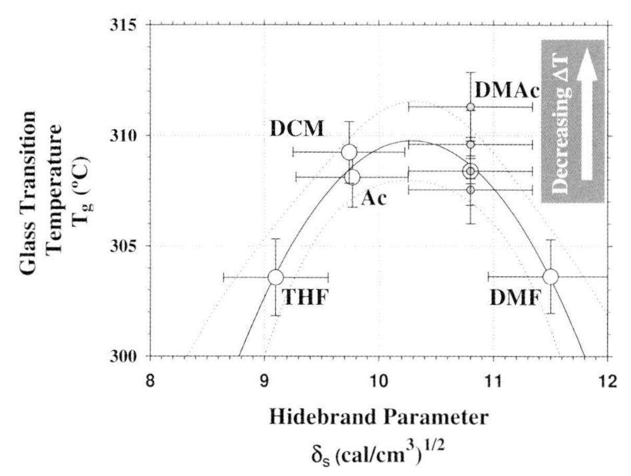


**Figure 5.9** Glass transition temperature versus the solvent solubility parameter for 6FDA-6FpDA membranes manufactured with different solvents or evaporated at different temperatures below the solvent boiling point, $\Delta T$.

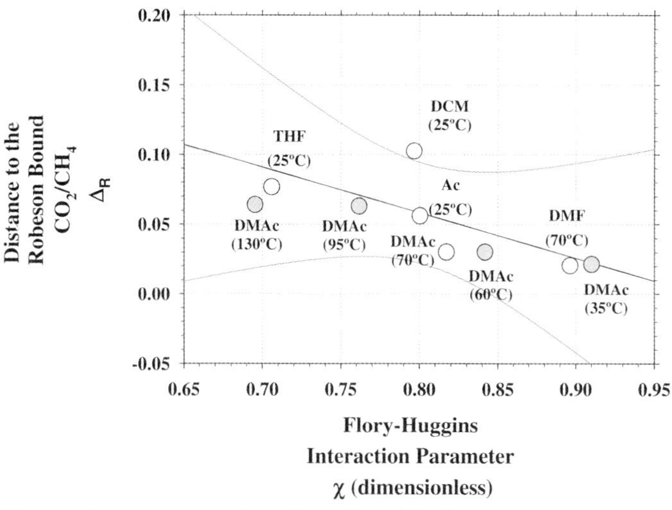


**Figure 5.10** Distance over the Robeson bound for the pair $CO_2/CH_4$ as a function of the Flory interaction parameter for the mebranes in Figures 5.8, 5.9.

versus permeability plot to the Robeson trade-off line, $\Delta_R$, can be used ($\Delta_R < 0$ if the point is below the Robeson bound and $\Delta_R > 0$ if the point is above it) [67]. In these terms, it can be shown that this distance to the Robeson's curve is a decreasing function of the solvent–polymer interaction energy, as shown in

Figure 5.10. This shows that the membrane quality increases when the solvent is chosen to have low interaction energy with the polymer.

### 5.4.3
### Asymmetric Polymeric Membranes

The more important factors from an industrial point of view are a high flux or productivity and a high selectivity or separation effectiveness. It is here that asymmetric membranes find more application, due to their high flux. When the same material forms two layers differing in their structure, with a thin active dense skin layer associated with another layer that acts as a mechanical support and has no significant resistance to mass flux, the resulting membranes are called integral.

In our case, these membranes were made of the same material and prepared by phase inversion process [72, 73]. The dense selective skin layer was possible because of the evaporation of solvent during the initial period and the macroporous layer sticking to the skin layer was formed due to the exchange between the solvent and nonsolvent systems inside the precipitation bath [74].

The polymeric material used to prepare integral asymmetric gas separation membranes was Matrimid 5218. This is a commercial polyimide with good properties for gas separation. A mixture of tetrahydrofuran (THF) and $\gamma$-butil-lactone (GBL) was used as solvent, while *n*-butanol (*n*-BuOH) was used as nonsolvent. The precipitation media was a bath with ethanol and water. Taking into account the corresponding ternary phase diagram, the compositions selected were 14% of Matrimid and 10% or 12% of *n*-BuOH in a 50:50 mixture of THF/GBL. Films were cast on a leveled glass plate and the solvent was left to evaporate at ambient temperature during a range from 2 s to 40 s.

In order to avoid or to minimize the influence of micropores appearing on the surface of the dense layer, the membranes were recovered by a thin layer of

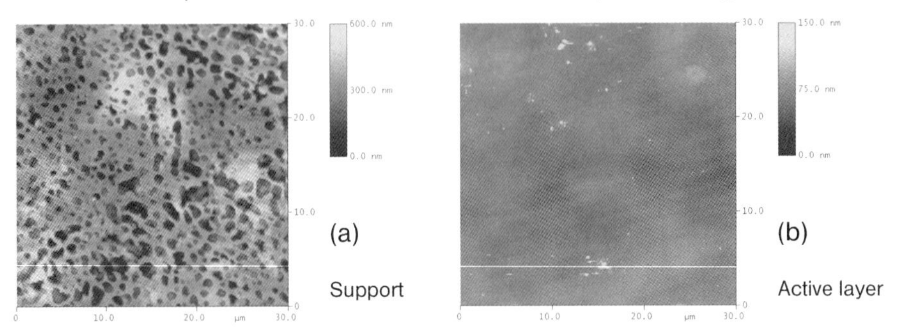


**Figure 5.11** AFM pictures of active and support layer of a Matrimid membrane 14/10 (14% of Matrimid, 10% of *n*-BuOH, in a 50:50 mixture of THF/GBL) evaporated during 10 s.

silicone rubber. This was chosen because it has a high permeability with a relatively low selectivity for all gases. The membrane was submerged in a solution prepared with 2% silicone in iso-octane during some seconds.

The structural characterization of these materials is difficult. The main factor here should be the thickness of the several dense or porous layers. In effect, the width of the dense skin layer determines the flux. Because of this, it is interesting to obtain thin membranes [75–78], with good mechanical, thermal and chemical resistances [79–81]. Actually, the borders within the several layers of the membrane are not easy to detect as far as they correspond to the same material.

In this sense we put special emphasis on the study of AFM and SEM pictures. Figure 5.11 shows the support and active layers as obtained by tapping mode AFM. The borders between layers are better revealed by phase contrast and force modulation techniques, as shown in Figure 5.12. The corresponding thicknesses for the Matrimid dense layer are shown in Figure 5.13 as a function of the time of evaporation used in order to eliminate the solvent once phase inversion took place.

Figure 5.13 also shows the corresponding thickness for the Matrimid dense layer as obtained from SEM. The thickness obtained by assuming that only the dense active layer of Matrimid has a significant resistance to the flux of gas according to

$$\delta'_{m} = \frac{P_{m}\Delta p}{J} \tag{5.6}$$

are also shown in Figure 5.13 [72, 73]. If the two thin layers of silicone rubber deposited on the active layer and on the porous layer are taken into account, the

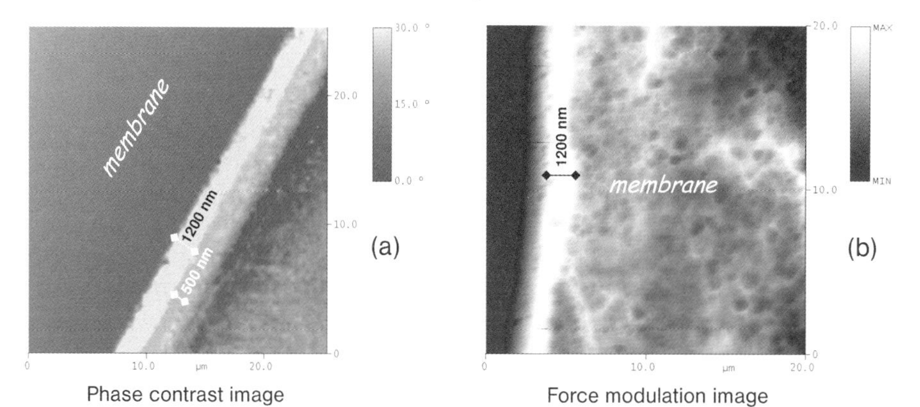


**Figure 5.12** Phase contrast and force modulation images of a Matrimid membrane 14/10 (14% of Matrimid, 10% of *n*-BuOH, in a 50:50 mixture of THF/GBL) evaporated during 40 s.

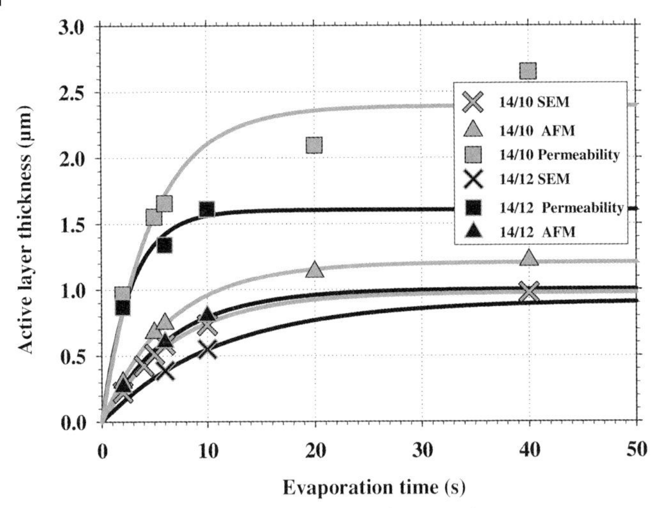


**Figure 5.13** Thickness of the active layer of Matrimid membranes 14/10 and 14/12 as a function of the evaporation time obtained by different methods.

width of the active layer of dense Matrimid could be evaluated by

$$\delta_{\mathrm{m}} = \frac{\delta'_{\mathrm{m}}}{1 + \frac{P_{\mathrm{m}}}{P_{\mathrm{S}}}} - 2\delta_{\mathrm{s}} \tag{5.7}$$

where $\delta_S$ (thickness of each of the two films of silicone rubber) is found to be $0.48 \pm 0.12\,\mu m$ as measured by AFM and SEM (note that $\delta_S$ does not depend on the evaporation time, nor on the composition of the initial polymer solution, but rather on the method of silicone deposition, which is the same for all cases). This procedure leads to values for the thickness of the Matrimid dense layer that, as shown in Figure 5.14, are in good accordance with those directly measured by AFM. The values for $P_m$ and $P_S$ are taken from the literature [82, 83].

## 5.4.4 Mixed Matrix Membranes

Activated carbons (with high permeability and selectivity but with very inconvenient mechanic properties when used alone) are proposed as inorganic fillers in order to profit from their different adsorption capacities for polar and unsaturated compounds against nonpolar and saturated chemicals. Some characteristics of the used activated carbons are shown in Table 5.1. ABS has been selected as the polymeric matrix thus far because it is a copolymer that combines the good selectivity of glassy polymers with the high permeability of rubbery polymers [84, 85].

Both sides of the membranes have been studied by AFM and areas from $50 \times 50\,\mu m$ to $0.5 \times 0.5\,\mu m$ have been scanned. The obtained roughness are always higher for the upper side than for the down side. Moreover, $R_q$ of both sides of the

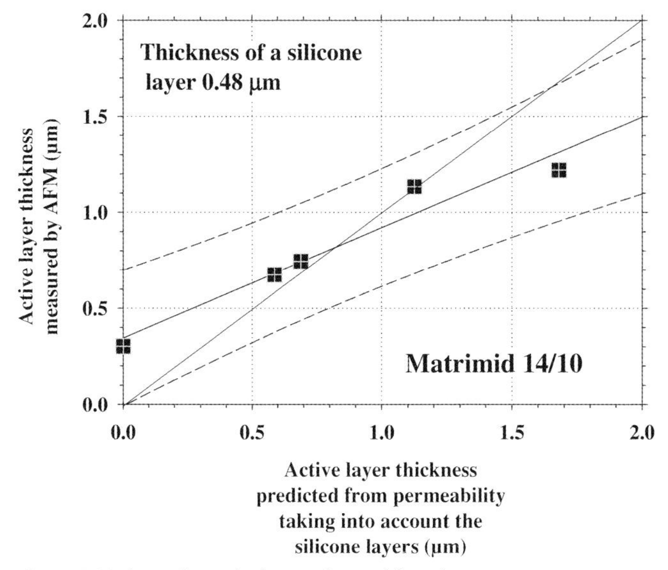


**Figure 5.14** Active layer thickness obtained by taking into account the silicone layers as a function of the corresponding thickness as measured by AFM for the 14/10 Matrimid membrane.

**Table 5.1** Some characteristics of the activated carbons used.

| | AC1 | AC2 |
|---|---|---|
| BET surface, $S_w$ ($m^2$/g) | 3272 | 818 |
| Monolayer capacity (BET), $V_m$ ($cm^3$(STP)/g) | 752 | 188 |
| Apparent density, $\rho_f$ (g/$cm^3$) | 0.28 | 0.42 |
| Average pore diameter $d_p$ (Å) | Bimodal | Bimodal |
| | 9.0 \|22.4 | 8.6 \|70.0 |
| Mean particle diameter, $d$ (μm) | $0.90 \pm 0.08$ | $4.5 \pm 2.8$ |

membrane increase with the content of active carbon when big areas are scanned. For scanned areas below 2.5 × 2.5 μm, roughness becomes independent of the AC content, which seems to confirm that there is always a layer of polymer over the activated carbon particles. This is also confirmed by phase contrast AFM, which shows that no heterogeneous viscoelasticity can be detected on the membrane surfaces. The constant roughnesses so reached are: 6.40 nm for the upper side and 3.50 nm for the under side of membranes containing AC2; and 4.89 nm for the upper side and 3.65 nm for the under side of membranes containing AC1.

The different roughness for both sides of each membrane can be attributed to the presence of pores ranging from 40 nm to 150 nm that should, probably, be due to the evacuation of solvent and appear more frequently, as can be observed in

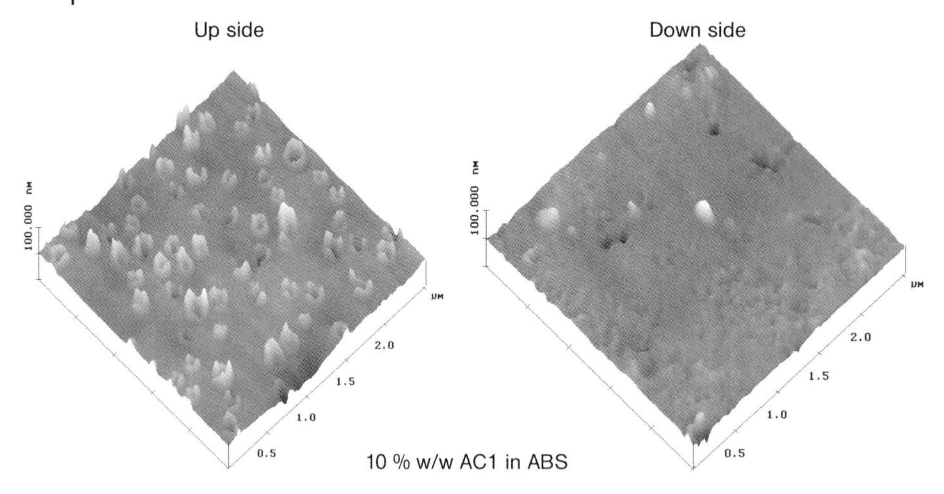


**Figure 5.15** Up and down side topographic AFM pictures of a MMCM membrane containing a 10% of the AC1 active carbon in ABS.

Figure 5.15, in the side open to air during casting. The surface density of these pores is 4.64 pores/$\mu m^2$ in the upper side and 1.92 pores/$\mu m^2$ in the under side for membranes containing AC2 and 7.52 pores/$\mu m^2$ and 2.20 pores/$\mu m^2$ respectively for membranes containing AC1. These pores appear more for membranes containing AC1 because AC1 particles have more porous volume (higher BET area, lower apparent density, as can be seen in Table 5.1). They are purely superficial as far as otherwise permeability should be much higher than actually found and selectivity much lower. Roughness also depends on the size of activated carbon particles, as far as AC1 has a mean diameter of 0.9 µm whereas AC2 has a mean diameter of 4.4 µm.

Transverse sections show carbon particles and agglomerates from 1 µm to 30 µm for AC1 membranes and 3 µm to 20 µm for AC2 membranes, when big scanned areas are studied. These sizes should correspond to aggregates up to 30 particles of AC1 and five particles of AC2. These sizes are within the optical microscopy range and can actually be seen by optical microscopy.

In higher resolution images, corresponding to small scanned areas, much smaller particles are revealed: ranging from 60 nm to 750 nm (with an average of 140 nm) for membranes containing AC1 and from 60 nm and 900 nm (with an average of 300 nm) for those containing AC2. It is worth noting that also these small activated carbon particles are bigger for AC2 than for AC1, as is the case with those shown in Table 5.1.

The biggest agglomerates detected are visualized directly in topography due to the process of fracture, while the small particles are detectable by using the phase contrast technique. Examples of the corresponding AFM pictures are shown in Figures 5.16 and 5.17. Higher carbon densities are found for AC2 containing membranes.

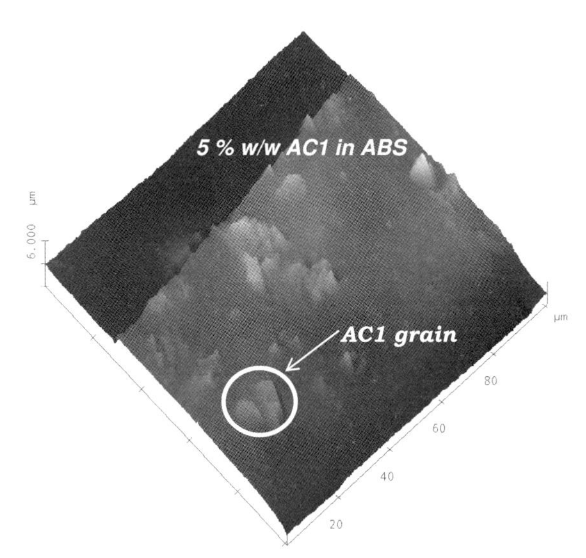


**Figure 5.16** Topographic AFM picture of a transversal section of an MMCM containing a 5% of the AC1 active carbon in ABS.

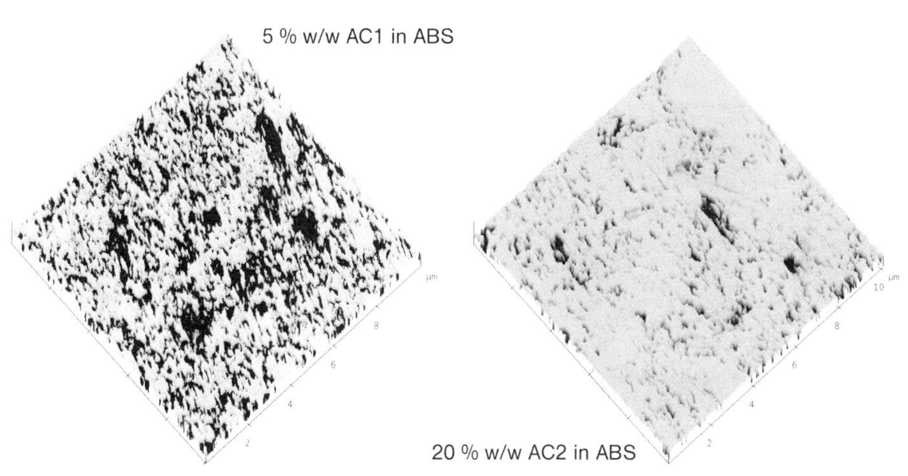


**Figure 5.17** Phase contrast AFM pictures of MMCM containing a 5% AC1 and a 20% AC2 active carbons in ABS.

All transverse sections always show a good adherence and contact between the polymer continuous matrix and the activated carbon particles. No changes are seen in the viscoelastic properties of the polymeric phases in contact with carbon disperse phase.

Permeability and selectivity clearly increase with the content of active carbon, as shown in Figure 5.18. According to the analysis of Moore and Koros [55], such

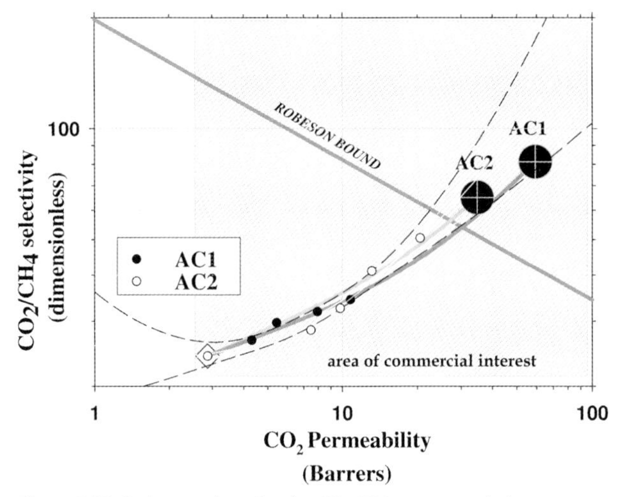


**Figure 5.18** Robeson chart for the $CO_2/CH_4$ gases and the MMCM membranes obtained from ABS including AC1 and AC2 active carbon as filler (2–10% of AC1; 20–40% of AC2) showing the best fitting Maxwell predictions for the pure active carbons.

simultaneous increase in permeability and selectivity should correspond to an absence of modifications of the polymer properties in the interface with the inorganic filler, with no clogging of transport path though the filler particles and with a good adhesion of filler to polymer. This good adhesion agrees with the absence of voids around particle fillers seen in SEM and AFM pictures; and it allows an application of the Maxwell theory.

## Acknowledgments

The authors would like to thank CICYT Plan Nacional de I+D+I (projects MAT2005-04976, MAT2004-01946, CTQ2006-01685). We also would like to thank: V. Peinemann (in whose laboratory in the GKSS in Geestaacht, Germany, the asymmetric Matrimid membranes were made) and J. Marchese (who manufactured the MMCM studied here).

## References


1 S.T. Hwang, K. Kammermeyer, *Membranes in Separations*, Krieger, Malabar, **1984**.
2 M. Mulder, *Basic Principles of Membrane Technology*, Kluwer, Dordrecht, **1991**.
3 W.S.W. Ho, K.K. Sirkar, *Membrane Handbook*, Van Nostrand Reinhold, New York, **1992**.
4 K. Kamide, S. Manabe, in: A.R. Cooper (ed.), *Ultrafiltration Membranes and Applications*, Plenum, New York, **1980**.
5 B. Rasneur, *Porosimetry (Characterization of Porous Membranes)*, Summer School on Membr. Sci. and Tech., Cadarache, **1984**.
6 R.E. Kesting, *Synthetic Polymeric Membranes, A Structural Perspective*, 2nd edn., John Wiley and Sons, New York, **1985**.
7 S. Lowell, J.E. Shields, *Powder Surface Area and Porosity*, John Wiley and Sons, New York, **1987**.
8 S.S. Kulkarni, E.W. Funk, N.N. Li, *Membranes*, in: W.S.W. Ho and K.K. Sirkar (eds), *Membrane Handbook*, Van Nostrand Reinhold, New York, **1992**.
9 S. Nakao, *J. Membr. Sci.* **1994**, *96*, 131–165.
10 G. Binnig, C.F. Quate, C. Gerber, *Phys. Rev. Lett.* **1986**, *56*, 930–933.
11 A.K. Fritzsche, A.R. Arevalo, M.D. Moore, C.J. Weber, V.B. Elings, K. Kjoller, C.M. Wu, *J. Membr. Sci.* **1992**, *68*, 65–78.
12 P. Dietz, P.K. Hansma, O. Inacker, H.D. Lehmann, K.H. Herrmann, *J. Membrane Sci.* **1992**, *65*, 101–111.
13 H. Bechhold, M. Schlesinger, K. Silbereisen, L. Maier, W. Nurnberger, *Kolloid Z.* **1931**, *55*, 172–198.
14 K. Schneider, W. Hölz, R. Wollbeck, S. Ripperger, *J. Membrane Sci.* **1988**, *39*, 25–42.
15 G. Reichelt, *J. Membrane Sci.* **1991**, *60*, 253–259.
16 L. Zeman, *J. Membrane Sci.* **1992**, *71*, 233–246.
17 ASTM, *Standard Test Method for Pore Size Characteristics of Membrane Filters by Bubble Point and Mean Flow Pore Test*, ASTM F316–03, **2003**.
18 ASTM, *Standard Test Methods for Pore Size Characteristics of Membrane Filters Using Automated Liquid Porosimeter*, ASTM E1294–89, **1999**.
19 E. Honold, E.L. Skau, *Science* **1954**, *120*, 805–806.
20 S.J. Gregg, K.S.W. Sing, *Adsorption, Surface Area and Porosity*, Academic, London, **1982**.
21 D. Dollimore, G.R. Heal, *J. Appl. Chem.* **1964**, *14*, 109.
22 A. Mey-Marom, M.G. Katz, *J. Membrane Sci.* **1986**, *27*, 119–130.
23 M. Brun, A. Lallemand, J.F. Quinson, C. Eyraud, *J. Chim. Phys.* **1973**, *6*, 979–989.
24 M. Sarbolouki, *Sep. Sci. Technol.* **1982**, *17*, 381–386.
25 R. Nobrega, H. de Balmann, P. Aimar, V. Sánchez, *J. Membrane Sci.* **1989**, *45*, 17–36.
26 C.L. Glaves, D.M. Smith, *J. Membrane Sci.* **1989**, *46*, 167–184.
27 K. Sakai, *J. Membrane Sci.* **1994**, *96*, 91–130.
28 K. Kaneko, *J. Membrane Sci.* **1994**, *96*, 59–89.
29 C.P. Bean, M.V. Doyle, G. Entine, *J. Appl. Phys.* **1970**, *41*, 1454.
30 H.B. Park, Ch.H. Jung, Young Moo Lee, A.J. Hill, S.J. Pas, S.T. Mudie, E. Van Wagner, B.D. Freeman, D.J. Cookson, *Science* **2007**, *318*, 254–258.
31 Y. Huang, D.R. Paul, *J. Membrane Sci.* **2004**, *244*, 167–178.
32 J.H. Kim, W.J. Koros, D.R. Paul, *Polymer* **2006**, *47*, 3104–3111.
33 F.P. Cuperus, *Characterization of Ultrafiltration Membranes*, Ph.D. Thesis, Twente, **1990**.
34 R. Riley, U. Merten, J.O. Gardner, *Science* **1964**, *143*, 801.
35 C. Riedel, R. Spohr, *J. Membrane Sci.* **1980**, *7*, 225–234.
36 T. Allen, *Particle Size Measurement*, Vol. 1, Chapman and Hall, London, **1997**.

**37** K.J. Kim, A.G. Fane, C.J.D. Fell, T. Suzuki, M.R. Dickson, *J. Membrane Sci.* **1990**, *54*, 89–102.

**38** P. Prádanos, M.L. Rodriguez, J.I. Calvo, A. Hernández, F. Tejerina, J.A. de Saja, *J. Membrane Sci.* **1996**, *117*, 291–302.

**39** J.I. Calvo, P. Prádanos, A. Hernández, W.R. Bowen, N. Hilal, R.W. Lovitt, P.M. Williams, *J. Membrane Sci.* **1997**, *128*, 7–21.

**40** L. Zeman, L. Denault, *J. Membrane Sci.* **1992**, *71*, 221–231.

**41** Digital Instruments, *Nanoscope Command Reference Manual*, Digital Instruments Veeco Metrology Group, Santa Barbara, **2001**.

**42** P.A. Shilyaev, D.A. Pavlov, *Fractal based approach to characterization of surface Geometry*, in: Proc. Int. Workshop "Scanning Probe Microscopy 2004", Institute for the Physics of Microstructures, Russian Academy of Science, N. Novgorod, **2004**.

**43** J. Macanas, L. Palacio, P. Prádanos, A. Hernández, M. Muñoz, *Appl. Phys. A* **2006**, *84*, 277–284.

**44** C.-C. Hu, K.R. Lee, R.C. Ruaan, Y.C. Jean, J.Y. Lai, *J. Membrane Sci.* **2006**, *274*, 192–199.

**45** L.M. Robeson, *J. Membrane Sci.* **1991**, *62*, 165–185.

**46** B.D. Freeman, *Macromolecules* **1999**, *32*, 375–380.

**47** M.L. Cecopieri-Gómez, J. Palacios-Alquisira, J.M. Domínguez, *J. Membrane Sci.* **2007**, *293*, 53–65.

**48** S.H. Huang, C.C. Hu, K.R. Lee, D.J. Liaw, J.Y. Lai, *Eur. Polym. J.* **2006**, *42*, 140–148.

**49** Z.F. Wang, B. Wang, Y.R. Yang, C.P. Hu, *Eur. Polym. J.* **2003**, *39*, 2345–2349.

**50** H. Lin, E. Van Wagner, J.S. Swinnea, B.D. Freeman, S.J. Pas, A.J. Hill, S. Kalakkunnath, D.S. Kalika, *J. Membrane Sci.* **2006**, *276*, 145–161.

**51** D.Q. Vu, W.J. Koros, S.J. Miller, *J. Membrane Sci.* **2003**, *11*, 335–348.

**52** L.M. Robeson, A. Noshay, M. Matzner, C.N. Merrian, *Angew. Makromol. Chem.* **1973**, *29/30*, 47–62.

**53** C. Zimmerman, A. Singh, W.J. Koros, *J. Membrane Sci.* **1997**, *137*, 145–154.

**54** E.E. Gonzo, M.L. Parentis, J.C. Gottifredi, *J. Membrane Sci.* **2006**, *277*, 46–54.

**55** T.T. Moore, W. Koros, *J. Mol. Struct.* **2005**, *739*, 87–98.

**56** A. Tena, R. Recio, L. Palacio, P. Prádanos, A. Hernández, A.E. Lozano, A. Marcos, J.G. de la Campa, J. de Abajo, *Poliimidas aromatic-alifáticas para la separación de CO2/N2. Permeabilidad y selectividad en function del polímero y de la temperature de procesado*, Proc. 10 Reunión GEP, Sevilla, **2007**, 191.

**57** R. Recio, P. Prádanos, A. Hernández, A.E. Lozano, A. Marcos, J.G. de la Campa, J. de Abajo, *Membranas poliméricas aromático-alifáticas para la separación de gases. Efecto del tratamiento térmico en las propiedades de separación*, Proc. 10 Reunión GEP, Sevilla, **2007**, 189.

**58** A. Marcos-Fernández, A.E. Lozano, J.G de la Campa, J. de Abajo, R. Recio, L. Palacio, P. Prádanos, A. Hernández, *Synthesis and characterization of polyethyleneoxide-containing copolyimides for gas separation membranes*, Proc. POLYMEX, Huatulco, **2006**.

**59** A. Marcos-Fernández, A.E. Lozano, J.G. de la Campa, J. de Abajo, R. Recio, A. Tena, L. Palacio, P. Prádanos, A. Hernández, *O-26, Phase-separation characterization of polyethyleneoxide-containing copolyimides for gas separation membranes*, Proc. 2 Workshop on Applications of Synchrotron Light to Non-Crystalline Diffraction in Materials and Life Science, Madrid, **2007**.

**60** A. Singh, K. Ghosal, B.D. Freeman, A.E. Lozano, J.G. de la Campa, J. de Abajo, *Polymer* **1999**, *40*, 5715–5722.

**61** C. Nagel, K. Günther-Schade, D. Fritsch, T. Strunskus, F. Faupel, *Macromolecules* **2002**, *35*, 2071–2077.

**62** R. Wang, C. Cao, T S Chung, *J. Membrane Sci.* **2002**, *198*, 259–271.

**63** S.A. Stern, Y. Mi, H. Yamamoto, A.K. StClair, *J. Polym. Sci. B. Polym. Phys.* **1989**, *27*, 1887–1909.

64 M.R. Coleman, W.J. Koros, *Macromolecules* **1999**, *32*, 3106–3113.

65 K.C. Khulbe, T. Matsuura, G. Lamarche, H.J. Kim, *J. Membrane Sci.* **1997**, *135*, 211–223.

66 C. Bas, R. Mercier, J. Sanchez-Marcano, S. Neyertz, N.D. Alberola, E. Pinel, *J. Polym. Sci. B. Polymer Phys.* **2005**, *43*, 2413–2426.

67 R. Recio, L. Palacio, P. Prádanos, A. Hernández, A.E. Lozano, A. Marcos, J.G. de la Campa, J. de Abajo, *J. Membrane Sci.* **2007**, *293*, 22–28.

68 R. Recio, L. Palacio, P. Prádanos, A. Hernández, A.E. Lozano, A. Marcos, J.G. de la Campa, J. de Abajo, *Desalination* **2006**, *200*, 225–226.

69 J.M. Prausnitz, R.N. Lichtenthaler, E. Gomes de Azevedo, *Molecular Thermodynamics of Fluid-Phase Equilibria*, 3rd edn., Prentice Hall, Upper Saddle River, **1999**.

70 D.W. Van Krevelen, *Properties of Polymers: Their Correlation with Chemical Structure: Their Numerical Estimation and Prediction from Additive Group Contributions*, Elsevier, Amsterdam, **1990**.

71 J. Fried, *Polymer Science and Technology*, 2nd edn., Prentice Hall, New York, **2003**.

72 R. Recio, P. Prádanos, A. Hernández, S. Shishatskiy, K.V. Peinemann, A. Bottino, G. Capannelli, *Asymmetric Membranes for Gas Separation. Different Techniques to Determine the Active Dense Skin Thickness*, Marie Curie Workshop on New Materials for Membranes, Geestacht, Germany, **2007**.

73 R. Recio, P. Prádanos, A. Hernández, F. Tejerina, K.V. Peinemann, S. Shishatskiy, *Caracterización con AFM de Membranas Mixtas de Inversión de Fase para Separación de Gases*, Proc. 5 Congreso Iberoamericano de Ciencia y Tecnología de Membranas (CITEM 2005), Valencia, **2005**.

74 S.P. Nunes, K.V. Peinemann, *Membrane Technology in the Chemical Industry*, Wiley-VCH, Weinheim, **2001**.

75 S.G. Kimura, *Preparation of Asymmetric Polymer Membranes*, US Patent 3709774, **1973**.

76 K.V. Peinemann, *Verfahren zur Herstellung einer Integral Asymmetrischen Membran*, German Patent DE 3420373, **1984**.

77 K.V. Peinemann, I. Pinnau, *Verfahren zur Herstellung und Erhöhung der Selektivitit einer Integral Asymmetrischen Membran*, German Patent DE 3525235, **1986**.

78 I. Pinnau, W.J. Koros, *Defect-free Ultrahigh Flux Asymmetric Membranes*, US Patent 4902422, **1990**.

79 D.T. Clausi, S.A. McKelvey, W.J. Koros, *J. Membrane Sci.* **1990**, *160*, 51–64.

80 H.J. Kim, A. Tabe-Mohammadi, A. Kumar, A.E. Fonda, *J. Membrane Sci.* **1999**, *161*, 229–238.

81 J.C. Jansen, M. Macchione, E. Drioli, *J. Membrane Sci.* **2005**, *255*, 167–180.

82 D.Q. Vu, W.J. Koros, S.J. Miller, *J. Membrane Sci.* **2003**, *211*, 311–334.

83 P. Jha, L.W. Mason, J.D. Way, *J. Membrane Sci.* **2006**, *272*, 125–136.

84 J. Marchese, E. Garis, M. Anson, N.A. Ochoa, C. Pagliero, *J. Membrane Sci.* **2003**, *221*, 185–197.

85 M. Anson, J. Marchese, E. Garis, N.A. Ochoa, C. Pagliero, *J. Membrane Sci.* **2004**, *243*, 19–28.

# 6
# Atomic Force Microscopy Investigations of Membranes and Membrane Processes

*W. Richard Bowen and Nidal Hilal*

## 6.1
## Introduction

The human imagination, including the scientific imagination, is highly visual. We are most easily convinced of the existence of phenomena and processes in the physical world if we can see them. The importance of seeing is deeply imbedded in language. Thus, when we finally understand some highly complex and abstract explanation we may comment, "I see that now". However, the size range of objects that the unaided human eye can directly observe is limited. We can use optical microscopes to extend the lower limit of this range down to objects of sizes comparable with the wavelength of light. Beyond this limit we need to use other means to "see".

Atomic force microscopy (AFM) is one means of imaging objects of dimensions from about the wavelength of light to those below a nanometer. The imaging principle is not optical. AFM images are generated by scanning a sharp tip across a surface, recording the movement of the tip and then using a computer to produce an image of the scanned surface. Thus, in the case of membranes it is possible to visualize the membrane surface properties, such as pores and morphology, using AFM. Fortuitously, the size range of objects that may be visualized by AFM corresponds closely to the size range of surface features that determine the separation characteristics of membranes.

However, the separation characteristics of membrane interfaces do not depend solely on the physical form of surface features. In liquids, surface electrical properties and the adhesion of solutes to membrane surfaces may also have profound effects on separation performance. It is thus exceedingly fortunate that an atomic force microscope may also be used to determine both of these additional controlling factors. Finally, means may be devised to quantify all of these controlling factors in liquid environments that match those of process streams.

The intention of the present chapter is to provide a concise review of the potential of AFM for the investigation of membranes and membrane processes using

*Monitoring and Visualizing Membrane-Based Processes*
Edited by Carme Güell, Montserrat Ferrando, and Francisco López

ISBN: 978-3-527-32006-6

examples from our own studies. The chapter begins with illustrative examples that outline the range of possibilities of AFM studies for membrane technology. Some more advanced topics are then considered: the correspondence between surface pore dimensions from AFM and molecular weight cut-off (MWCO), imaging in liquid and the determination of surface electrical properties, the effects of surface roughness on interactions with particles, "visualization" of the rejection of a colloid by a membrane pore and the use of AFM measurements in membrane development (particular applications of AFM to gas separation membranes can be found in Chapter 5). The written text and technical details are kept to a minimum throughout. Such details are described fully in the original publications. The intention here is to let the images "speak" for themselves!

## 6.2 The Range of Possibilities

It is helpful to understand a little about the nature and operation of an atomic force microscope. A schematic of some typical equipment is shown in Figure 6.1.

In this case, the position of the sample is controlled and varied by means of a piezoscanner and the corresponding movement of the tip, which is positioned at the end of a cantilever, is measured optically by means of a laser beam and a position-sensitive photodetector. There are many ways of controlling the relative movement of sample and tip; and physicists are continually developing increasingly sophisticated variations. Such complexities are dealt with in many research papers, reviews and books and need not concern us here. However, it is interesting to note the typical form of a cantilever and a tip, as shown in the electron microscope images of Figure 6.2.

Here the cantilever is ~100 μm long and the tip, which is located at the apex of the cantilever, a few microns long and just a nanometer or so in radius at its end.

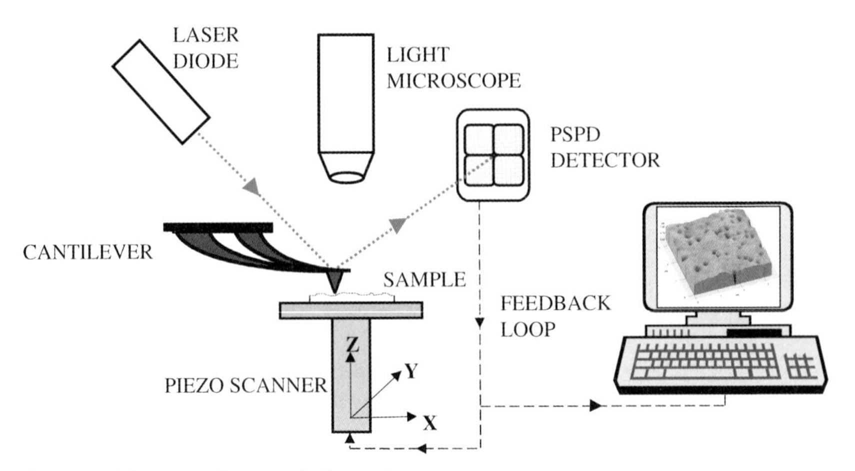


**Figure 6.1** Schematic of an atomic force microscope.

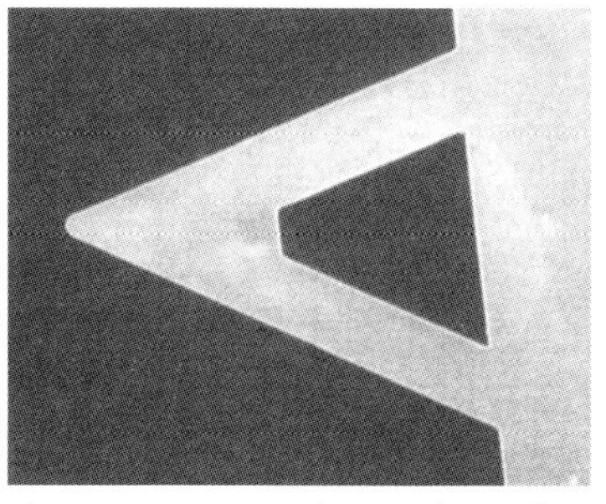

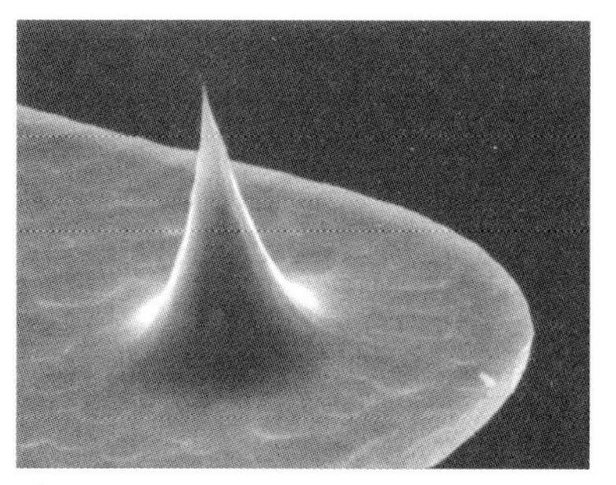

**Figure 6.2** AFM cantilever and tip. The cantilever is ~100 μm long and the tip is ~3 μm long.

Material scientists are constantly devising cantilevers of different geometries, materials and physical properties with tips of different shapes. For imaging very small features there is a special interest in fabricating very sharp tips (more details about AFM fundamentals can be found in Section 5.2.2).

A membrane technologist acquiring an atomic force microscope for the first time is best advised to begin with relatively simple measurements. A good starting point is to image some track-etch membranes in air. The pores in such membranes may also be imaged using a good optical microscope, which gives assurance about the images produced by AFM. As an example, Figure 6.3 show at AFM image of a Cyclopore membrane with specified pore dimensions of 0.2 μm [1].

Shown alongside the membrane image is the derived pore size distribution. Such distributions may be readily obtained using commercial image analysis software, either automatically or manually.

Once successful images of microfiltration membranes have been obtained it is a challenge to move downward in expected pore size or MWCO. Thus, Figure 6.4 shows a single pore in a PCI Membranes ES625 ultrafiltration membrane which has a specified MWCO of 25 000 [2].

The derived pore size distribution indicates a mean pore size of 5.1 nm, with a standard deviation of 1.1 nm. This is a pore of dimensions suitable for separating a

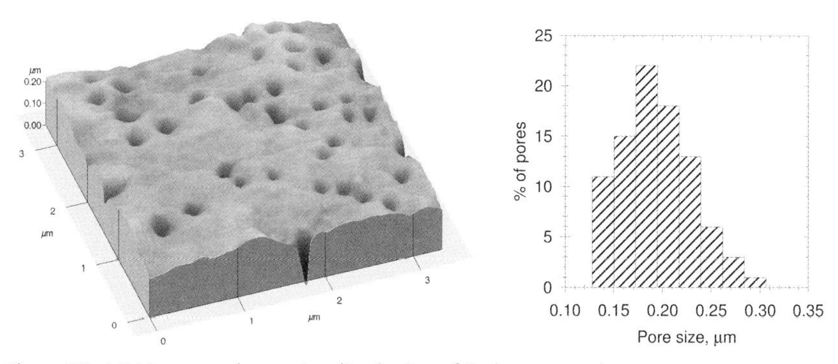


**Figure 6.3** AFM image and pore size distribution of Cyclopore membrane.

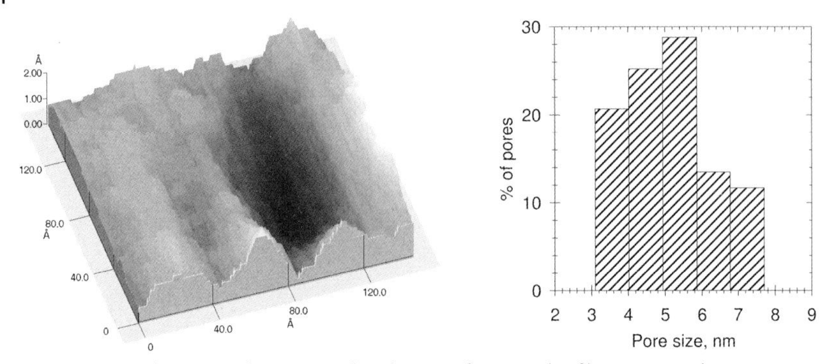


**Figure 6.4** Single pore and pore size distribution of ES625 ultrafiltration membrane.

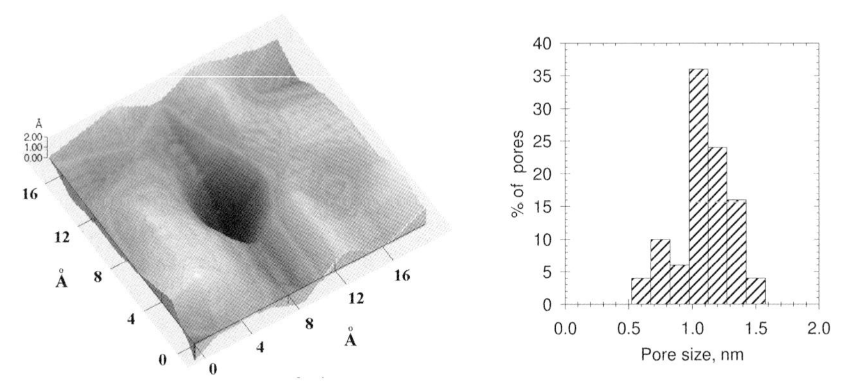


**Figure 6.5** Single pore and pore size distribution of XP117 nanofiltration membrane.

protein molecule. It should be noted that it is not always possible to image such small pores. In particular, it is necessary for the membrane surface roughness to have characteristic dimensions less than the pore dimensions for successful imaging.

An important challenge in science is "to boldly go" where no man (or woman) has gone before. Thus, Figure 6.5 shows an image and the corresponding pore size distribution of a single pore in a nanofiltration membrane, XP117 from PCI Membranes.

Caution is needed here. It is only in exceptional instances possible to obtain images of such small pores. Further, some scepticism is in order. The existence of pores in microfiltration membranes may be confirmed by optical images. It does not seem unreasonable to push the limit of belief in pores down to the ultrafiltration range. But in the nanofiltration range we need to be especially aware of the possibility of artefacts that look like pores and seeing what we wish to visualize. Whenever I show this image at conferences now I miss our much-lamented colleague and enthusiastic membrane proponent, the late Marcel Mulder, posing the first "question" by denying the existence of pores in nanofiltration membranes.

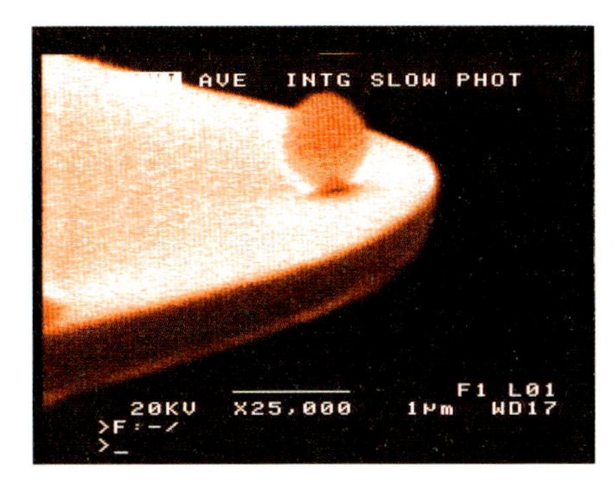


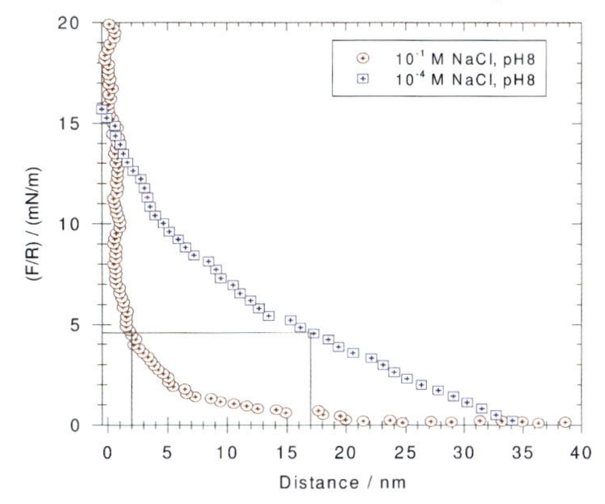


**Figure 6.6** SEM image of a 0.75 µm silica sphere attached to a tipless AFM cantilever and force distance curves for the approach of the probe to a Cyclopore microfiltration membrane in NaCl solutions at pH 8.0. F = force; R = sphere radius.

It was mentioned in Section 6.1 that an atomic force microscope can also quantify the other key properties controlling membrane performance. The crucial innovation in the determination of such properties is the development of *colloid probes*. Such probes are formed by attaching particles of dimensions of the order of ~1 µm to the end of *tipless* cantilevers. Such attachment may be carried out using the manipulation properties of an atomic force microscope, but greater success is achieved with the use of specially designed micromanipulation equipment. An example of a colloid probe is shown in the electron microscopy image of Figure 6.6. The silica colloid probed shown is at about the lower size limit for successful micromanipulation and subsequent measurement.

If such probes are manipulated to approach a single point on a membrane surface in a controlled manner in an electrolyte solution, it is possible to directly quantify the electrical double layer interactions between probe and membrane, also shown in Figure 6.6 for two solutions of differing ionic strengths. Such electrical interactions are very important in determining the rejection of colloids and biological macromolecules during ultrafiltration.

The adhesion of process stream components to membranes also has an important influence on membrane performance. Ideally, such adhesion should be avoided. As an example, Figure 6.7 shows a polystyrene colloid probe and data for the retraction of such a probe after it has been pushed into contact with membrane surfaces [3].

The depths of the depressions of the curves in Figure 6.7 give direct quantification of the adhesion of the probes to the surfaces. The ES404 membrane was an existing commercial membrane and the XP117 membrane a development membrane designed to have lower fouling properties (both membranes by PCI Membranes). The development membrane had significantly lower adhesion and hence significantly lower fouling potential that the existing membrane.

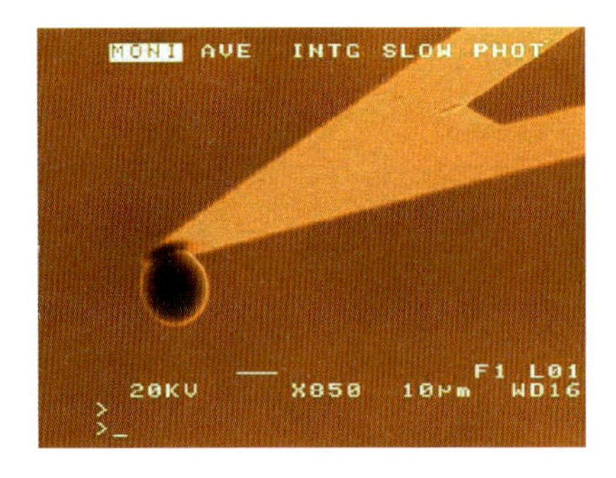


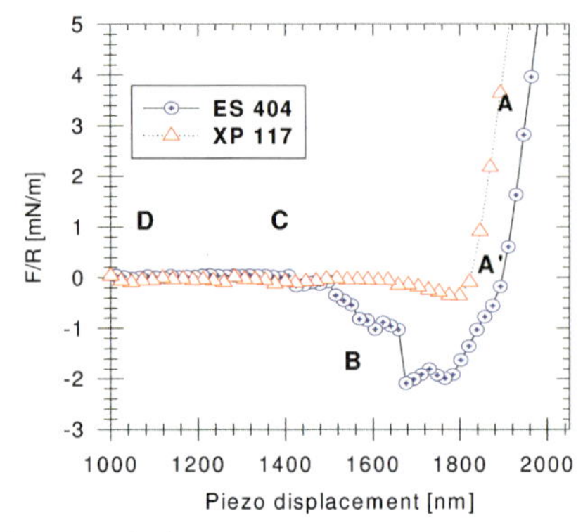


**Figure 6.7** SEM image of a 11 μm polystyrene sphere attached to a tipless AFM cantilever and force versus piezo displacement plot (retraction) for two membranes in $10^{-2}$ M NaCl solution at pH 8.0 (ES404, conventional; XP114, modified).

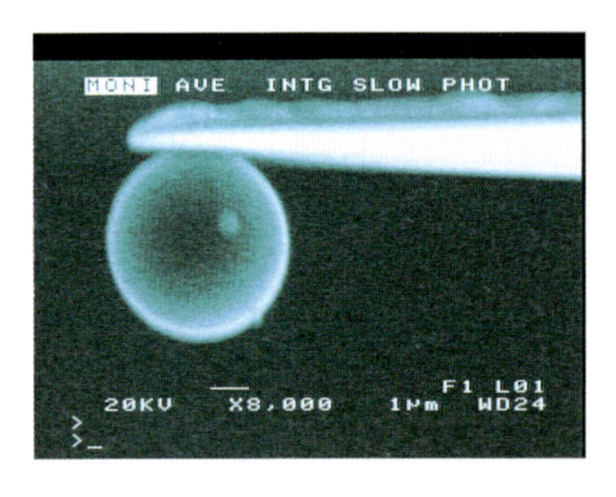


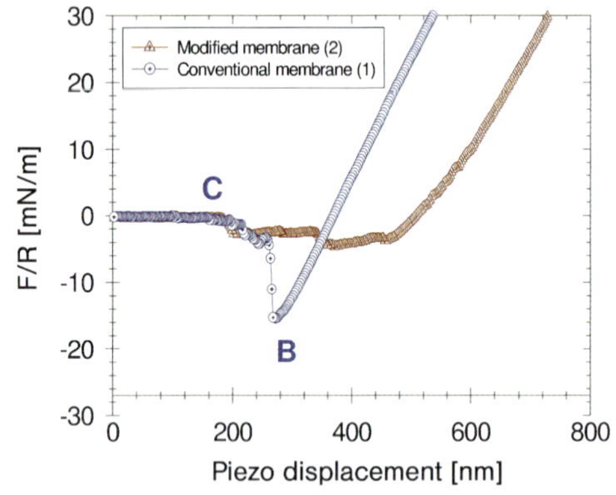


**Figure 6.8** SEM image of a BSA-coated silica sphere attached to a tipless AFM cantilever and force versus piezo displacement plot (retraction) for two membranes in $10^{-2}$ M NaCl solution at pH 5.0 (conventional, ES404; modified XP117).

Ultrafiltration membranes are used particularly for the processing of solutions of biological macromolecules. Such molecules are too small to immobilize singly on a cantilever. However, by adsorbing such molecules onto colloid probes it is possible to measure their adhesion to membrane surfaces. Such a protein-modified probe (using bovine serum albumin, BSA) and the corresponding adhesion data for two membranes is shown in Figure 6.8 [4].

The data show that the protein-modified probe has significantly lower adhesion to the development membrane (XP117) than to the existing commercial membrane (ES404); and hence the modified membrane has significantly lower fouling potential in the processing of such protein solutions.

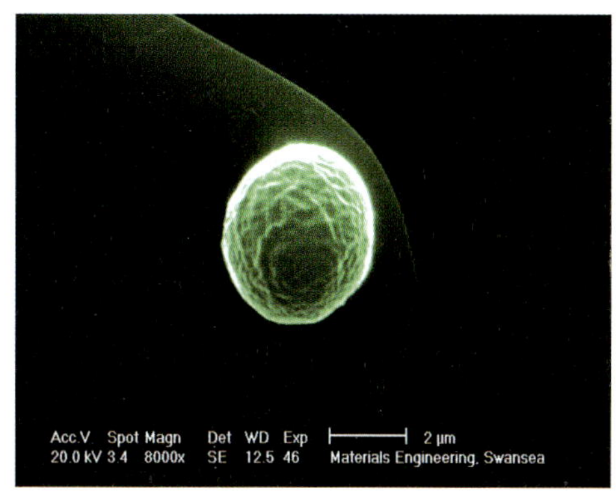


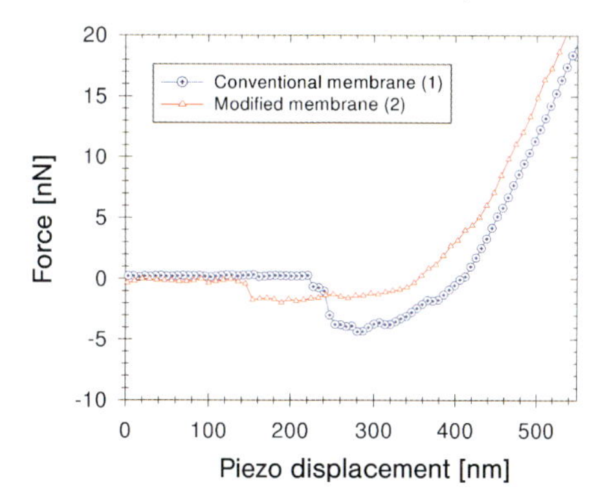


**Figure 6.9** SEM image of a yeast cell (*Saccharomyces cerevisiae*) attached to a tipless AFM cantilever – a cell probe – and force versus piezo displacement plot (retraction) for two membranes in $10^{-2}$ M NaCl solution at pH 5.0 (conventional, ES404; modified XP117).

Membranes are frequently used to process biological cell dispersions. In order to elucidate such processes, it has proved possible to immobilize single cells at tipless AFM cantilevers, creating *cell probes*, whilst maintaining the viability of such a cell (Figure 6.9) [4].

Such cell probes allow the direct measurement of the adhesion of biological cells to membranes. Interpretation of the data for such cell probes is more difficult than for colloid or modified colloid probes, as the cell can distort during the measurements. However, the data show that the XP117 development membrane has lower adhesion and hence lower fouling propensity, also for yeast cells.

## 6.3
## Correspondence Between Surface Pore Dimensions from AFM and MWCO

From a practical point of view, the choice of ultrafiltration membranes for a particular process is usually defined in terms of MWCO, defined such that solutes of such molecular weight would be 90% rejected by the membrane. Care is required in the use of MWCO, as rejection can depend on other factors such as charge and molecular shape, but it remains an extensively used parameter. It is possible to estimate the mean pore diameter of membranes from rejection data such as MWCO through the use of an appropriate mathematical expression. As MWCO is an historically important measure and AFM a relatively new technique it is pertinent to investigate the correspondence between the data respectively obtained.

Such an investigation has been carried out for a series of membranes with MWCO in the range 1000–10 000 (Desal-G, Osmonics) [5]. Images of membranes obtained in air at the extremes of this range, GE (1000) and GN (10 000), are shown in Figure 6.10. The higher MWCO membrane is noticeably rougher.

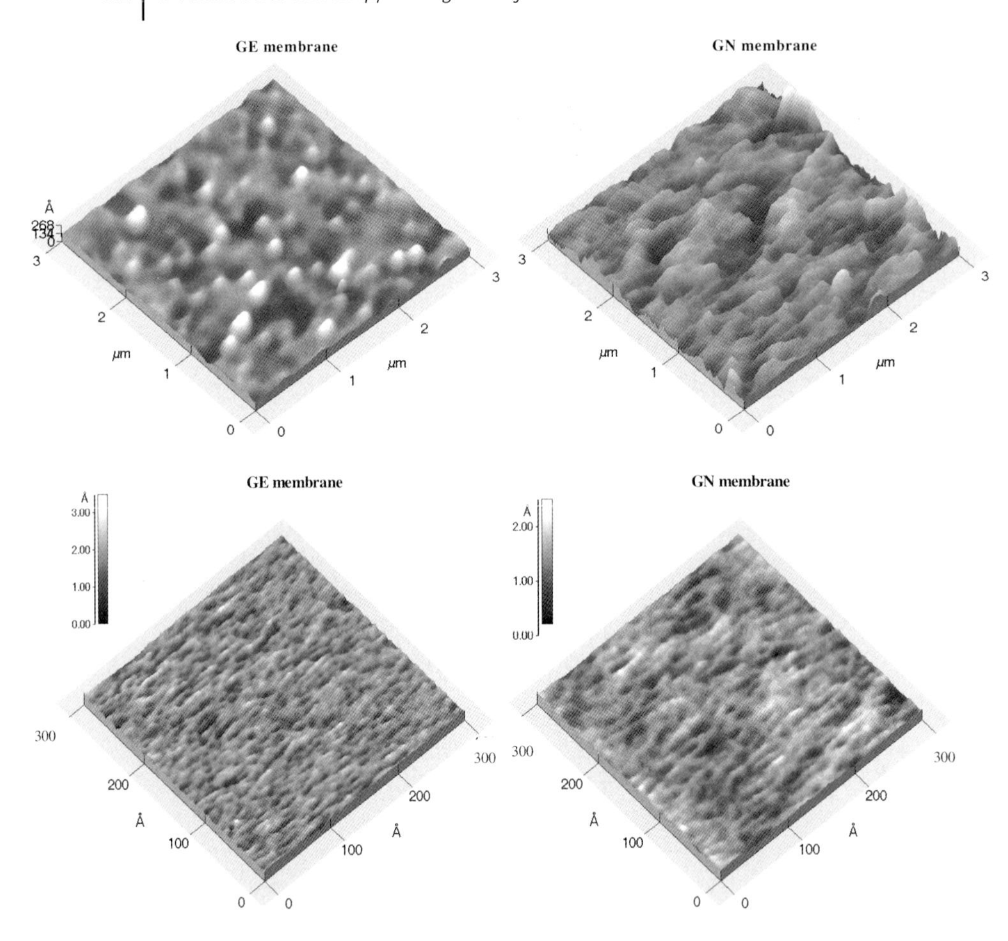


**Figure 6.10** AFM images of two Desal G membranes, GE (MWCO 1000) and GN (MWCO 10 000) at low resolution (above) and high resolution (below).

Indeed, there is a significant increase in surface roughness throughout the range, as shown by the data in Table 6.1.

From such images it is possible to determine the surface pore diameter distributions of the membranes. Table 6.2 gives data, together with the mean pore diameters from MWCO using three differing mathematical expressions. Depending on the expression used, the ratio of the mean pore diameters obtained from MWCO data and AFM measurements was in the range 1.04–2.42. Agreement between the diameters obtained in the two ways was best for the membranes with the lowest MWCO values. The two measurements are probing the membrane properties in different ways, but the overall correlation is encouraging.

**Table 6.1** Surface characteristics of Desal G membranes over areas of $3 \times 3\,\mu m$. $R_{p-v}$ = peak to valley distance; Rms = root mean square.

| Membrane | MWCO | $R_{p-v}$ (nm) | Rms roughness (nm) |
|---|---|---|---|
| GE | 1000 | 26.6 | 3.6 |
| GH | 2500 | 58.7 | 8.4 |
| GK | 3500 | 53.9 | 8.4 |
| GM | 8000 | 63.7 | 9.2 |
| GN | 10 000 | 88.1 | 11.7 |

**Table 6.2** Pore diameters for Desal G series membranes obtained from AFM measurements and MCWO. AFM values within brackets are standard deviations.

| Membrane | AFM mean pore diameter (nm) | MWCO mean pore diameter (nm) | | |
|---|---|---|---|---|
| | | Ex. 1 | Ex. 2 | Ex. 3 |
| GE | 1.8 (±0.3) | 1.9 | 2.4 | 2.0 |
| GH | 2.2 (±0.5) | 2.6 | 3.8 | 3.2 |
| GK | 2.5 (±0.5) | 3.0 | 4.4 | 4.2 |
| GM | 2.8 (±0.7) | 4.4 | 6.4 | 6.8 |
| GN | 3.1 (±0.9) | 4.8 | 7.2 | 7.6 |

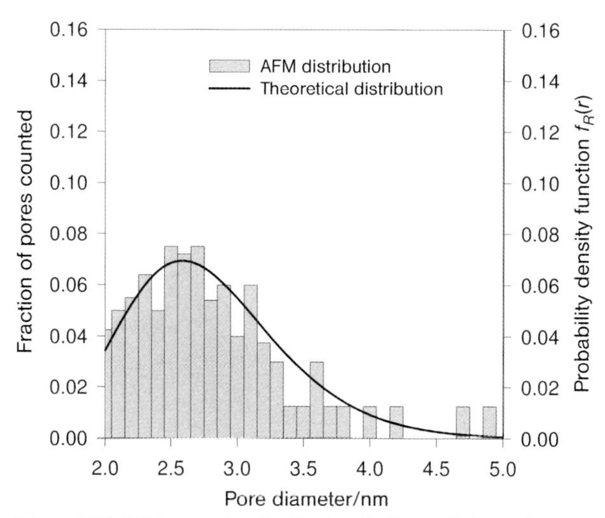


**Figure 6.11** AFM pore size distribution for Desal GN membrane and theoretical log normal distribution.

However, AFM has the important advantage of providing a measure of pore size distribution. This is indicated in Table 6.2 by the standard deviations. A full distribution for one of the membranes is given in Figure 6.11.

Knowledge of such distributions is very important if the membrane is to be selected for a fractionation process, where a narrow distribution is highly desirable. Further, theoretical modelling to predict membrane process performance often assumes a log normal distribution of pore sizes. Figure 6.11 shows that this is a reasonable assumption in this case.

## 6.4
## Imaging in Liquid and the Determination of Surface Electrical Properties

For membranes that are used in liquid systems it is very useful to have the possibility of imaging in a solution corresponding to the processing conditions of pH and ionic content. An atomic force microscope gives great control of imaging protocols, allowing investigation of the procedures that give the best membrane images, in particular the imaging force. Figure 6.12 shows the force of interaction between an AFM tip and a Cyclopore membrane with a specified pore diameter of 0.1 μm in solutions at constant pH but varying ionic strength [6].

It may be seen that the range of the interaction increases greatly as the ionic strength decreases in accordance with electrical double layer theory. It is then possible to image the membrane using a force at any point on the curves in Figure 6.12. Two series of such images, one at various forces at constant ionic strength

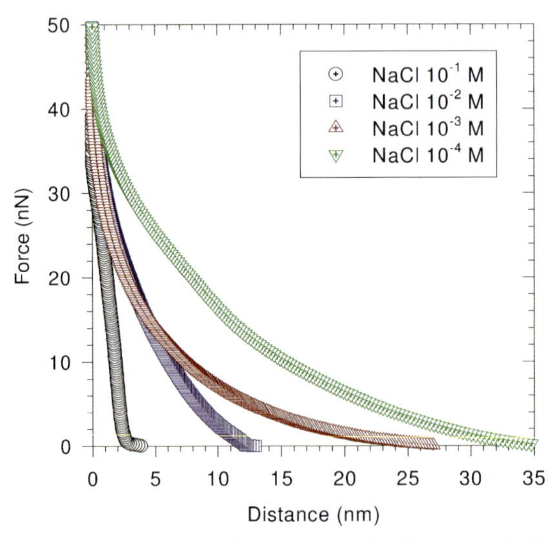


**Figure 6.12** Force vesus distance curves for the approach of an AFM tip to a Cyclopore microfiltration membrane in NaCl solutions of various ionic strengths at pH 6.5.

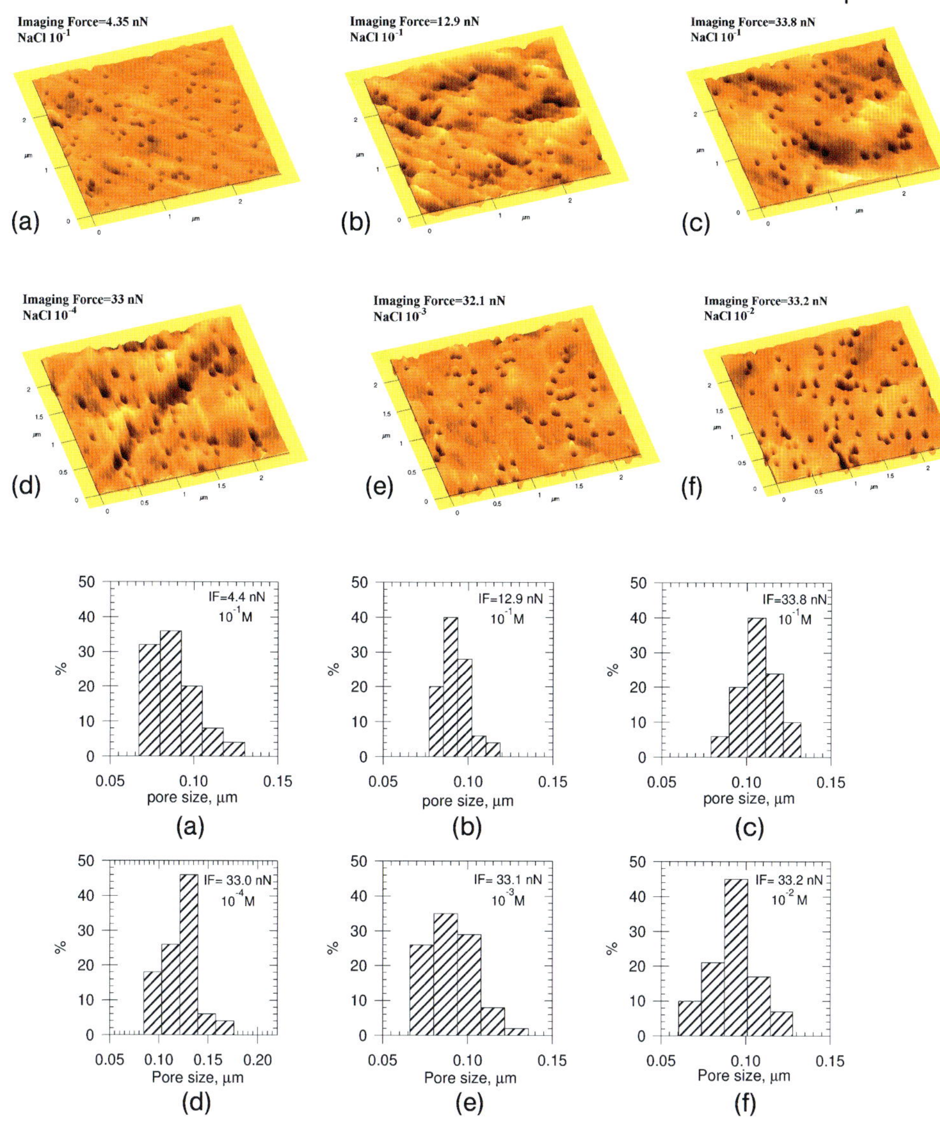


**Figure 6.13** AFM images of a Cyclopore membrane in electrolyte solutions: (a–c) at various forces in $10^{-1}$ M NaCl solution; (d–f) at approximately constant force at various ionic strengths. All data at pH 6.5. IF = imaging force.

and the other at approximately constant force at various ionic strengths are shown in Figure 6.13. It may be seen that the quality of the images improves with increasing imaging force and with increasing ionic strength. The derived pore diameter distributions also vary with the imaging conditions, as shown in the corresponding graphs in Figure 6.13 [6].

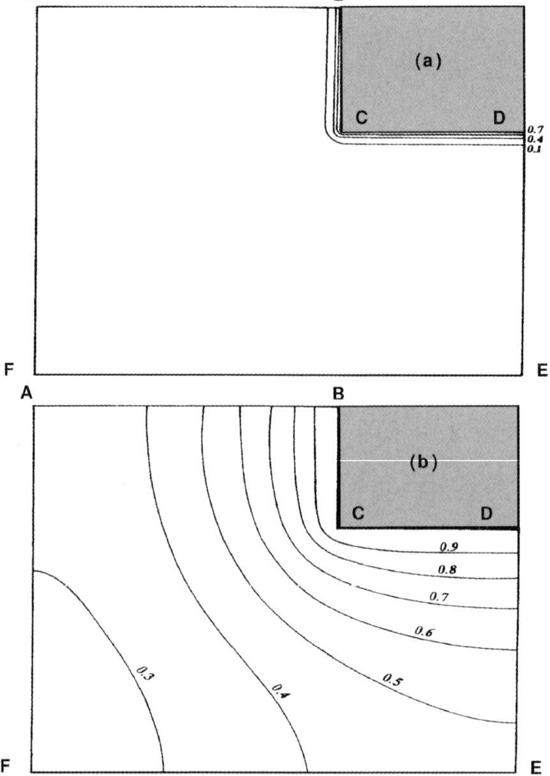


**Figure 6.14** Calculated isopotential lines at the entrance to a membrane pore of diameter 0.1 μm. (a) $10^{-1}$ M solution, (b) $10^{-2}$ M solution.

The reason for this variation may be understood from Figure 6.14 which shows calculated isopotential lines for a $10^{-1}$ M solution and a $10^{-4}$ M solution respectively.

Roughly understood, the imaging tip will follow an isopotential line during the imaging process. At the high ionic strength, all of the calculated lines lie close to the membrane surface (shaded, spinning the image 360° out of the plane of the paper would generate a pore). At the lower ionic strength, some of the isopotential lines lie far from the surface, so a clear, veracious pore image would not be obtained when such a line is followed, as would be the case at low imaging forces. In conclusion, the best imaging conditions in ionic solutions are at high imaging force and if possible and appropriate at high ionic strength.

Although most membrane technologists now recognize the important contribution of membrane surface electrical properties in defining the separation characteristics, there remains confusion as to how best to describe and quantify such interactions. There is an unfortunate tendency in the applied membrane literature to use the terms "charge" and "potential" loosely, almost interchangeably. There is also a regrettable lack of precision in the interpretation of membrane streaming potentials, which are the basic data most commonly used to quantify membrane surface electrical properties. The interpretation of streaming potential data can be complex, even for perfectly smooth and chemically uniform planar surfaces. Most membrane surfaces have roughness comparable with electrical double layer dimensions, so *along-the-surface* membrane streaming potential data give only some average property. *Through-the-membrane* streaming potential data may only be usefully interpreted in comparison with numerical double layer calculations as pore diameters are often less than characteristic electrical double layer dimensions.

Fortunately, AFM in conjunction with the colloid probe technique offers an alternative means of membrane surface electrical properties characterization. If a colloid probe is approached towards a surface it is possible to quantify the force of interaction. Figure 6.15 shows typical data for a Desal DK membrane, which is one of the least rough membranes [7].

Moreover, if the surface potential or surface charge of the colloid has been determined and the solution is of defined ionic content, it is possible to calculate the potential or charge of the surface under investigation by matching the experimentally obtained curves to theoretical calculations based on electrical double layer theory. In the example shown, the best-fit membrane surface charge was $-0.00114\,\mathrm{C\,m^{-2}}$ and the best-fit membrane surface potential was $-64\,\mathrm{mV}$. Furthermore, an important advantage of the colloid probe technique is that it allows exploration of variations in surface electrical interactions at different points on the membrane surface, as the following section shows.

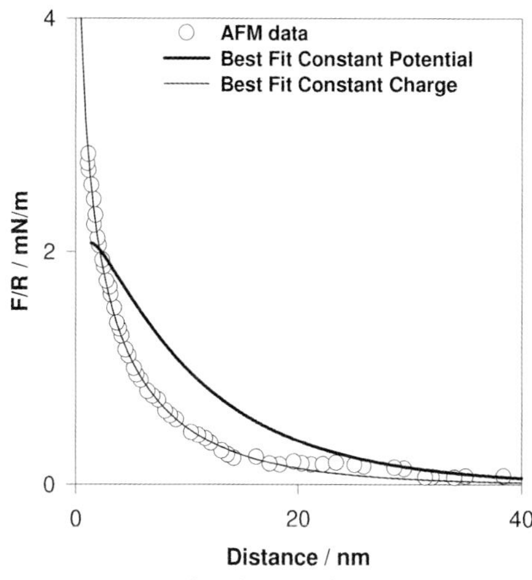


**Figure 6.15** Best fit solutions of theoretical force–distance curves to AFM experimental force–distance data for a Desal membrane in $10^{-3}$ M NaCl.

## 6.5
## Effects of Surface Roughness on Interactions with Particles

Surfaces are usually imaged using sharp tips in order to produce images with the highest possible definition. However, from a processing viewpoint an important factor is how process stream components such as solutes and colloidal particles interact with the surface. If a surface has feature variations that have dimensions comparable with those of the process stream components, then such interactions may show variation at different locations on the surface. An effective way of gauging such effects is to image the surface with an appropriate stream component, for example a particle that is immobilized to form a colloid probe. Figure 6.16 shows results for a reverse osmosis membrane (AFC99, PCI Membranes) imaged in saline solution, first with a sharp tip and then with a 4.2 μm silica sphere [8].

The membrane surface features are still apparent but less well defined in the colloid probe image. This is a rather rough membrane showing clear peaks and valleys. Following imaging, it is possible to position the colloid probe at any point on the surface to quantify colloid–surface interactions. Figure 6.17 shows long-range electrostatic interactions quantified at peaks and valleys respectively; and Table 6.3 presents some analysis of the data.

Both the magnitude and range of the electrical double layer interactions on the peaks are greatly reduced compared with those in the valleys. In both cases the range is very different from that at a planar surface, both experimentally (data for mica are shown for comparison) and theoretically (the Debye length). It is also possible to quantify the adhesive interaction between colloid probe and membrane surface at different locations, as shown in Figure 6.18 and Table 6.4.

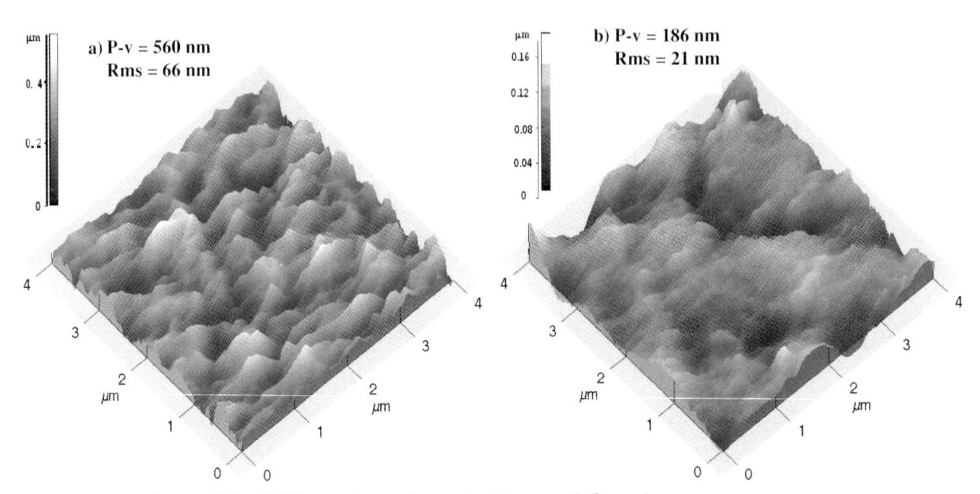


**Figure 6.16** AFC99 membrane imaged with a tip (left) and with a 4.2 μm colloid probe (right). P–v = peak to valley; Rms = root mean square.

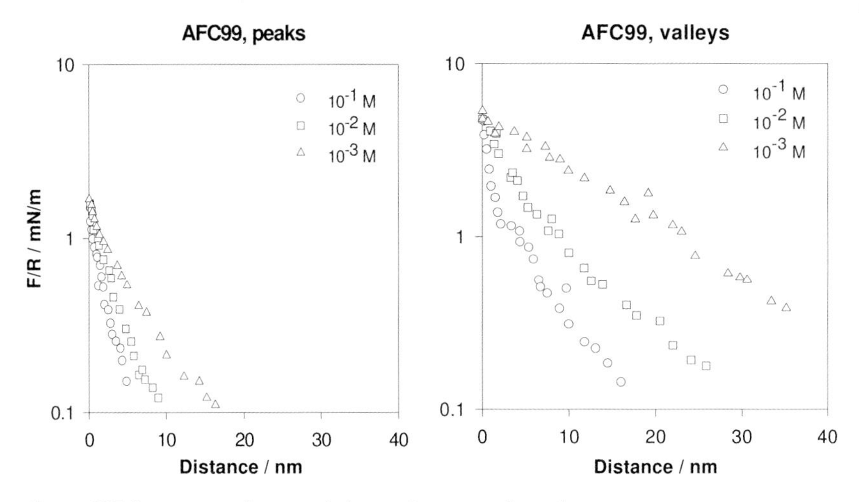


**Figure 6.17** Long-range electrostatic interactions at peaks and valleys for an AFC99 membrane at various ionic strengths in NaCl solutions. F = force; R = sphere radius.

**Table 6.3** Effective experimental decay lengths for approach curves between silica colloid probe and AFC99 membrane and mica, with Debye length.

| NaCl concentration (M) | Effective decay length (nm) | | | Debye length (nm) |
|---|---|---|---|---|
| | Membrane peak | Membrane valley | Mica | |
| $10^{-3}$ | 7.7 (±1.4) | 12.1 (±1.5) | 9.2 (±0.5) | 9.6 |
| $10^{-2}$ | 3.2 (±0.6) | 6.8 (±1.2) | 3.5 (±0.3) | 3.1 |
| $10^{-1}$ | 1.6 (±0.4) | 4.9 (±1.3) | 1.8 (±0.2) | 0.96 |

The adhesion of the colloid probe is markedly lower at the peaks on the membrane surface that in the valleys, with the difference increasing with decreasing salt concentration and reaching a factor of more than 20 in $10^{-3}$ M solution. The wide variation in interactions shows that theoretical descriptions of membrane fouling need to take explicit account of surface morphology. Further, the results show that the selection of membranes for the filtration of specific process streams would benefit from an assesment of the size of likely foulants and the membrane roughness, especially the periodicity of the roughness. Fouling could be minimized by using membranes with roughness such that only adhesion at peaks is possible – where the effects of crossflow are also maximum.

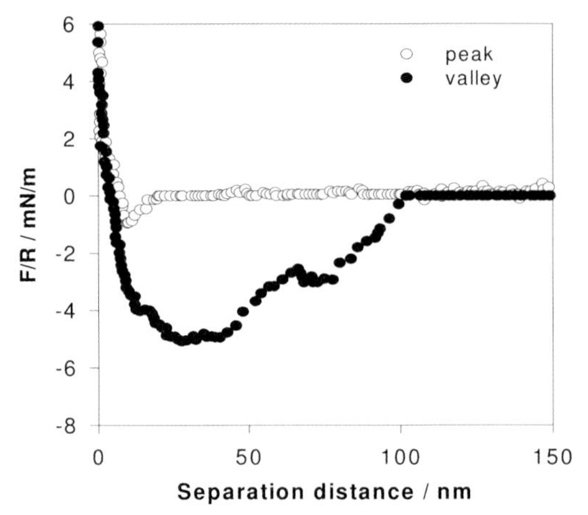


**Figure 6.18** Adhesion of a silica colloid probe at a peak and a valley on an AFC99 membrane in $10^{-3}$ M NaCl solution (pull-off force).

**Table 6.4** Normalized adhesion forces for silica colloid probes and AFC99 membrane.

| Surface type | NaCl concentration (M) | F/R (mN/m) |
|---|---|---|
| Membrane peak | $10^{-3}$ | −0.3 (±0.17) |
| | $10^{-2}$ | −1.4 (±0.43) |
| | $10^{-1}$ | −2.3 (±0.48) |
| Membrane valley | $10^{-3}$ | −6.5 (±2.2) |
| | $10^{-2}$ | −8.1 (±3.5) |
| | $10^{-1}$ | −8.7 (±4.0) |

## 6.6 "Visualization" of the Rejection of a Colloid by a Membrane Pore and Critical Flux

One of the most useful practical operating concepts for membrane processes is that of a critical filtration flux or critical operating pressure. These critical parameters are such that below such critical values rejection will occur and fouling will be minimum, while above these critical values both transmission and fouling may take place. For colloidal particles, the critical values may arise as a balance between the hydrodynamic force driving solutes toward a membrane pore and an electrostatic (electrical double layer) force opposing this motion.

Science fiction writers have imagined tiny probes, travelling for example through the human body, which would allow us to visualize hidden microscopic phenomena. Such probes remain figments of the imagination. However, a colloid

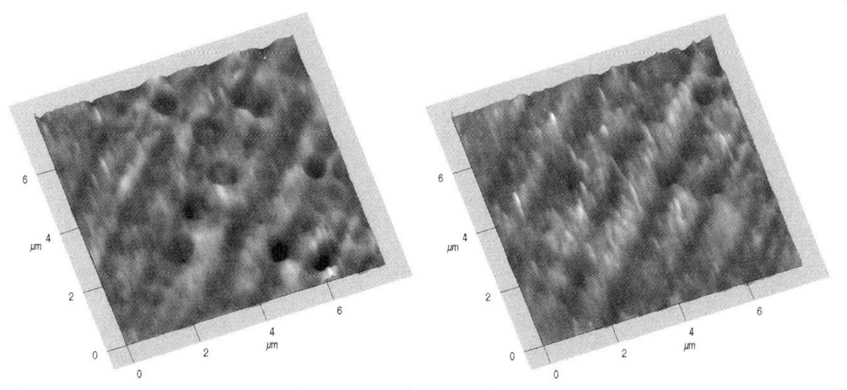

**Figure 6.19** Imaging a 1.0 µm Cyclopore membrane with a 0.75 µm colloid probe: in 0.1 M NaCl, pH 8 (left); in 0.0001 M NaCl, pH 8 (right). Normalized imaging force 4.6 mN/m.

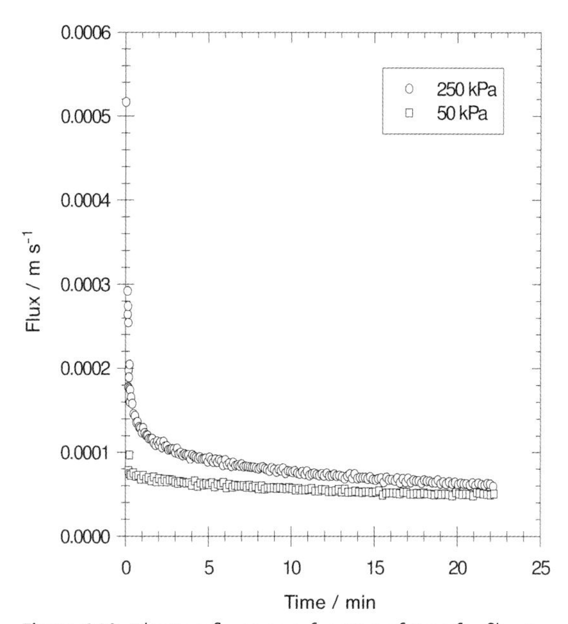


**Figure 6.20** Filtration fluxes as a function of time for filtration of particles very close in dimensions to the pore dimensions. Silica particles, 0.001 M NaCl, pH 6.0, mean particle diameter 86 nm, mean pore diameter 85 nm, calculated critical pressure = 130 kPa.

probe moving along a membrane surface can allow us to visualize how a colloidal particle would "see" such a surface, Figure 6.19 [9].

The figure shows how a 0.75 µm silica colloid probe "sees" the pores in a 1.0 µm Cyclopore membrane in solutions of two ionic strengths when imaged in each case with an applied force of 4.6 mN $m^{-2}$. It is only at the highest ionic strength that the pores are clearly apparent, for at the lowest ionic strength such a force is experienced too far from the membrane surface for the electric field to still show sufficient evidence of membrane porosity. Cases where the process stream component dimensions are close to those of the membrane pores should preferably be avoided as rapid pore blocking may occur if the critical flux or pressure is exceeded. Such conditions may, however, be theoretically predicted. Thus, Figure 6.20 shows the time dependence of flux for such a case where the critical pressure is calculated to be 130 kPa. Above this pressure there is a rapid decrease in flux with time, but below this pressure only a very gradual flux decrease occurs.

## 6.7
## The Use of AFM in Membrane Development

The ability of AFM to guide the development of improved membranes has been shown in the development of polysulfone-sulfonated poly(ether ether) ketone (PSU/SPEEK) blend membranes. The aim of the work was to develop highly charged membranes with correspondingly low adhesion characteristics and high critical fluxes [10]. The SPEEK provides negative charges, as shown by the structure in Figure 6.21.

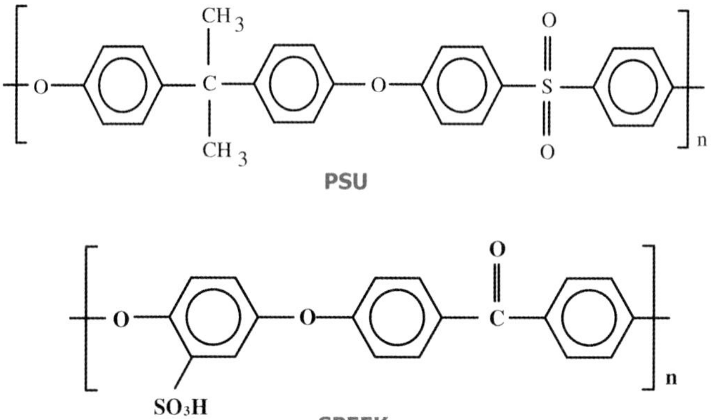

PSU

SPEEK

**Figure 6.21** Structures of polysulfone and sulfonated poly (ether ether) ketone.

One great benefit of SPEEK is that it acts as a pore formation-promoting agent. This gives membranes with high porosity and high fluxes, as shown by the increase in water flux with increasing SPEEK content shown in Figure 6.22.

The increase in porosity may be directly confirmed by high-resolution images of an unmodified PSU membrane and a PSU/SPEEK membrane with 5% of the charged polymer (Figure 6.23).

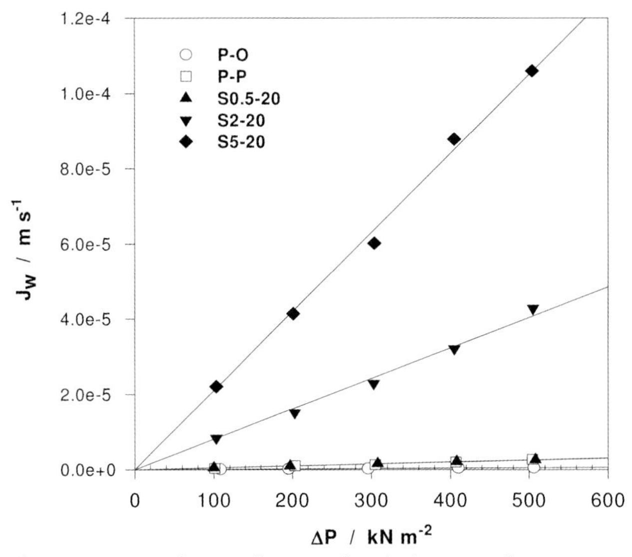

**Figure 6.22** Water flux as a function of applied pressure for five membranes of various percentages of SPEEK. P–O and P–P: 0%. S0.5–20: 0.5%. S2–20: 2.0%. S5–20: 5%.

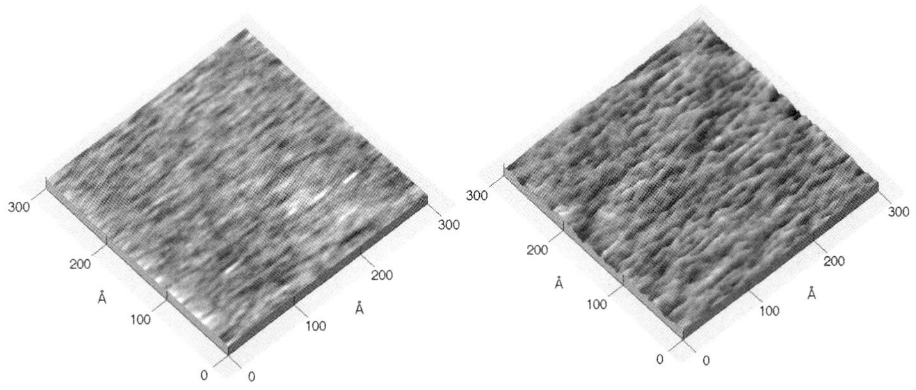

**Figure 6.23** High-resolution images of unmodified membrane (left) and membrane modified with 5% SPEEK (right, S5-20).

The adhesion of a range of colloid probes, both inorganic and biological, is greatly reduced at the PSU/SPEEK membranes, as shown by the data in Table 6.5.

Such low adhesion forces show that the membranes are well suited to many types of separation. The removal of humic acid was investigated as an example of a challenging separation. The blend membranes gave very good separation with little fouling. By quantification of the fraction of humic acids deposited during filtration it was possible to show that both planar (S5-20) and tubular (T5-20) versions of the membranes showed high critical flux values, whereas membranes

**Table 6.5** Normalized adhesion forces for a range of colloid probes at a PSU/SPEEK membrane (S5-20).

| Materials | Mean F/R (mN/m) | |
|---|---|---|
| | SPEEK/PSU | PSU |
| Silica | 0.84 (±0.35) | 6.5 (±0.8) |
| Cellulose | 0.51 (±0.23) | 2.2 (±0.32) |
| Latex | 2.1 (±0.52) | 14.3 (±0.8) |
| BSA | 1.4 (±0.64) | 12.3 (±1.4) |
| Yeast | 3.2 (±0.85) | 9.5 (±1.5) |
| Spores | 2.7 (±1.9) | 8.6 (±2.7) |

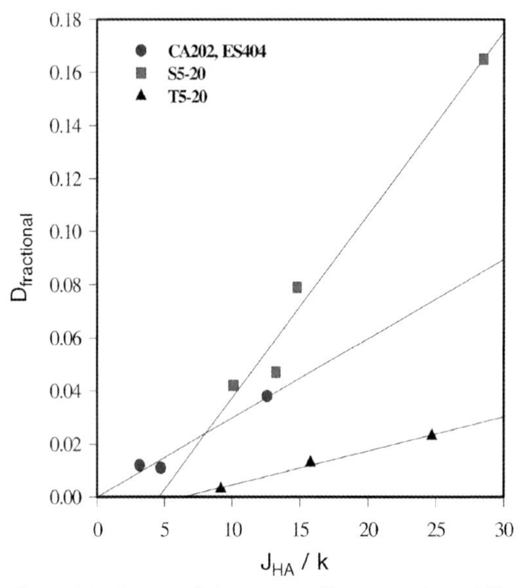


**Figure 6.24** Fractional deposition of humic acids at different operating fluxes. The intercepts on the *x*-axis indicate the critical fluxes.

already in commercial use for such separations (CA202, ES404; PCI Membranes) did not show such a favorable property (Figure 6.24) [11].

To allow for the different membrane geometries the data in Figure 6.24 are expressed as critical Peclet numbers ($J/k$, where $J$ is the flux, $k$ is the mass transfer coefficient). The critical Peclet numbers were 4.6 and 6.4 for the planar and tubular membranes respectively.

## 6.8 Conclusions

The many benefits of AFM in investigations of membranes and membrane processes may be summarized as:

1. Atomic force microscopy can determine the key properties of synthetic membranes: pore size distribution, surface morphology and surface roughness, surface electrical properties, surface adhesion.
2. Correspondence between surface pore dimensions from AFM and MWCO is good. In addition, AFM gives surface pore size distribution.
3. Operation in liquid and colloid probe techniques are particular advantage of AFM.
4. AFM can establish the effects of changes in interactions over the surface of membranes, for example, due to local morphology.
5. AFM allows the visualization of solute/membrane interactions.
6. AFM is a very useful asset in assessing the properties of membranes during their development.

Indeed, it is apparent that the capabilities of AFM are very closely matched to the knowledge requirements of membrane scientists and engineers.

### Acknowledgments

We thank all who have contributed to this research, especially Teodora Doneva, Bob Lovitt, Martin Peer, Peter Williams, Chris Wright and Huabing Yin.

### References

1 W.R. Bowen, N. Hilal, R.W. Lovitt, P.M. Williams, *J. Membrane Sci.* **1996**, *110*, 233–238.
2 W.R. Bowen, N. Hilal, R.W. Lovitt, P.M. Williams, *J. Membrane Sci.* **1996**, *110*, 229–232.
3 W.R. Bowen, N. Hilal, R.W. Lovitt, C.J. Wright, *J. Membrane Sci.* **1998**, *139*, 269–274.
4 W.R. Bowen, N. Hilal, R.W. Lovitt, C.J. Wright, *J. Membrane Sci.* **1999**, *154*, 205–212.
5 W.R. Bowen, T.A. Doneva, *Surf. Interface Anal.* **2000**, *29*, 544–547.
6 W.R. Bowen, N. Hilal, R.W. Lovitt, A.O. Sharif, P.M. Williams, *J. Membrane Sci.* **1997**, *126*, 77–89.
7 W.R. Bowen, T.A. Doneva, J.A.G. Stoton, *Colloid. Surface. A.* **2002**, *201*, 73–83.
8 W.R. Bowen, T.A. Doneva, *Desalination* **2000**, *129*, 163–172.
9 W.R. Bowen, N. Hilal, M. Jain, R.W. Lovitt, A.O. Sharif, C.J. Wright, *Chem. Eng. Sci.* **1999**, *54*, 369–375.
10 W.R. Bowen, T.A. Doneva, H.B. Yin, *J. Membrane Sci.* **2001**, *181*, 253–263.
11 W.R. Bowen, T.A. Doneva, H.B. Yin, *J. Membrane Sci.* **2002**, *206*, 417–429.

# 7
# Confocal Raman Microscopy for Membrane Content Visualization

*Philippe Sistat, Patrice Huguet, and Stefano Deabate*

## 7.1
## Introduction

This chapter deals with microRaman confocal microscopy as a useful tool for visualizing membrane processes. Section 7.2 is devoted to a short presentation of the Raman effect theoretical background. Sections 7.3 to 7.5 describe the Raman microscopy technique and main data processing methods currently used to gain quantitative information from spectral data obtained in this way. Section 7.6 deals with Raman imaging and visualization of chemical contents in transparent materials. Finally, a few examples of in situ visualization of molecular species migrating through a polymer membrane are reported (Section 7.7). This chapter ends with a critical review of ex situ and in situ Raman microscopy characterization of membrane processes and Raman imagery.

This summary account for a very large subject necessarily depends on a number of more detailed articles that record the primary sources for particular areas. The reader will find a wider presentation of the theory and practice of Raman microscopy in [1–4].

## 7.2
## The Raman Effect

The inelastic scattering of light, named the Raman effect after its experimental discoverer [5, 6], provides a complementary technique to infrared spectrometry for the study of vibrational and rotational spectra. In Raman spectrometry, the sample (a solid, a transparent liquid, a gas) is illuminated with a monochromatic light beam of wavenumber $\tilde{\nu}_0$ (typically from some specific type of laser). Most of the incident light is transmitted without change while a small portion is deflected (scattered) from the original direction, within the whole solid angle. In addition to the elastic scattering without change of wavenumber $\tilde{\nu}_0$ (Rayleigh scattering), the

*Monitoring and Visualizing Membrane-Based Processes*
Edited by Carme Güell, Montserrat Ferrando, and Francisco López

ISBN: 978-3-527-32006-6

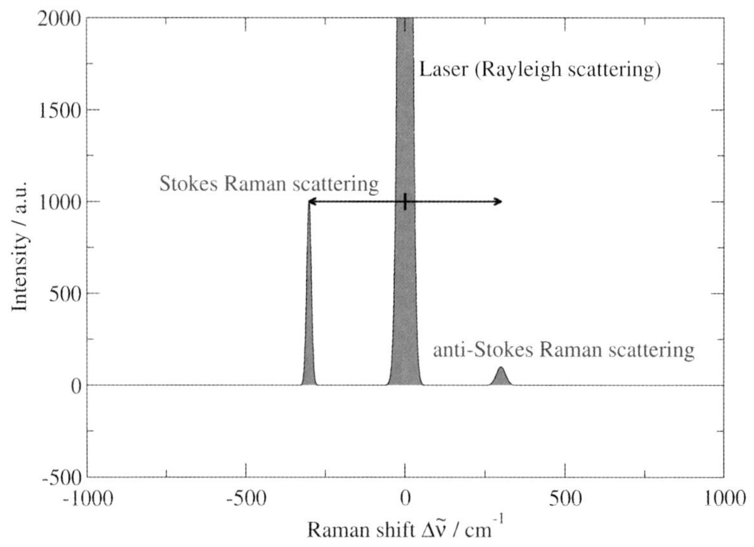


**Figure 7.1** Stokes and anti-Stokes scattering.

deflected light also exhibits discrete components of altered wavenumber (Raman scattering). Pairs of new lines appear in the spectrum which are symmetrically positioned with respect to the Rayleigh line, i.e. at wavenumbers $\tilde{\nu}' = \tilde{\nu}_o \pm \tilde{\nu}_M$. The wavenumber $\tilde{\nu}_M$ corresponds to a transition between vibrational or rotational energy levels of molecular systems. More rarely, electronic transitions are involved. In this case, the energy of the incident and scattered photons are different and the scattering is inelastic.

Raman scattering is a very weak effect, with an intensity usually three to five orders of magnitude lower than Rayleigh scattering. The Raman bands whose wavenumber is shifted towards lower values than the exciting light wavenumber (i.e. $\tilde{\nu}' = \tilde{\nu}_S = \tilde{\nu}_o - \tilde{\nu}_M$) are referred to Stokes lines, whereas those appearing at higher values (i.e. $\tilde{\nu}' = \tilde{\nu}_{aS} = \tilde{\nu}_o + \tilde{\nu}_M$) are referred to anti-Stokes bands (Figure 7.1). Usually, only the most easily observable Stokes side (see Section 7.2.1) of the Raman spectrum is reported, plotted as a relative wavenumber scale $\tilde{\nu}_o - \tilde{\nu}' = \tilde{\nu}_M$ (thus, the abscissa is preferentially referred to "wavenumber shift, cm$^{-1}$"), in the range from about 0 cm$^{-1}$ to 4000 cm$^{-1}$. In terms of absorption spectroscopy, this corresponds to the medium and far-infrared ranges, including vibrational and rotational molecular transitions as well as the lattice vibrations of crystals.

### 7.2.1
### Partial Quantum Mechanical Treatment of the Raman Effect

At the beginning of the twentieth century a general understanding of the origins of molecular spectra was achieved with the advent of the quantum theory. According to the basic principle of quantum mechanics, the energy associated with the vibrational and rotational degrees of freedom of a molecule is quantized, i.e. it can only take a discrete set of values characterized by the quantum number $\nu$ or $J$ (Figure 7.2). Radiation absorbed or emitted by a molecular system results from an upward or downward transition between two energy levels.

The molecular energy gain or loss, $\Delta E$, corresponds to the energy of the absorbed or emitted radiation which is, in turn, directly proportional to the radiation frequency or wavenumber:

$$\Delta E = h\nu = hc\tilde{\nu} \tag{7.1}$$

where $h$ is the Planck constant ($h = 6.62608 \times 10^{-34}$ J.s), $c$ is the light speed (in vacuum $c = c_0 = 2.99792458 \times 10^8\ \mathrm{m\,s^{-1}}$), $\nu$ is the frequency and $\tilde{\nu}$ is the wavenumber.

For condensed phases, infrared absorption/emission is a single photon process involving direct transition between two vibrational levels (usually the ground state, $\nu$=0, and the first excited state, $\nu = 1$). Differently, Raman scattering involves two quasi-simultaneous transitions consisting in the annihilation of the incident photon (whose $E = h\nu_o \gg h\nu_M$), followed by the creation of a new one. The photons of the incident radiation interacting with the sample induces the potential energy of the molecules to be raised from the ground to a virtual state (transition not corresponding to an eigenvalue of the system energy; Figure 7.1). Almost immediately, most molecules return to the ground state ($\nu = 0$) through the emission of a photon of the same frequency $\nu_o$ (Rayleigh scattering). A small fraction of photons come back to the first excited vibrational state ($\nu = 1$), so the energy of the created (scattered) photon is $h(\nu_o - \nu_M)$, observed at the wavenumber $\tilde{\nu}' = \tilde{\nu}_o - \tilde{\nu}_M$

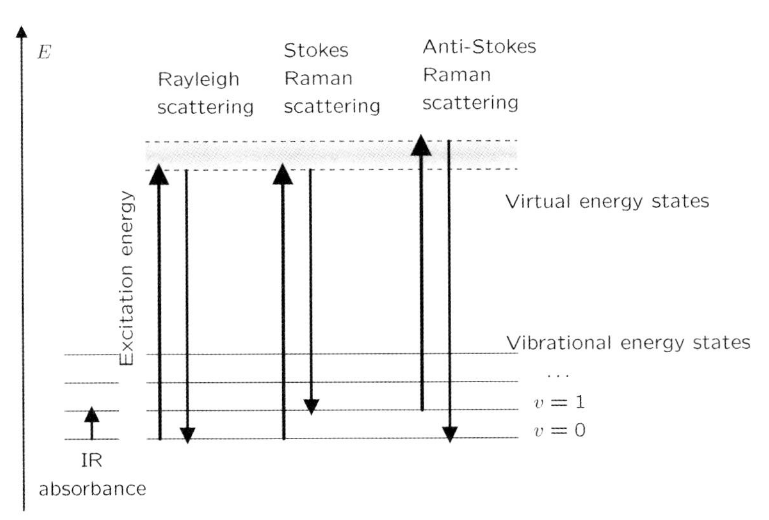


**Figure 7.2** Transitions between vibrational energy levels involved in the processes of infrared absorption, Rayleigh and Raman scattering.

(Stokes Raman scattering). Molecules that are in an excited state ($v = 1$) can undergo analogous effects when illuminated by a laser. Again, most of these molecules give rise to Rayleigh scattering but a small number comes back from the virtual to the ground state ($v = 0$). In this case, the energy radiation $h(v_o + v_M)$, corresponds to a Raman line at the wavenumber $\tilde{v}' = \tilde{v}_o + \tilde{v}_M$ (anti-Stokes scattering). Because of the strong decrease of the Boltzmann population of the excited states with $v$, the anti-Stokes lines intensity is lower than the Stokes bands and decreases faster with the Raman shift.

A given normal vibration exhibits spectral activity if the electric dipole moment $\mu$ of the molecule changes during this vibration, i.e. when the value of the dipole moment derivative, taken at equilibrium, differs from zero:

$$\left(\frac{\partial \mu}{\partial Q}\right) \neq 0 \tag{7.2}$$

where $Q$ is the normal coordinate associated to the vibration. In quantum theoretical terms, a transition between two energy levels may appear in the spectrum if the related transition moment $\mu_{fi}$ is nonzero:

$$\mu_{fi} = \left\langle \psi_f | \hat{\mu} | \psi_i \right\rangle \neq 0 \tag{7.3}$$

where $\psi_i$ and $\psi_f$ are the wave functions of the initial and final states, respectively, and $\hat{\mu}$ is the dipole moment operator. For direct absorption or emission of radiation (infrared spectroscopy), $\hat{\mu}$ corresponds to the permanent electric dipole operator. For light scattering (Raman spectroscopy), $\hat{\mu}$ is the induced dipole moment operator.

## 7.3 Raman Microspectrometry

Originally, Raman microspectrometry was developed to improve the Raman detection sensibility to small samples containing a limited number of molecules scattering the light. The intensity of the signal delivered by a spectrometer detector excited by a Raman line at the wavelength λ can be expressed as:

$$I_\lambda = I_0 \sigma_\lambda N \Omega S_d s_\lambda \tag{7.4}$$

where $I_0$ ($W\,m^{-2}$) is the laser irradiance at the sample, $\sigma_\lambda$ ($m^2\,sr^{-1}\,mol^{-1}$) the differential cross-section for the Raman band, $N$ (mol) is the number of molecules in the probed volume $V$ ($m^3$), $\Omega$ (sr) the solid angle of collection of the Raman light, $S_d$ ($m^2$) the surface of the detector heated by the scattered light and $s_\lambda$ the sensitivity of the detector at the wavelength λ.

When a small volume of matter is examined, compensation for the decreased number of molecules $N$ in the probed volume and increase of the detection sensitivity can be obtained by modifying the $I_0$ and $\Omega$ parameters. Some authors [7, 8] have explored a number of techniques for developing microRaman instruments and so demonstrated that the best way to develop microRaman analysis is to use a microscope for both exciting the sample and collecting the back-scattered light. Indeed, a microscope objective is able to focus the laser beam into a very small volume (a few cubic microns) and therefore to greatly increase the local radiance $I_0$. At the same time, the objective can collect the scattered light under a wide solid angle, increasing $\Omega$. Moreover, the use of a microscope allows to select any part of interest of a sample and, thus, to develop Raman mapping (see Section 7.7).

## 7.4 Confocal Raman Microscopy

Application of the confocal technique to Raman microprobing is an efficient way to obtain spectra without interference and to carry out 2D or 3D maps of small samples. Recently optical microscopes have been developed with a highly focused laser source (i.e. the laser spot size is only limited by the diffraction phenomena related to the optical system) that drastically enhances the spatial resolution. Whereas in a conventional microscope, the entire field of view is uniformly illuminated and observed, the confocal setup allows spatial filtering by optically conjugated pinhole diaphragms (Figure 7.3). The confocal arrangement cuts off the light coming from the small region of the sample coinciding with the laser focalization point and eliminates the contributions due to light scattered from domains beyond the focal plane. In other words, it allows a sensible decrease in the sampling volume without losing too much detected intensity.

Figure 7.3 depicts a sketch of the geometric configuration normally used to focus the laser beam, visualize the sample and collect the scattered light. Raman microspectrometers were originally designed to work with a confocal microscope.

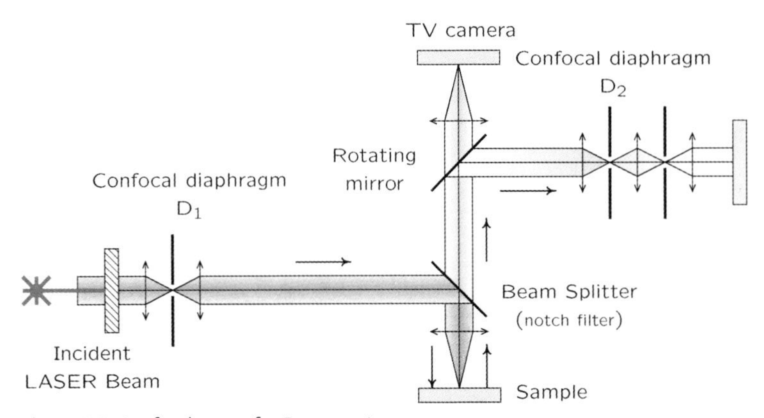


**Figure 7.3** Confocal setup for Raman microspectroscopy. Holes D1 and D2 are optically conjugated.

In order to obtain a real confocal configuration, a very accurate optical alignment is required as well as a high degree of stability and reproducibility of the mechanical and optical alignment.

The main advantages of the confocal microscopy are:

- A slight enhancement of the lateral resolution. Theoretically, the confocal setup may decrease the spreading function, but it should be mentioned that this requires very small diaphragms to be used, which in turn may decrease the amount of detected intensity.
- A very high enhancement of the axial resolution (depth discrimination), allowing the "optical sectioning" of a sample (if transparent). The axial resolution can be defined as the full width at half maximum (FWHM) of the intensity profile along the $z$ optical axis (see Figure 7.4). In confocal microscopy, a good approximation of the axial resolution is given by the expression:

$$\Delta z \geq \frac{\pm 4.4 n\lambda}{2\pi(\mathrm{NA})^2} \tag{7.5}$$

where $n$ is the refractive index of the analyzed material, $\lambda$ the laser wavelength and NA the numerical aperture of the microscope objective. For a $\times 100$ objective with $\mathrm{NA} = 0.9$, an incident radiation with $\lambda = 632\,\mathrm{nm}$ (red line of a He/Ne laser) and a sample with $n = 1.5$, the depth discrimination can be estimated at $\Delta z = 0.8\,\mu\mathrm{m}$.

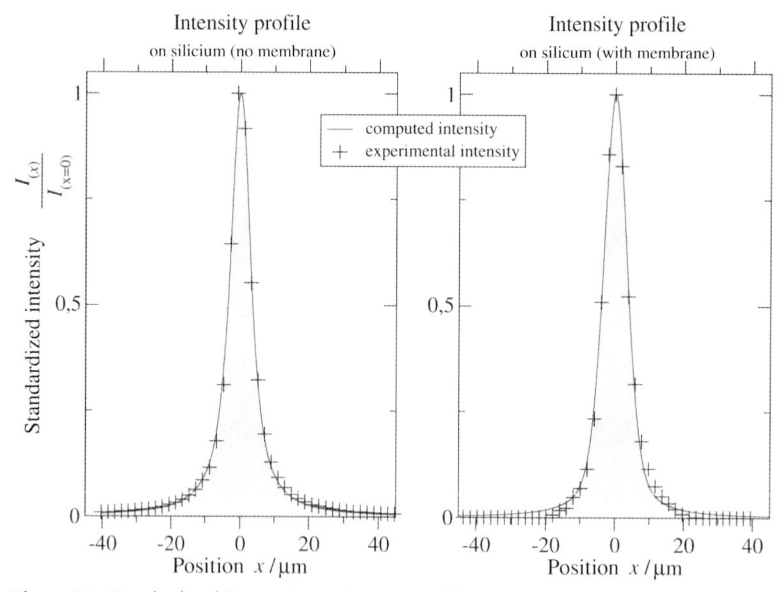


**Figure 7.4** Standardized Raman intensity scattered from a silicon sample as a function of the position. The focus is moved along the normal to the sample.

A complete calculation of the depth of focus for confocal microscopes is given in several books [9–11] and papers [12–14] Theoretical and experimental studies on the confocal depth resolution of immersion objectives were carried out by Wilson [15, 16]. The real resolution depth of a confocal Raman setup depends on the working conditions and can be experimentally obtained by measuring the intensity of the Raman lines of a silicon target, as a function of the penetration depth. Indeed, the silicon surface acts as a mirror plane. Figure 7.4 shows the example of the depth resolution determination within a dialysis cell containing a polymeric anion exchange membrane. A test was carried out with a $\times 60$ water immersion objective, with a long working distance (w.d. = 0.23 mm, NA = 1.2) and with compensation of the optical glass cover slide thickness.

Thanks to all the enhancements developed in the past decades, confocal Raman spectroscopy is now a technique of choice for the characterization of synthetics membranes [17–19] and chemical heterogeneous systems [20].

## 7.5 Specific Data Processing

Raman data collected into spectra contains a lot of information. However, this information is not usually directly available and the data must be processed to get qualitative and quantitative chemical informations. The main fact is that Raman line intensities are proportional to the concentration of chemical components and the Raman line positions are characteristics of the chemical bonding. Nevertheless, some phenomena like undesirable fluorescence, band overlapping, noise and spikes (outlier points) complicate the experimenter's task.

### 7.5.1 Baseline and Fluorescence

Most often the Raman spectra are spoilt due to a residual intensity linked with the fluorescence phenomenon. This phenomenon is undesirable in conventional Raman spectroscopy and it is necessary to remove the fluorescence background from the spectrum. For this purpose, several algorithms can be used, consisting in the subtraction of either a constant intensity value, or a linear or a polynomial baseline.

The fluorescence phenomenon is a severe drawback for the quantitative processing of spectroscopic data. In the case of polymeric membranes, their autofluorescence is known to decrease with the following treatments [21]:

- Membrane washing with an acid (hydrochloric acid for instance).
- Membrane cleaning by applying an ionic transport through the membrane (electrodialysis is best in this case).
- Increase of the direct laser illumination time.

## 7.5.2 Raman Lines Assignment

Raman spectroscopy is particularly attractive since it allows both qualitative and quantitative measurements at the same time. Each spectrum is the fingerprint of given species through the positioning of the different Raman lines. Indeed, the position of each line is directly related to the energy involved in the molecular vibrational motion. This energy depends at the same time on the nature of the atomic constituents (i.e. their mass) and on the strength of the interatomic chemical bonds. Schematically speaking, spectra result from the superpositioning of the different vibrational modes exhibited by each chemical group constituting the larger molecular entity.

Several methods exist to extract information from a spectrum. In the simplest case, it is possible to locate a specific band of a constituent which will allow identification. Some tables and sample spectra of simple compounds can be found in the literature [22–24]. If the different constituents of the studied sample do not overlap, it is easy to carry out identification and the following quantitative processing of spectra. In the case where there is overlapping, some mathematical techniques must be used to isolate the different contributions.

## 7.5.3 Quantitative Processing

Three principal approaches can be developed in order to extract quantitative information from Raman spectra. The first one is based on intensity ratio, which is a straightforward approach. The second relies on nonlinear least square fitting of the spectra by known functions. The third uses the arsenal of chemometric techniques, such as factor analysis.

### 7.5.3.1 Intensity Ratio

The most elementary data processing that can be implemented is based on intensity ratios. An intensity ratio $r_{\tilde{\nu}_1,\tilde{\nu}_2}$ can be defined as the relative intensities of two bands appearing at wavenumbers $\tilde{\nu}_1$ and $\tilde{\nu}_2$, so $r_{\tilde{\nu}_1,\tilde{\nu}_2} = I_{\tilde{\nu}_1}/I_{\tilde{\nu}_2}$. Since the intensity of the Raman scattering is linear with respect to concentrations, it is possible to compare respective intensities recorded at different wavenumbers. In the absence of bands overlapping, the intensity ratio $r_{\tilde{\nu}_1,\tilde{\nu}_2}$ is proportional to the concentration ratio of the corresponding species and then $r_{\tilde{\nu}_1,\tilde{\nu}_2} \propto C_1/C_2$, where $C_1$ and $C_2$ $\propto$ are the respective concentrations of the two species involved. When Raman bands are overlapped some additional precaution must be taken.

Often, it is possible to find a band featuring a specific molecular group whose concentration remains constant throughout the whole medium investigated. In this case, and without overlapping, it can be used as reference to compute intensity ratios. This reference band has to be strong enough in order to minimize the errors on the computed ratios.

When a solution is investigated, an internal reference can be added as a new solute whose concentration is known exactly. Obviously, every interaction of the added solute with the other species previously present in the solution (in terms of chemical reaction, coupled transport or any other effect) has to be avoided.

#### 7.5.3.2 Nonlinear Least Squares Fitting

Usually, Raman spectra can be seen as the sum of single specific bands, each centered on a wavenumber $\tilde{\nu}_o$. Raman bands are also characterized by some spreading around the central wavenumber $\tilde{\nu}_o$, defined by the full width at half maximum (FWHM), $w$, of the band. Unlike intensity, FWHM does not generally change with the concentration of the related chemical species. Theoretically, Raman bands can be described by two kinds of dispersion function: Gaussian or Lorentzian. The most general case is a linear combination of both functions.

**Gaussian** In a first attempt, the intensity of a Raman band can be modeled by the Gaussian function:

$$I_G(\tilde{\nu}) = I_{max} G(\tilde{\nu}) = I_{max} e^{-4(\tilde{\nu}-\tilde{\nu}_o)^2 \ln(2)/w^2} \tag{7.6}$$

where $\tilde{\nu}_o$ is the wavenumber corresponding to the maximum intensity, $I_{max}$. When integrating the intensity to obtain the spectral band area $A$, it follows:

$$A_G = \int I_{max} G(\tilde{\nu}) d\tilde{\nu} = I_{max} w \sqrt{\frac{\pi}{4\ln(2)}} \tag{7.7}$$

**Lorentzian** The other model for the Raman band shape is the Lorentzian function:

$$I_L(\tilde{\nu}) = I_{max} L(\tilde{\nu}) = I_{max} \frac{1}{1 + 4\frac{(\tilde{\nu}-\tilde{\nu}_o)^2}{w^2}} \tag{7.8}$$

Integration gives:

$$A_L = \int I_{max} L(\tilde{\nu}) d\tilde{\nu} \tag{7.9}$$

$$A_L = I_{max} \frac{\pi w}{2} \tag{7.10}$$

**General Case** The most general case of a Raman band shape is the linear combination of Gaussian and Lorentzian functions:

$$I_{GL}(\tilde{\nu}) = g I_G(\tilde{\nu}) + (1-g) I_L(\tilde{\nu}) \tag{7.11}$$

where $g$ is the Gaussian ratio, often written as a percentage. Figure 7.5 depicts the shape of both functions computed with the same parameters ($\tilde{\nu}_0 = 2000\,\text{cm}^{-1}$; $w = 20\,\text{cm}^{-1}$; $I_0 = 10$ a.u.).

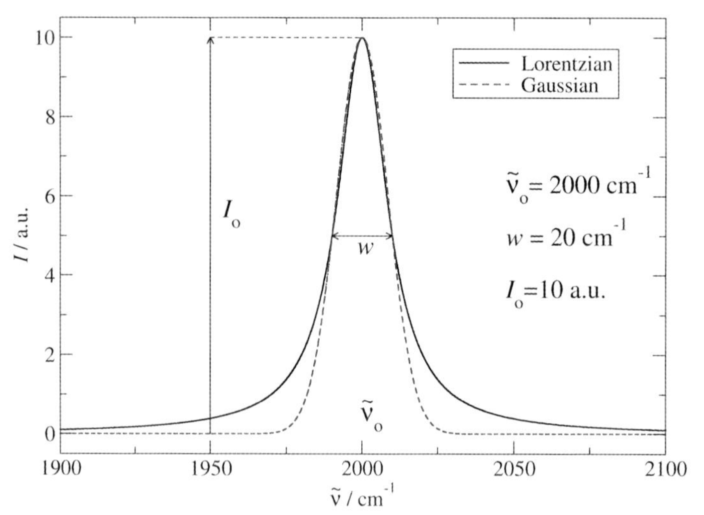


**Figure 7.5** Gaussian and Lorentzian functions computed with $\tilde{\nu}_0 = 2000\,\text{cm}^{-1}$, $w = 20\,\text{cm}^{-1}$, $I_0 = 10$ a.u.

Hence, the number of parameters necessary to describe a Raman band with mixed Gaussian and Lorentzian contributions is four, namely $\tilde{\nu}_o$, $w$, $I_{max}$ and $g$.

**Fitting** When dealing with real media, Raman spectra are the weighted sum of several contributions (bands), each of them described by a specific Gaussian–Lorentzian function. Thus, the whole experimental spectrum can be theoretically recomposed by a linear combination of Gaussian–Lorentzian functions, according to the formula:

$$I_{fit}(\tilde{\nu}) = \sum_{i=1}^{n} I_{GL}\left(\tilde{\nu}; < \tilde{\nu}_o^{(i)}, w^{(i)}, I_{max}^{(i)}, g^{(i)} >\right) \tag{7.12}$$

where $n$ is the number of scattering species and $< \tilde{\nu}_o^{(i)}, w^{(i)}, I_{max}^{(i)}, g^{(i)} >$ is the set of specific parameters related to the $i$-th Gaussian–Lorentzian contribution:

$$I_{fit}(\tilde{\nu}) = \sum_{i=1}^{n} I_{max}^{(i)} \left( g^{(i)} e^{-4\left(\tilde{\nu}-\tilde{\nu}_o^{(i)}\right)^2 \ln(2)/w^{(i)2}} + \left(1 - g^{(i)}\right) \frac{1}{1 + 4\left(\tilde{\nu} - \tilde{\nu}_o^{(i)}\right)^2 / w^{(i)2}} \right) \tag{7.13}$$

A given spectrum is fitted by minimizing the following sum:

$$S = \sum_{j=1}^{m} \left[ I_{\exp}\left(\tilde{\nu}_j\right) - I_{fit}\left(\tilde{\nu}_j; < I_{max}^{(i)}, \nu_o^{(i)}, w^{(i)}, g^{(i)} >\right) \right]^2 \tag{7.14}$$

where $j$ is the wavenumber index and $m$ is the total number of wavenumbers for which the intensity is recorded. The process is known as nonlinear least squares fitting and is often implemented as the Levenberg–Marquardt method which works very well in practice and is actually the standard of nonlinear least squares routines [25–27]. The Levenberg–Marquardt method is implemented in the majority of data processing software for spectroscopy.

The main drawback of this fitting method is that a great number of bands has to be used for complex spectra, this leads to a huge set of fitting parameters. Besides, some parameters should be kept fixed during analysis in order to minimize the risk to obtain incoherent results. Thus, some minimal knowledge of the system under study must be taken into account to help in the choice of the parameters to be fixed. Briefly, the higher the Raman bands overlapping in a spectrum the lower the information extracted from the fit. In the case of Raman visualization experiments (see Section 7.6), the huge number of parameters to be fitted forces the experimenter either to decrease the width of the investigated spectral domain or to refer to other techniques, such as principal component analysis.

#### 7.5.3.3 Principal Component Analysis and Related Methods

One definition of factor analysis is given by Malinowsky in [28]: "*Factor analysis is a multivariate technique for reducing matrices of data to their lowest dimensionality by the use of orthogonal factor space and transformations that yield predictions and/or recognizable factor*".

By this technique, a full set of recorded spectra is stored in a $m \times n$ intensity matrix $\mathbf{I}$, in which each line corresponds to a specific wavenumber value and each column to a specific spectral acquisition. Usually such a data matrix corresponds to the spatial or temporal accumulation of spectra. The matrix element $I_{i,j}$ corresponds to the $i$-th wavenumber $\tilde{\nu}_i$ recorded during the $j$-th acquisition. Since the Raman intensity is linearly proportional to the concentration of the scattering species, the data matrix corresponds to $\mathbf{I} = \mathbf{A}_{\text{purespectra}}\ \mathbf{B}_{\text{concentration}} + \mathbf{E}$, where $\mathbf{A}_{\text{purespectra}}$ is a $m \times p$ matrix containing as columns the $p$ spectra of each individual constituent of the compound, $\mathbf{B}_{concentration}$ is a $p \times n$ matrix containing as columns the concentration factors for each acquisition and $\mathbf{E}$ is the $m \times n$ experimental error matrix. From a mathematical point of view, the data matrix $\mathbf{I}$ can be decomposed into a matrix product, for instance through the singular value decomposition algorithm [27], giving the three matrices $\mathbf{U}$, $\mathbf{S}$ and $\mathbf{V}$:

$$\mathbf{I} = \mathbf{USV}' \tag{7.15}$$

In this expression $\mathbf{S}$ is a $p \times p$ diagonal matrix whose elements are the square roots of the eigenvalues, sorted in decreasing order of magnitude.

Usually, not all of the $p$ factors are needed to explain the features of the spectra and only $n_f$ factors can report data variation other than noise. Then the factor model can be compressed into:

$$\underbrace{\tilde{\mathbf{I}}}_{m\times n} = \underbrace{\tilde{\mathbf{A}}}_{m\times n_f}\underbrace{\tilde{\mathbf{B}}}_{n_f\times n} = \underbrace{\tilde{\mathbf{U}}}_{m\times n_f}\underbrace{\tilde{\mathbf{S}}}_{n_f\times n_f}\underbrace{\tilde{\mathbf{V}}'}_{n_f\times n} \tag{7.16}$$

by only taking into account the first $n_f$ factors where $n_f < p$.

$n_f$ must be high enough for that $\tilde{\mathbf{I}}$ provides an acceptable approximations of the original data matrix, i.e $\tilde{\mathbf{I}} \approx \mathbf{I}$.

Once the number of meaningful factors, $n_f$, is determined, a transformation step can allow the recognition of chemical factors and in turn find the concentration of the different chemical components. For this to be accomplished, some different techniques are available and discussed in the literature [28].

## 7.6
## Visualization Using Raman Spectroscopy

Raman scattering allows the identification of chemical bonding in a sample, while usually providing an accurate quantification of the scattering species concentration; this also applies to membrane systems [29–31]. Coupled with an optical microscope, Raman spectroscopy is a very useful tool for producing molecular chemical images of materials. The main Raman microspectroscopy drawbacks are: (i) the requirement to work with transparent samples and (ii) the intrinsic weakness of the Raman effect which needs long integration times and accurate data processing. Recently, the great improvement in weak signal detection systems has allowed quantitative information about the molecular spatial distribution inside heterogeneous material to be obtained, with a resolution limited only by the diffraction phenomena of the optical system. By judiciously selecting non-overlapping Raman lines featuring the various molecular constituents of the sample, it is possible to draw up a map that gives the relative abundance (2D or 3D distribution) of each chemical component inside the material [20, 32].

Three main kinds of visualization can be carried out by confocal Raman microscopy: line, plane or parallelepiped scanning. All of them can be carried out by point-to-point illumination with a microcontrol translation table. Alternatives techniques to the point-scanning technique are line imaging and widefield imaging. Both of them allow a distribution of the laser energy along a line or a plane, while the spatially resolved spectra are recorded by 2D CCD detectors [2]. However, even if they present some interesting features, like a lower integration time, point-scan imaging is still the most used technique.

### 7.6.1
### Line Scanning

In point-by-point scanning Raman microscopy, the laser spot is focused onto the surface or along the axial axis of a sample and a spectrum of each spatial position is collected. Scans of microvolumes can be carried out subsequently along a line, as shown in Figure 7.6a. A controlled microdisplacement device allows a very high mechanical resolution of the position. The sample is sequentially moved from position to position and stays at rest for a given time to take the Raman spectrum. The integration time corresponds to the time delay taken at each point.

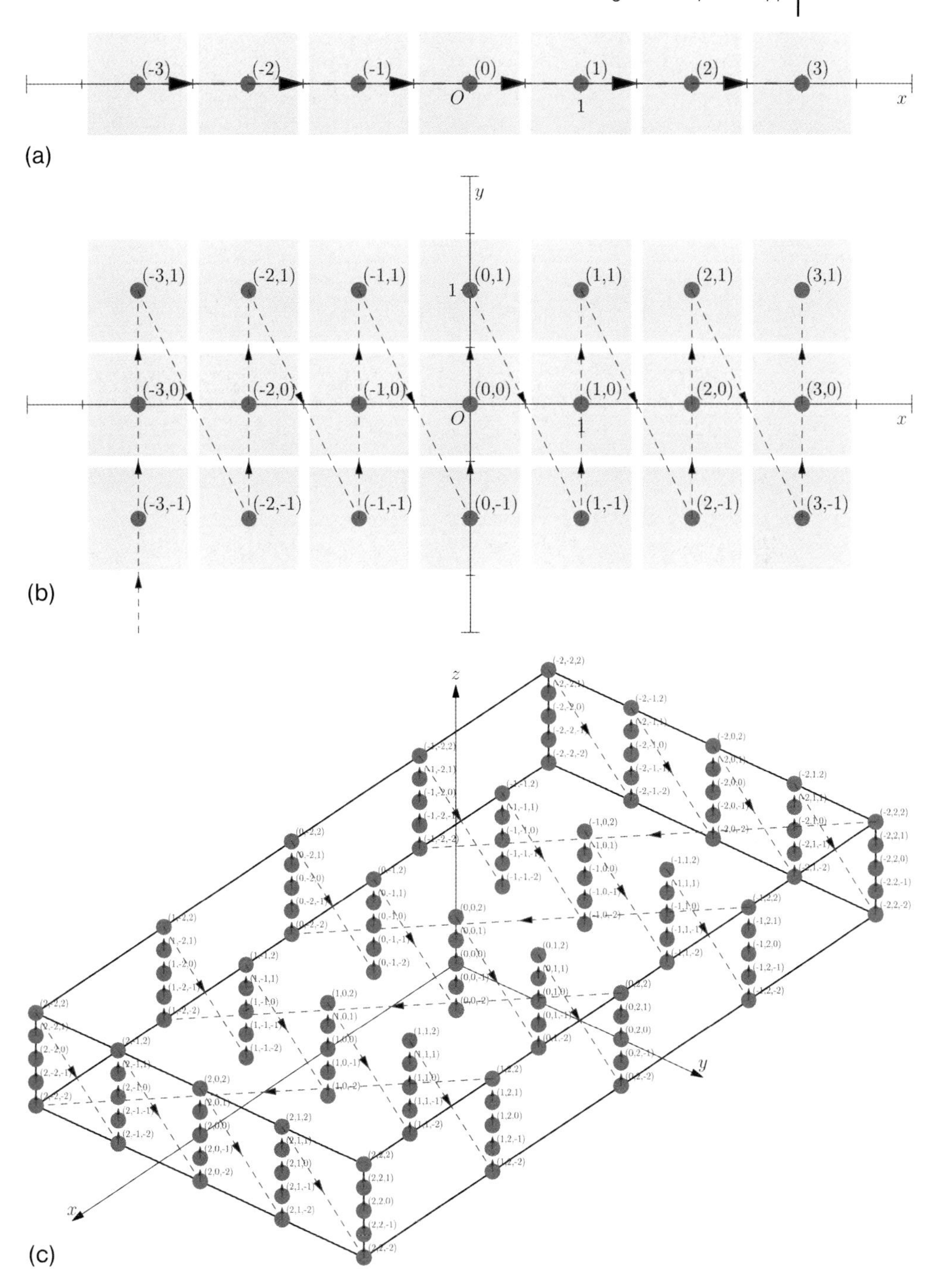


**Figure 7.6** Point-to-point illumination for scanning along a line (a), a rectangular domain plane within one plane (b) and a rectangular box/cuboid (c).

Another linear scan technique involves a mirror which continuously moves the focus point along the line. The detector is then continuously hit by photons emitted from different points along the scan line. In this case, detection from all the points along the scanned line is carried out during the same cycle. As a consequence, the laser energy is dispersed along this line and the confocal mode cannot be used.

### 7.6.2 Plane

Automatic positioning devices can be used to scan a plane in confocal mode. Usually, the surface exploration consists in a succession of line scans, as shown in Figure 7.6b. The kind of sequence used for an experiment is important for transient state studies. Let us be interested in a $(m,n)$ grid. If $\Delta t$ is the time taken between two successive positions, and if we denote $(i,j)$ as the focus point position located at the $i$-th row and the $j$-th column on the grid, the instant for recording the Raman intensity at this point falls between $t$ and $t + \Delta t$, where $t = \Delta t[(i-1)+(j-1)n]$. The real data must be taken into account in the study of transient phenomena.

### 7.6.3 Volume

The most complete space exploration within a sample consists in probing the whole volume, as shown in the Figure 7.6c. When working with a micro-positioning device, an obvious limitation of this kind of experiment is related to the time needed to scan all points. If we used a $m \times n \times p$ grid, the time required to check all points is $m \times n \times p \times \Delta t$. This formula clearly shows that, even if the applied integration time is of a few seconds, it is not realistic to investigate transient phenomena by this way: the spatial resolution must be low or the scanned volume dimension must be small.

## 7.7 Membrane Systems Applications

Raman microspectrometry can be applied to the study of membrane systems in contact with liquid solutions or gases; and it can allow the visualization of ionic species, solvent, or membrane structure [20, 33–44]. However, the cell used for investigation must be specifically designed in order to obtain the best optical coupling between the system under study and the spectrometer.

### 7.7.1 Cell Design

Experimental studies of the membrane transport by confocal Raman microspectrometry can be carried out only in some specific cases. Amongst various

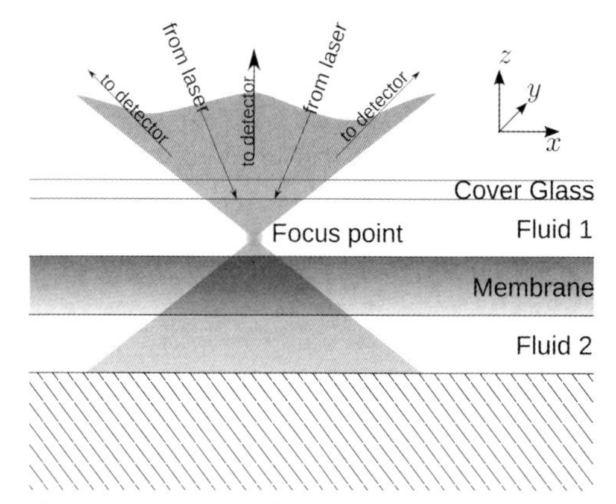


**Figure 7.7** Experimental setup to study translucent membranes clamped between two bathing solutions by confocal Raman spectroscopy.

reasons the most relevant one is the necessity of a translucent medium. Thus, only membranes that weakly absorb the visible light can be used an this is mainly linked to the case of polymeric membranes.

Another point deals with the constraints on the physical dimension of the optical setup. In confocal mode, the membrane system must be very close to the microscope objective since its working distance is usually of the order of a few millimeters. Therefore, the design of a cell depends on these considerations. For membrane transport studies, the system must contain the fluid channels, the membrane and an optical window in a strictly limited thickness. Figure 7.7 depicts such a cell with two channel and a membrane. Fluids can be gases [40] or liquids [38, 41]. From a practical point of view the use of an immersion objective can enhance the solid angle collecting the emitted light.

### 7.7.2 Visualization of Ionic Species Distribution

Raman spectroscopy is sensitive to molecular ions; monoatomic ions like $Na^+$, $Cl^-$ cannot been "seen". Some preliminary works have been done on the visualization of molecular ions under transport conditions with the help of Raman spectroscopy [35, 39]. The membrane sample is for instance an anion exchange membrane with grafted ionic sites, homogeneous at the microscale and sufficiently translucent to allow the transmission of the laser beam. The membrane is put under the microspectrometer in the $xy$ plane orthogonally to the laser beam axis $z$; and the membrane/solution system is then scanned point by point along the $z$ axis with a resolution of 1 μm. Chemical imaging is carried out by measuring simultaneously the spectra (chemical information) and the spatial information. A typical spatially resolved spectra is plotted as both function of wavenumber shift and position in Figure 7.8. In this example the membrane is clamped between two solutions: on

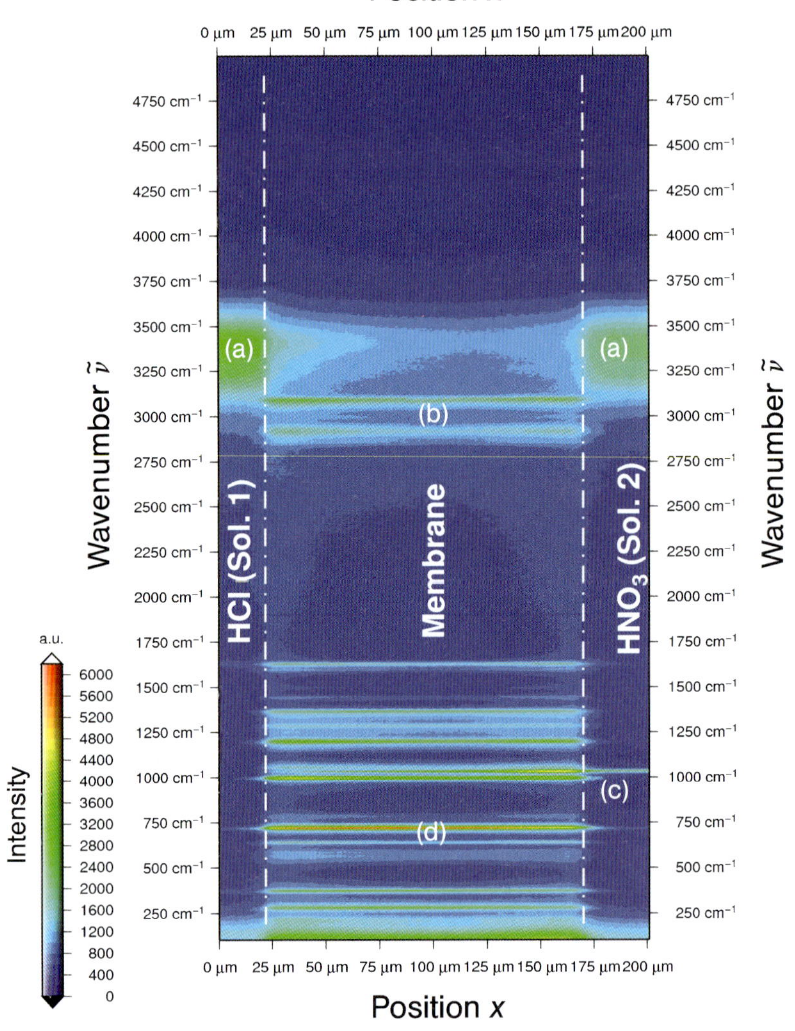


**Figure 7.8** Intensity map recorded on an anion exchange membrane (ARA from Solvay) during dialysis between hydrochloric acid and nitric acid. Position refers to the membrane normal axis: (a) water (OH stretching), (b) aliphatic $CH_2$ stretching and aromatic CH stretching (c) nitrates symmetric stretching, (d) $CF_2$ symmetric stretching (membrane matrix).

the left a hydrochloric acid solutions ($1\,mol\,l^{-1}$), on the right a nitric acid solution ($1\,mol\,l^{-1}$). The wavenumber shift ranges from $200\,cm^{-1}$ to $4900\,cm^{-1}$. As can be seen, although only one spatial dimension is explored, a lot of information is obtained. The spatial and spectral domains marked (a) are related to the presence of water at high concentration (solvent in the bathing solution); within the membrane the water-specific band is weak, which means a low concentration of water. Some specific bands (clearly two) related to the membrane structure appears in domain (b) and do not exist in the solutions, these bands are characteristic of the C–H bond stretching in an aliphatic (cf. aromatic) structure in the membrane for the lowest wavenumber, $\sim 2910\,cm^{-1}$ (cf. the highest, $\sim 3095\,cm^{-1}$). Another band near $1035\,cm^{-1}$ is characteristic of nitrate ions even if there is some overlapping with membrane matrix specific band ($\nu_{C-C}$). The nitrate line obviously appears in the left solution (nitric acid). The intensity of this band can be used to visualize the concentration profile of nitrate ions within the system under dialysis conditions. Finally, chemical information about $CF_2$ distribution within the matrix is readable at (d), Raman line at $\sim 720\,cm^{-1}$.

The visual information can be related to very accurate spectroscopic information, as can be seen in Figure 7.9 when the dataset is plotted as an accumulation of spectra in the spectral domain only. The nitrate band appears in the nitric acid solution (A) and in the membrane (overlapped with a membrane matrix Raman line), whereas there are no nitrate ions in the other side of the membrane (B). The wide water band (stretching $\nu_{OH}$) can also give some structural properties of water, both inside and outside the membrane.

A typical concentration profile of ions is obtained after some mathematical processing of data and shown in Figure 7.10.

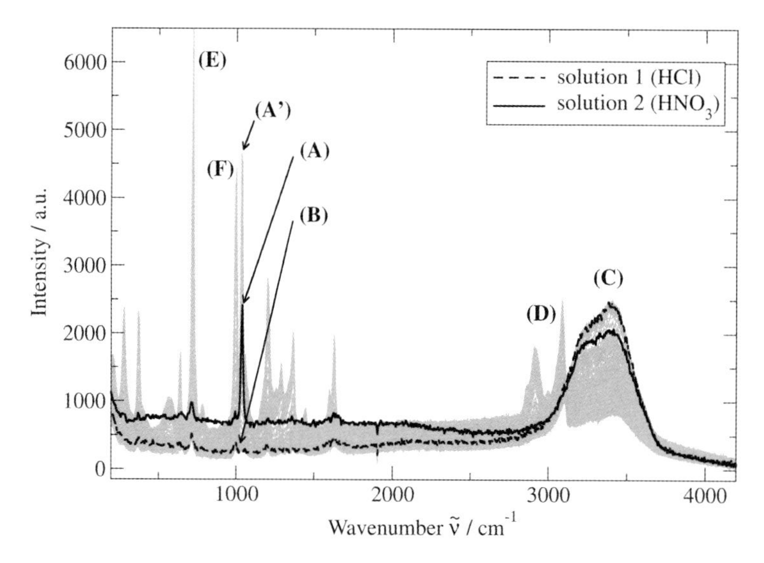


**Figure 7.9** Superimposed spectra during the depth scan of an ARA membrane: (A) and (A′) intensity of the nitrate band in the nitric acid solution and within the membrane respectively, (B) nitrate absence in the hydrochloric solution, (C) water intensity, (D–F) membrane matrix intensities.

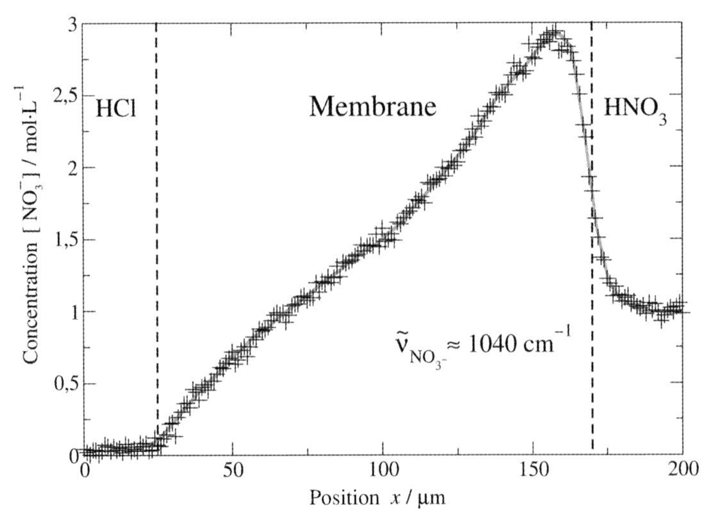


**Figure 7.10** Concentration profile for the nitrate ion through the membrane solution system computed from integration of the nitrate band ($1040\,cm^{-1}$).

The same approach can be applied to the study of solvent mixture (see Section 7.7.3).

Others setups have been applied to the study of concentrational changes in a polymer gel electrolyte submitted to a current flow [45], in fuel cells [46, 47] or in a lithium battery [48]. Ions gradients can be visualized, while structural evolution can be pointed out.

FT-Raman spectroscopy is also a well suited method for adsorption studies of anions in various anion exchange membranes [49]. It can probe the competitive adsorption and coordination of anions in anion exchange membranes.

### 7.7.3
### Visualization of Solvent Distribution

With almost the same design as in Section 7.7.2, hyperspectral data for chemical imaging of a membrane system can be collected by confocal Raman microspectroscopy in order to study solvent distribution within a membrane [38–41]. Figure 7.11 depicts the spatial and spectral information measured on a proton-conducting membrane clamped between methanol and water solutions.

In this figure, domain (a), around $3300\,cm^{-1}$, is related to the water content of the medium but also to some extent takes account of the methanol presence. Indeed, this band is characteristic of –OH stretching.

More specific to methanol are the two bands in domain (b): $\nu_S$ and $\nu_{AS}$ for $CH_3$. Discrimination can be carried out in solution with the (c) band ($\nu$ for CO) that belongs only to the methanol molecule. Visualization can be completed by plotting

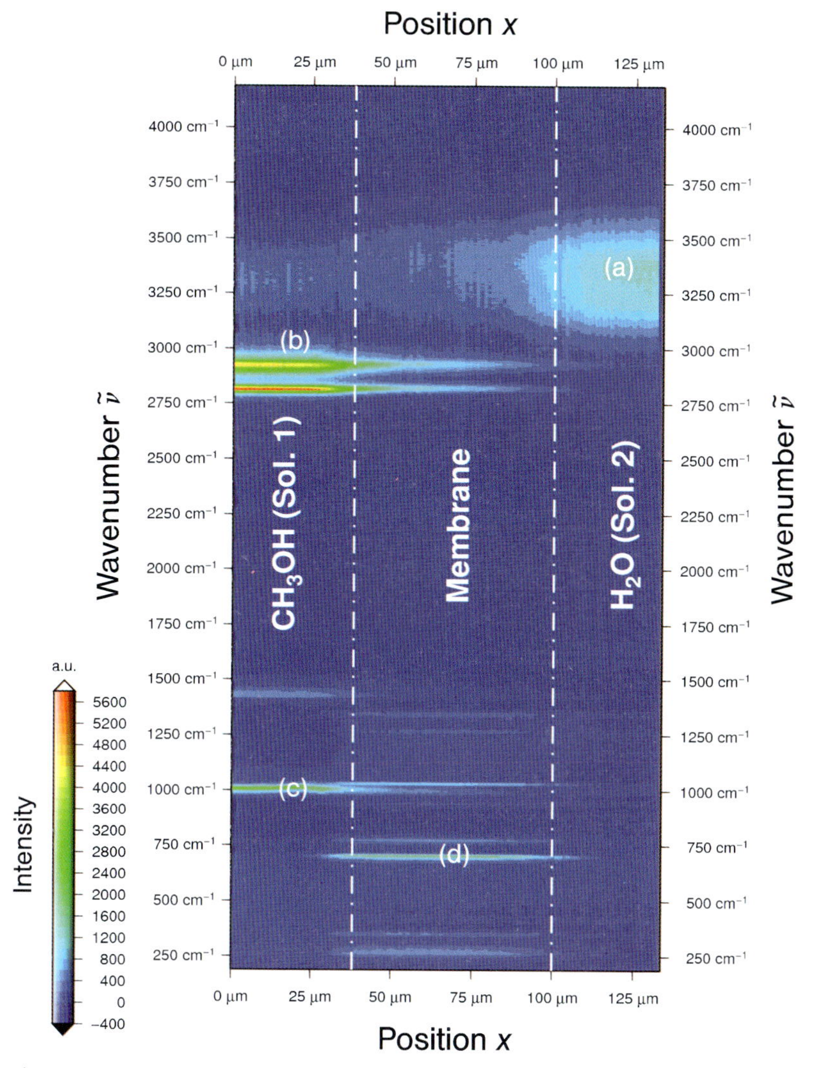


**Figure 7.11** Intensity map recorded on a cation exchange membrane (Nafion) during cross diffusion between water and methanol. Position refers to the membrane normal axis: (a) water (OH stretching), (b) symmetric and anti-symmetric stretching $CH_3$, (c) CO stretching (methanol), (d) $CF_2$ symmetric stretching (membrane matrix).

the spectra on the same picture (Figure 7.12) in order to extract information about structures or evolution.

This kind of work can be useful for a better understanding of the methanol crossover in methanol fuel cells [40, 41]. Very accurate measurement of local concentration gradient of solvent can be estimated.

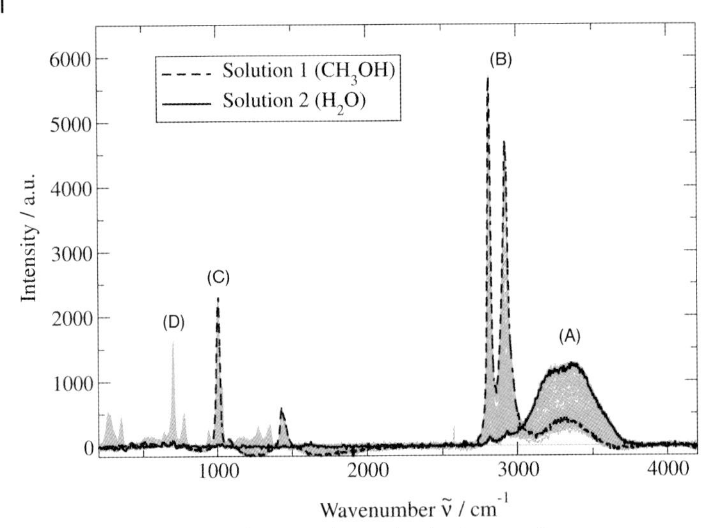


**Figure 7.12** Superimposed spectra during the depth scan of a Nafion membrane: (A) water (OH stretching), (B) symmetric and anti-symmetric stretching $CH_3$, (C) CO stretching (methanol), (D) $CF_2$ symmetric stretching (membrane matrix).

Some calculations can be carried out to give corrected depth profiles. Usually, this approach needs some modeling of the diffraction and refraction effects that arise in confocal Raman spectroscopy [41, 50]. Indeed, the depth profiles obtained in a conventional way contain, on the one hand, an intrinsic error due to the diffraction phenomena of the optical system, and on the other hand, an error due to the refractive index difference between the membrane (polymer sample) and its environment (bathing solutions or gas).

### 7.7.4 Gas Permeation

Raman scattering by gases is not a very strong effect and gives only weak signals. So gas permeation is usually not measurable by Raman spectroscopy. However, degradation of some chemical components consecutive to gas permeation can be detected in situ by Raman spectroscopy: any formation of a deposited solid (carbon for instance) allows its identification and evaluation of membrane degradation [51].

### 7.7.5 Membrane Synthesis

Confocal Raman spectroscopy is also a powerful tool for the monitoring of membrane synthesis. It has been used for instance to demonstrate in situ the formation of a surface liquid layer on the top of membrane-forming systems [52] during a vapor-induced phase separation (VIPS) process. The kinetic aspects of the phase inversion process for asymmetric membrane formation can also be reached [53] by this in situ technique; and any concentration change in the reacting species can easily be measured at any spot during membrane formation. In the same way, the extent of polymer cross-linking that results following in situ photo-polymerization on a membrane assembly has been realized by near infrared Raman spectroscopy [54].

### 7.7.6
### Biological Media

The nondestructive behavior of Raman spectroscopy, its ability to probe samples under in vivo conditions and its spatial resolution in the low micron scale allow new insights into biological media (living cells, biological membranes). The mapping of single cells by near infrared Raman microspectroscopy is now possible [55]: after attribution of different Raman band to proteins, lipids, cholesterol and nucleic acid and after a principal component analysis of the Raman data, the main cellular constituents of cells can be identified. The transmembrane ion gradients of sealed, hemoglobin-free erythrocyte membrane vesicles can be measured through intensity ratios [56]; also, the influence of a transmembrane potential on the potassium ions distribution can be pointed out. However, such cations cannot be directly detected and only their effect on the intensity of Raman scattering of $CH_3$ stretching is measured. Artificial membranes like planar supported lipid bilayers can also be studied in situ thanks to Raman scattering techniques [57].

Another specific membrane has been studied by confocal Raman spectroscopy: human skin. The Raman effect has been used to visualize water concentration profiles in human skin in vivo [58, 59] and transdermal drug delivery. Optical sections can be obtained without the needs for physically dissecting the tissue.

## 7.8
## Conclusion

The utility of Raman spectroscopic methods for extracting position-dependent chemical information from a variety of membrane system is well established. Confocal Raman spectroscopy is a very powerful, nondestructive technique that can work in situ at the micron scale. In the future, Raman imagery will continue to open new possibilities for monitoring membrane processes. For this to be achieved, emphasis must be put, on the one hand, on the development of new design for cells, allowing online recording of spectroscopic data within a working system, and on the other hand, on the improvement of spectroscopic data processing.

## References


1 J.M. Chalmers, P.R. Griffiths (Eds.), *Handbook of Vibrational Spectroscopy, vol. 1: Theory and Instrumentation*, John Wiley and Sons, Chichester, **2002**.

2 J.M. Chalmers, P.R. Griffiths (Eds.), *Handbook of Vibrational Spectroscopy, vol. 2: Sampling Techniques*, John Wiley and Sons, Chichester, **2002**.

3 J.M. Chalmers, P.R. Griffiths (Eds.), *Handbook of vibrational spectroscopy, vol. 3: Sample Characterization and Spectral Data Processing*, John Wiley and Sons, Chichester, **2002**.

4 R. Lewis, G.M. Edwards (Eds.), *Handbook of Raman Spectroscopy*, Dekker, New York, **2001**.

5 C.V. Raman, *Indian J. Phys.* **1928**, *2*, 387.

6 C.V. Raman, K.S. Krishnan, *Indian J. Phys.* **1928**, *3*, 399.

7 M. Delhaye, P. Dhamelincourt, *J. Raman Spectrosc.* **1975**, *3*, 33.

8 P. Dhamelincourt, *Étude et réalisation d'une Microsonde Moléculaire à Effet Raman*, PhD dissertation, Université de Lille I, Lille, **1979**.

9 T. Wilson, C. Sheppard, *Theory and Practice of Scanning Optical Microscopy*, Academic, New York, **1984**.

10 S. Kimura, C. Munakata, *Appl. Optics* **1990**, *29*, 489.

11 T. Wilson, *Confocal Microscopy*, Academic, London, **1990**.

12 G.J. Brakenhoff, H.T. van der Voort, N. Nanninga, *Chem. Act.* **1984**, *163*, 231.

13 C. Sheppard, *J. Phys. D.* **1986**, *19*, 2077.

14 W. Zhao, J. Tan, L. Qiu, P. Jin, *Sensor Actuator A Phys.* **2005**, *120*, 17.

15 M. Schwertner, M.J. Booth, T. Wilson, *J. Microsc.* **2005**, *217*, 184.

16 E.J. Botcherby, R. Juskaitis, T. Wilson, *Optic. Comm.* **2006**, *268*, 253.

17 K.C. Khulbe, T. Matsuura, *Polymer* **2000**, *41*, 1917.

18 J. Ethève, P. Huguet, C. Innocent, J.-L. Bribes, G. Pourcelly, *J. Phys. Chem. B* **2001**, *105*, 4151.

19 C. Innocent, P. Huguet, J.-L. Bribes, G. Pourcelly, M. Kameche, *Phys. Chem. Chem. Phys.* **2001**, *3*, 1481.

20 M. Pastorczak, M. Wiatrowski, M. Kozanecki, M. Lodzinski, J. Ulanski, *J. Mol. Struct.* **2005**, *744/747*, 997.

21 M. Chaouki, P. Huguet, J.-L. Bribes, *J. Mol. Struct.* **1996**, *379*, 219.

22 D. Lin-Vien, N.B. Colthup, W.G. Fateley, J.G. Grasselli (Eds.), *The Handbook of Infrared and Raman Characteristic Frequencies of Organic Molecules*, Academic, San Diego, **1991**.

23 K. Nakamoto (ed.), *Infrared and Raman Spectra of Inorganic and Coordination Compounds. Part A: Theory and applications in Inorganic Chemistry*, John Wiley and Sons, New York, **1997**.

24 K. Nakamoto (ed.), *Infrared and Raman Spectra of Inorganic and Coordination Compounds. Part. B: Applications in Coordination, Organometallic, and Bioinorganic Chemistry*, John Wiley and Sons, New York, **1997**.

25 K. Levenberg, *Q. Appl. Math.* **1944**, *2*, 164.

26 D. Marquardt, *SIAM J. Appl. Math.* **1963**, *11*, 431.

27 W.H. Press, S.A. Teukolsky, W.T. Vetterling, B.P. Flannery, *Numerical Recipes: The Art of Scientific Computing*, 3rd edn, Cambridge University Press, Cambridge, **2007**.

28 E.R. Malinowski, *Factor Analysis in Chemistry*, 3rd edn, Wiley-Interscience, New York, **2002**.

29 M. Chaouki, P. Huguet, F. Persin, J.-L. Bribes, *New J. Chem.* **1998**, *22*, 233.

30 M.E. Vallejo, P. Huguet, C. Innocent, F. Persin, J.-L. Bribes, G. Pourcelly, *J. Phys. Chem. B* **1999**, *103*, 11366.

31 H. Ericson, T. Kallio, T. Lehtinen, B. Mattson, G. Sundholm, F. Sundholm, P. Jacobsson, *J. Electrochem. Soc.* **2002**, *149*, A206.

32 P.D.A. Pudney, T.M. Hancewicz, D.G. Cunningham, M.C. Brown, *Vib. Spectrosc.* **2004**, *34*, 123.

33 B. Mattsson, H. Ericson, L.M. Torell, F. Sundholm, *J. Polym. Sci. Pol. Chem.* **1999**, *37*, 3317.

34 B. Mattson, H. Ericson, L.M. Torell, F. Sundholm, *Electrochim. Acta*, **2000**, *45*, 1405.

35 C. Thibault, P. Huguet, P. Sistat, G. Pourcelly, *Desalination*, **2002**, *149*, 429.
36 M. Schmitt, B. Leimeister, L. Baia, B. Weh, I. Zimmermann, W. Kiefer, J. Popp, *ChemPhysChem*, **2003**, *4*, 296.
37 P. Huguet, T. Kiva, O. Noguera, P. Sistat, V. Nikonenko, *New J. Chem.* **2005**, *29*, 955.
38 H. Matic, A. Lundblad, G. Lindbergh, P. Jacobsson, *Electrochem. Solid State Lett.* **2005**, *8*, A5.
39 P. Huguet, P. Sistat, S. Deabate, E. Petit, *Desalination*, **2006**, *200*, 173.
40 P. Scharfer, W. Schabel, M. Kind, *J. Membrane Sci.* **2007**, *303*, 37.
41 S. Deabate, R. Fatnassi, P. Sistat, P. Huguet, *J. Power Sour.* **2008**, *176*, 39.
42 P. Georén, J. Adebahr, P. Jacobsson, G. Lindbergh, *J. Electrochem. Soc.* **2002**, *149*, A1015.
43 A. Martinelli, A. Matic, P. Jacobsson, L. Börjesson, M.A. Navarra, A. Fernicola, B. Scrosati, *Solid State Ionics*, **2006**, *177*, 2431.
44 A. Martinelli, A. Matic, P. Jacobsson, L. Börjesson, M.A. Navarra, D. Munaò, S. Panero, B. Scrosati, *Solid State Ionics*, **2007**, *178*, 527.
45 D. Ostrovskii, P. Jacobsson, *J. Power Sources* **2001**, *97/98*, 667.
46 L.S. Sarma, C.-H. Chen, G.-R. Wang, K.-L. Hsueh, C.-P. Huang, H.-S. Sheu, D.-G. Liu, J.-F. Lee, B.-J. Hwang, *J. Power Sour.* **2007**, *167*, 358.
47 H. Ekström, B. Laffite, J. Ihonen, H. Markusson, P. Jacobsson, A. Lundblad, P. Jannasch, G. Lindbergh, *Solid State Ionics*, **2007**, *178*, 959.
48 I. Rey, J.C. Lassègues, P. Baudry, H. Majastre, *Electrochim. Acta*, **1998**, *43*, 1539.
49 J. Matulionytė, R. Ragauskas, O. Eicher-Lorka, G. Niaura, *Chemija*, **2006**, *17*, 1.
50 A. Gallardo, S. Spells, R. Navarro, H. Reinecke, *Macromol. Rapid Commun.* **2006**, *27*, 529.
51 F. Iguchi, N. Sata, H. Yugami, H. Takamura, *Solid State Ionics*, **2006**, *177*, 2281.
52 P. Menut, Y.S. Su, W. Chinpa, C. Pochat-Bohatier, A. Deratani, D.M. Wang, P. Huguet, C.Y. Kuo, J.Y. Lai, C. Dupuy, *J. Membrane Sci,.* **2007**, DOI:10.1016/j.memsci.2007.11.016.
53 H.J. Kim, A.E. Fouda, K. Jonasson, *J. Appl. Polym. Sci.* **2000**, *75*, 135.
54 M. Murphy, K. Faucher, X.-L. Sun, E. Chaikof, R. Dluhy, *Colloid. Surface. B* **2005**, *46*, 226.
55 C. Krafft, T. Knetschke, A. Siegner, R. Funk, R. Salzer, *Vib. Spectrosc.* **2003**, *32*, 75.
56 R. Mikkelsen, S. Verma, D. Wallach, *Proc. Natl Acad. Sci. USA* **1978**, *75*, 5478.
57 C. Lee, C. Bain, *Biochim. Biophys. Acta* **2005**, *1711*, 59.
58 P. Caspers, G. Lucassen, H. Bruining, G. Puppels, *J. Raman Spectrosc.* **2000**, *31*, 813.
59 P. Caspers, G. Lucassen, G. Puppels, *Biophys. J.* **2003**, *85*, 572.

# 8
# In Situ Characterization of Membrane Fouling and Cleaning Using a Multiphoton Microscope

*Robert Field, David Hughes, Zhanfeng Cui, and Uday Tirlapur*

## 8.1
## Overview

Ultra- and microfiltration are pressure-driven membrane processes that have found applications in the diary, food, chemical, biotechnology and water treatment industries. Now fouling of membranes (e.g. rejected species plugging membrane pores or the formation of a cake on the membrane surface) reduces filtration performance and eventually results in the need for cleaning the membranes or even replacing them. Fouling remains a significant problem in many applications. Much of the current understanding relating to membrane fouling has been gleaned from the collection and analysis of macroscopic parameters such as flux and transmembrane pressure from which fouling mechanisms have been inferred. Now multiphoton microscopy (MPM) offers the opportunity for us to gain greater insights as it generates images at submicron resolution. The aim of this chapter is to introduce the technique of MPM, to present some background information on fouling and then present the results of some initial studies on fouling and cleaning.

## 8.2
## Optical Techniques for Characterization of Membrane Fouling

### 8.2.1
### Introduction

Optical techniques are principally used to visualize the membrane surface and the deposits thereon. The resolution of these images depends upon the illumination source. Systems using what might be termed "standard" visible light based microscopy are capable of resolutions of 0.5 μm [1]. This makes them suitable for the imaging of filtration cakes of micron-sized particles. To achieve higher resolutions

*Monitoring and Visualizing Membrane-Based Processes*
Edited by Carme Güell, Montserrat Ferrando, and Francisco López

ISBN: 978-3-527-32006-6

fluorescence-based microscopy is required, which uses laser light or electron microscopy where illumination is provided by a beam of high energy electrons. Electron microscopy has the potential to achieve nanometer resolutions [2].

Early systems based on visible light used high-zoom video cameras positioned to view the membrane from the side. The ability to continuously record, in real time, activity at the membrane surface made this type of system ideal for in situ imaging of cake deposition, provided the particles were sufficiently large. A 15 × zoom video camera was used to record in situ particle motion of large polyethylene particles (125–180 μm) close to a stainless steel mesh filter [3]. The particles in this study were much larger than those typically found in microfiltration applications. Another study [4] used a similar system to record the deposition of more realistic calcite (2.6 μm, 25 μm) and anatase (0.5 μm) particles. Individual particles could not be resolved and only a cake thickness was quoted. In both studies the cakes thicknesses were in the millimeter range.

A significant advance in optical characterization was made with the introduction of an approach now known as *direct observation through the membrane* (DOTM) [5, 6]. The microscope objective is positioned on the permeate side and focused through the permeate channel and the membrane onto the membrane surface on the feed side. Thus the membranes used must be transparent. Although this is a major limitation, significant findings have been made, as discussed elsewhere.

Using technology similar to DOTM but viewing the membrane from the feed side, others [7, 8] investigate foulant removal by backpulsing and validate models for biofoulant deposition [9]. These studies feature an expanded range of membranes but the feed concentration must be limited as high feed turbidity makes imaging impossible.

DOTM type systems when used with flat sheet membranes provide a top-down or birds-eye view of the membrane, so coverage of the surface maybe assessed, but the thickness of the cake cannot be determined. Side views of cake layer formation on the outside of hollow fibers under outside-to-inside filtration of yeast suspensions have been achieved by modification of the DOTM technique [10, 11]. From the images presented, cake thicknesses could potentially have been measured but the authors did not address this possibility. The interested reader is also referred to Chapter 2, which encompasses an extended review of such techniques.

To achieve resolutions below 0.5 μm electron microscopy must be employed. Scanning electron microscopy (SEM) provides high-resolution 2D surface images and is frequently used for ex situ imaging of virgin and fouled membranes. Before imaging, fouled membranes must be removed from the module and undergo some form of preparation. This usually involves dehydration and spattering with nanometer-sized gold particles; however this process can vary (see Chapter 3).

An alternative to SEM is the use of fluorescence microscopy. The most common technique is confocal laser scanning microscopy (CLSM). CLSM offers two significant advantages over SEM, namely: the ability to produce 3D images in situ and the discrimination of species based on fluorescence. Both the $x,y$ resolution and the $z$ penetration depth are to some degree system- and sample-dependant (further discussion of this point can be found in Chapter 4). With the exception of the studies reported here, all other fluorescence microscopy on membrane systems has been undertaken using CLSM. This technique requires that the foulant species must be labeled with fluorophores. CLSM uses either UV or visible lasers and single photon excitation from three separate laser sources to evoke red, green and blue fluorescence.

Optical techniques for the analysis of membrane fouling are still in their infancy and there exists much room for new techniques which can provide one or more of the following: in situ imaging, improved resolution and the ability to differentiate between foulant species. MPM has the potential to address all three needs.

### 8.2.2 Principles of Multiphoton Microscopy

Multiphoton microscopy is a near infrared nonlinear optical imaging system that is generally used in the form of a two-photon laser scanning fluorescence microscopy. It is a relatively new technique [12]. The technique works by exciting fluorescence in a sample through the near simultaneous absorption of two photons. The use of a two-photon absorption process as opposed to a single photon absorption process as found in confocal microscopy offers the advantages of increased penetration and reduced sample damage, thus MPM has become the premier method for 3D imaging of thick biological samples [13].

Fluorescence maybe excited by a single- or two-photon process. In a single-photon process, one photon provides all the energy required to move the atom to a higher energy level and prime the fluorescence process. A similar process can be achieved with two photons, if they are absorbed near simultaneously. Absorption of the first photon raises the atom to a "virtual" energy level which is not normally an allowed energy level. The second photon which moves the atom up to an allowed energy level must be absorbed within the timescales for molecular energy fluctuations, as dictated by the Heisenberg uncertainty principle, namely $10^{-16}$ s [13]. The processes are illustrated in Figure 8.1 on a simplified Jablonski diagram. Actually two-photon absorption is a very rare natural process. In bright daylight a good photon absorber would undergo a single photon interaction roughly every second but a two-photon simultaneous interaction would only occur once every 10 million years or so [14].

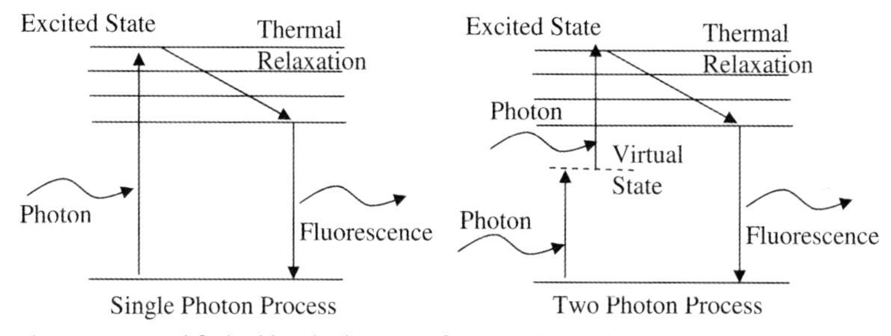


**Figure 8.1** Simplified Jablonski diagram of one- and two-photon processes.

Thus to utilize the two-photon process for MPM a powerful tightly focused femto or pico second pulsed laser is required to supply a high density of photons for a short time in a very small area.

In principle, one could move from two photons to three, as illustrated in Figure 8.2, but in practice this would be difficult to achieve. Thus the designation MPM is a slight misnomer. For two-photon excitation to occur it is necessary that the photons be focused to give a high density. They are focused upon a subfemtoliter volume for a very short time. Through sequential scanning, the illumination of small portions is achieved. The intensity of the excited fluorescence decreases with distance from the focal point to the fourth power. Consequently significant fluorescence is only excited in a very small volume around the focal point. Thus all of the fluorescence essentially comes from the focal point. In a single-photon process, such as confocal microscopy, scanning is also employed but fluorescence is excited in an hourglass shape around the focal point, due to considerable excitation of fluorescence even by unfocused illumination.

With MPM, fluorescence is only generated in the area being imaged and this has two advantages. First, the processing of the fluorescence generated is much simpler and secondly the absence of the additional fluorescence reduces photobleaching and phototoxicity, which can be a particular problem when imaging living samples [15]. Further, with MPM, no pinhole is required to reject fluorescence from axially out of focus areas, which allows all the fluorescence generated to be collected.

As the energy of a photon is inversely proportional to its wavelength, lower energy, longer wavelength near infrared photons, with wavelengths between 750 nm and 1050 nm might, at first glance, be expected to penetrate a sample less

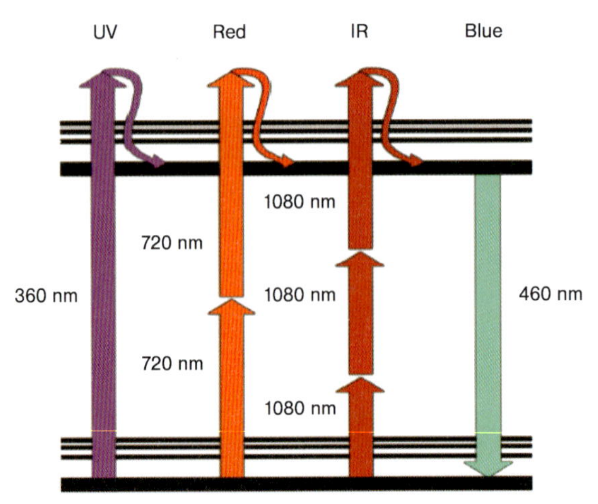


**Figure 8.2** Three ways of achieving an emission wavelength of 460 nm. The processes illustrated are: single photon excitation at UV, two-photon excitation with red light, multiphoton excitation with infrared light. (Adapted by Clarence Yapp from [17]).

well than higher-energy photons in the UV and visible range, with wavelengths of 351 nm to 568 nm. However such photons are more prone to scattering and absorption, which limits the depth at which blur-free images can be obtained [14, 16]. For example, the Rayleigh scattering of UV wavelength photons is a particular problem in single-photon confocal microscopy [17]. Furthermore, another point to note is that the emission wavelength for MPM is substantially shorter than the excitation wavelength, which makes for easier differentiation and hence detection. In contrast, the emission wavelength for confocal microscopy is longer, but usually by only 50–200 nm. This results in more difficult detection in confocal systems.

The fluorescence observed by MPM results from having either a naturally fluorescent material, from the addition of fluorescent labels or from second harmonic generation (SHG). Collagen is a material that gives SHG; the input of two photons at 900 nm results in UV output at 450 nm. The SHG property of collagen has been exploited in some tissue engineering work but SHG has yet to be used in membrane fouling studies. Fluorophore labeling of nonfluorescent species to give measurable fluorescence is a prerequisite in the absence of SHG.

To generate 3D images, the $xy$ plane is scanned at different focal depths (positions in the $z$ direction). The resulting set of images is referred to as a stack. Scanning in the $xy$ plane is achieved by movable mirrors which shift the focal point in lines across the sample whilst the $z$ position is controlled by a nanopositioner connected to the microscope stage. At Oxford, all images have been generated using a modified BioRad multiphoton microscope, as described in Section 8.4.

## 8.3 Partial Review of Fouling and Cleaning

This review is included to provide some context for the work described below. It briefly examines protein fouling, cell–protein fouling and membrane cleaning.

Single proteins have molecular weights in the 10–100 kDa range and dimensions in the nanometer range. Thus when filtering protein suspensions through microfiltration membranes, whose pores are hundreds of nanometers in diameter, limited fouling and high protein transmission might be expected. However many studies have shown proteins to be potent foulants of microfiltration membranes, with protein aggregates (a number of single proteins jointed together) playing an important role. The fouling may be either internal or external and the location of the fouling depends upon the protein, membrane and operating conditions. A two-stage fouling mechanism is now generally adopted to model the protein fouling of track-etched membranes; initial internal fouling of the pores or blockage of them by aggregates is followed by the formation of an external cake of aggregated material. These cakes of aggregates are very thin, in comparison with the thick cakes formed when filtering particulate suspensions, but they are very densely packed and have a high resistance. This combination of internal fouling followed by the development of a high resistance cake leads to large rapid declines in flux when filtering proteins suspensions under constant transmembrane pressure (TMP). The fouling layers are also found to be irreversible, requiring chemical cleaning to remove them from the membrane.

Industrial applications of crossflow microfiltration often involve multicomponent feeds. One such process is the separation of proteins from a fermentation broth. There exist many studies on the fouling of both real and model broths, where the fouling is characterized in terms of flux decline under constant TMP, or TMP rise under constant flux. Numerous real fermentation broths have been studied [e.g. [18–20]]. A popular model protein–cell system is the mixture of washed yeast cells and bovine serum albumin (BSA) [e.g. [21, 22]] and this has been used in our work. The more complex scenario of washed yeast cells and a mixture of proteins has also been studied [23]. In all cases a cake of yeast cells was found to form on the membrane surface and under conditions where the cake acts as a filter aid an enhancement in the flux and protein transmission above that for filtration of the protein alone maybe achieved [23]. A cake may be considered a filter aid if the cake captures foulant species, stopping these species reaching the membrane; and the consequent reduction in fouling at the membrane outweighs any increase in specific resistance due to the captured species blocking the interstices in the cake [24]. For the yeast and protein case, the yeast cell cake captures protein aggregates whilst allowing single native protein molecules to pass through. Consequently protein fouling of the membrane is reduced as this is primarily caused, particularly at the initiation stage by aggregates rather than single native protein molecules [25–27].

## 8.4 Materials and Methods

### 8.4.1 Outline of the Application of MPM to Membrane Filtration

At Oxford, all images have been generated using a modified BioRad Radiance 2100 MP multiphoton microscope (Zeiss; Jena, Germany). Brief details of the system are given here; it is thoroughly detailed elsewhere [28, 29]. Near infrared (NIR) laser beams ($\lambda = 800$ nm) are obtained from a tunable 76 MHz femtosecond pulsed Ti:sapphire laser (Mira 900-F; Coherent, Ely, UK) pumped by a 7 W multiline argon ion laser (Verdi; Coherent). Mean laser powers within the system are electronically regulated via the pockol cell of the BioRad beam conditioning unit (BCU). The NIR femtosecond pulsed laser beams are focused to a diffraction-limited spot using a high-number-aperture objective. At this spot fluorescence is excited. The fluorescence from the sample returns via the objective, either to the fluorescence detectors for imaging, or to the SpectraCube (Applied Spectral Imaging, Israel) for spectral analysis. Image processing is considered below.

The Bio-Rad scanning system is attached to a TE300 Microscope (Nikon; Surrey, UK), with a $60 \times 1.2$ NA plan apochromat water immersion lens (Nikon). For in situ 3D imaging the standard configuration is combined with a $1.5\times$ digital zoom (applied via the computer). The 60× objective is used as it gives good resolution and the required working distance of approximately 1 mm. It will immediately be noted that this provides a real constraint; the lens is external to the cell containing the sample but the surface being imaged has to be just 1 mm away from it. In the work described later the cell is a membrane module, details of which are given in the next section.

Digital zooms ranging from 1× to 10× can be applied; these reduce the area scanned to some fraction of the maximum possible scanable area for a given objective. This increases resolution but can lead to problems with sample damage, such as photobleaching and sample burning. A 1.5× digital zoom gives acceptable resolution whilst minimizing sample damage.

In the $xy$ plane the image is typically 137 μm by 137 μm, giving an $xy$ imaged area of $1.877 \times 10^{-2}\,\text{mm}^2$. Both the $x$ and $y$ directions are split into 512 pixels and this results in a pixel area of $7.160 \times 10^{-2}\,\mu\text{m}^2$. Therefore the $xy$ resolution is typically 0.27 μm. The standard step size in the $z$ direction is 1.5 μm. This results in a typical voxel volume (i.e. an $xyz$ cuboid) of $0.107\,\mu\text{m}^3$.

In the work on protein fouling the SpectraCube is used for spectral analysis. This instrument allows precise determination of the wavelength of the emitted fluorescence from the sample. For spectral analysis, fluorescence is excited as for imaging but the fluorescence is directed to the SpectraCube rather than the imaging fluorescence detectors. The SpectraCube has a resolution of under 5 nm at a wavelength of 400 nm and under 16 nm at 700 nm [30].

The multiphoton microscope must scan the sample to obtain an image, thus the time taken for imaging is an important consideration. Depending upon the thickness of the sample, resolution in the $z$ plane and the number of images that are averaged to create a final image, it takes between 1 min and 20 min. If the target moves during the time taken to complete the imaging, the resulting 3D image contains blurring and/or streaks. Clearly this is another restriction which needs to be taken into account when planning experiments.

### 8.4.2 Module Design

The design of a membrane module for use with an MPM presented some considerable challenges. The basic constraints imposed by the MPM on the design of the module were as follows:

- The maximum dimensions of the module are dictated by the dimensions of the microscope stage, which is 115 mm by 115 mm.
- The working depth, defined as the distance between the tip of the objective and the point of imaging must not exceed approximately 1 mm.
- The path between the tip of the objective and the point of imaging must contain only materials of known refractive index, be transparent and free from defects.

To assist in the analysis of results and allowing for the constraints imposed by the microscope, the module was designed to be geometrically similar to a rectangular channel crossflow module [31]. The module consisted of two Perspex plates, with the lower plate having a circular recess into which a 1 mm thick rubber gasket was placed. The gasket had a slot 10 mm by 37 mm cut out of it to form the permeate channel. A second 0.5 mm thick rubber gasket was placed against the circular upper plate and secured with a small amount of silicone grease. A slot 10 mm by 45 mm was cut from the gasket to form the feed channel. A 47 mm diameter membrane disc was then placed between the two halves. Other details can be found in [31, 32]. Here it is noted that the channel height of 0.5 mm was chosen to ensure the membrane was within the working distance of the objective.

### 8.4.3 Filtration Circuit

A schematic diagram of the filtration circuit together with the MPM is depicted in Figure 8.3 and details can be found in [31, 32]. As mentioned above there must be no motion during image acquisition. This is an exacting requirement as even movements of a fraction of a micron result in a blurring of the image. To ensure that the membrane and any foulant species present are suitably still, the filtration process is temporarily stopped to facilitate imaging. Stopping of the filtration process has no effect on the filtration data. The standard experimental procedure is as follows:

1. The module is integrated into the microscope stage and connected to filtration loop.
2. The system is charged with buffer, the TMP and crossflow velocity for the run are set and the buffer (pure water) flux taken at experimental conditions.
3. The buffer reservoir is replaced with the feed reservoir and the feed pump is started.
4. The feed pump is stopped after the desired filtration time and the membrane surface is imaged at the desired locations.
5. After imaging the feed pump is restarted and collection of filtration data begins immediately.
6. Steps 4 and 5 are then repeated for each subsequent time point.

### 8.4.4 Image Analysis

In the work described below, the images from the MPM were processed and analyzed using LaserSharpe (Zeiss), Image J (NiH downloaded from http://rsb.info.nih.gov/ij/), and Imaris (Bitplane, Zurich, Switzerland). The MPM generates separate stacks of images for each of the red (R), green (G) and blue (B) channels. The programs listed enable the separate color channels from the microscope to be

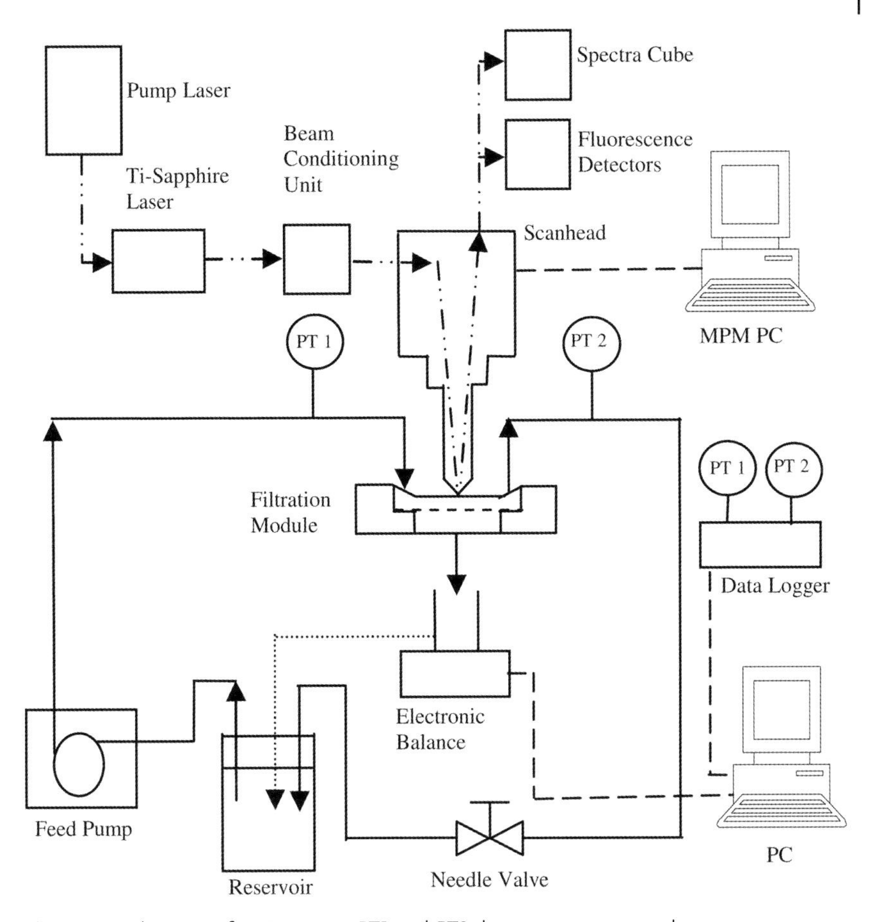


**Figure 8.3** Schematic of MPM set-up. PT1 and PT2 denote pressure transducers.

merged to form a single RGB stack. The RGB stack can then be projected from any angle.

LaserSharpe is software which interfaces the microscope with the controlling computer. It also incorporates some image processing and analysis functions. The program is mainly used whilst actually imaging to quickly determine the quality of images. The major tool used for image processing is Imaris. The program is a versatile processing tool which allows for complete control of all image properties and will generate projections of RGB stacks at any angle. Image J has more limited processing features but has a number of very useful analysis tools, which include measurement and particle counting. Further details can be found in [31].

The particle counting tool can be used to count and size individual particles, or to determine the fraction of an image which is covered by a given species. The procedure for determining either parameter is essentially the same, varying only at the final step. The methodology used is very similar to that used previously [9] and is summarized and illustrated in [32]. Briefly the RGB stack is projected to give a 2D top-down view. The top down view is then converted from an RGB color image to an 8-bit grayscale image which has 256 gray levels. The grayscale image must then be thresholded to produce a black and white image. The process of thresholding involves defining the cut-off point, the gray levels above which become black and below which they become white. This task may be performed either automatically by the software or manually by the user. It is preferable to allow the software to set the threshold as this removes any user bias; however this is not always possible as the software occasionally returns nonsensical thresholds.

## 8.5 Imaging Cake Fouling

### 8.5.1 Introduction

There exists no single system which can fully characterize a cake either in situ or ex situ. Ex situ techniques provide information on cake structure with a high resolution but their application is fraught with difficulty as removal and preparation of the membrane and cake may cause unknown changes to both. In general, but in particular when characterizing filtration cakes, in situ techniques are preferable. Numerous in situ techniques exist for measuring cake thickness but these do not give information on cake structure. Conversely the DOTM system and others like it give structural information, but are limited to "cakes" only one particle thick, in other words monolayers. To provide information simultaneously on both cake structure and thickness a system capable of gathering data in three dimensions is required.

CSLM with its ability to provide 3D images offers an obvious opportunity to gather information simultaneously on cake structure and thickness. Limited attempts have been made to characterize membrane fouling using CLSM but for reasons given previously MPM is a better option. When imaging biological samples, penetrations of greater than 400 μm may be achieved [13] but unfortunately, with yeast cells, our experience suggests that the penetration achievable is an order of magnitude lower.

An advantage of MPM over techniques based on visible light is that greater resolution can be achieved using fluorescence. Thus fine details and small structural features which have not previously been visualized with other in situ techniques may be revealed. The use of different fluorophores to label different particles also aids identification of the species. A disadvantage of the requirement to label is the need to limit the study to "washed" yeast cells because fluorophore labeling does affect the filtration performance of unwashed cells, as described in the next section.

### 8.5.2
### Effect of Fluorophore Labeling

Fluorophore labeling of nonfluorescent species is a prerequisite in the use of MPM. Fast dried bakers yeast (YSC2; Sigma Aldrich, UK) was rehydrated using 10 mM phosphate buffer. Yeast has no autofluorescence under 800 nm excitation, so fluorophores (also termed fluorescent probes or dyes) must be added so the yeast can be imaged. Before filtration the yeast suspension is centrifuged to remove an excess fluorophores and the extracellular polymeric substances (EPS), a process known as washing. The omission of this stage was considered as discussed below.

Yeast was suspended in 90 ml of 10 mM phosphate buffer at pH 7.5 and well stirred. Fluorophores were then added. Table 8.1 gives details of the fluorophores. The suspension (now 100 ml in volume) was gently stirred for 30 min to allow the yeast cells to absorb the fluorophores at $20 \pm 2\,^{\circ}C$. The suspension was then washed to remove excess fluorophores and EPS. A washing procedure was used similar to that reported by others [7]. The suspension was placed into two 50 ml centrifuge tubes, and centrifuged at 2500 rpm for 10 min. The supernatant was removed and the yeast pellet was resuspended with buffer. The process was performed three times after which the yeast suspension was further diluted to the required concentration using the same buffer. In all experiments fresh yeast suspensions were prepared on the same day and used within a few hours.

On average unwashed yeast has been found to be 73.6% yeast cells (by mass). This is close to values previously reported in literature [7, 33]. Sizing of the washed yeast was performed using both a particle sizer (Malvern Loc Mastersizer) and the multiphoton microscope. The equivalent spherical diameter found using the particle sizer was 5.04 μm. The diameter of the yeast cells found with the MPM was 4.11 μm. It should be noted that the multiphoton images were analyzed assuming the yeast cells to be spherical. Figure 8.4 shows the filtration behavior of labeled and unlabeled yeast cells and the filtration behavior of labeled and unlabeled washed yeast cells. The labeling was performed with all three of the fluorophores detailed above.

It is readily seen that the unwashed fluorophore labeled suspension has a much sharper flux decline and a significantly lower flux after 60 min than an unwashed nonfluorophore labeled suspension. The excess fluorophores in the suspension, although only at low concentrations in the final suspension (CW 10 mg $L^{-1}$, AO 0.66 mg $L^{-1}$, PI 0.1 mg $L^{-1}$) have a significant effect on filtration performance. However once the suspension has been washed to remove excess fluorophores the difference is small, with the fluorophore and nonfluorophore labeled washed yeast suspensions having fluxes after 60 min of 285.5 L $m^{-2}$ $h^{-1}$ and 267.7 L $m^{-2}$ $h^{-1}$, respectively. The washing procedure is effective in removing the excess fluorophores from the suspensions, leaving a suspension of fluorophore labeled washed yeast with filtration performance similar to nonfluorophore labeled washed yeast suspension.

As the washing procedure removes extracellular material such as EPS, one is only able to use MPM to investigate, with confidence, the behavior of washed cells.

**Table 8.1** Details of fluorophores added to yeast suspensions.
Emission and excitation data from http://www.olympusmicro.com.

| Fluorophore | Color | Peak excitation (nm) | Peak emission (nm) | Item(s) labeled | Stock conc. ($g\,L^{-1}$) | Volume added | Comment |
|---|---|---|---|---|---|---|---|
| Calcofluor white (CW) | Blue | 440 | 500–520 | Cell wall | 0.5 | 10 mL | Used in all experiments |
| Acridine orange (AO) | Alkaline cell: green<br>Acid cell: red/purple | 502 | 526 | Cell acidity | 10.0 | 33 μL | Not used in yeast/protein mixtures. Slightly labels the membrane green |
| Propidium iodide (PI) | Red | 536 | 617 | Nuclei of dead cells | 1.0 | 50 μL | Not used in yeast/protein mixtures |

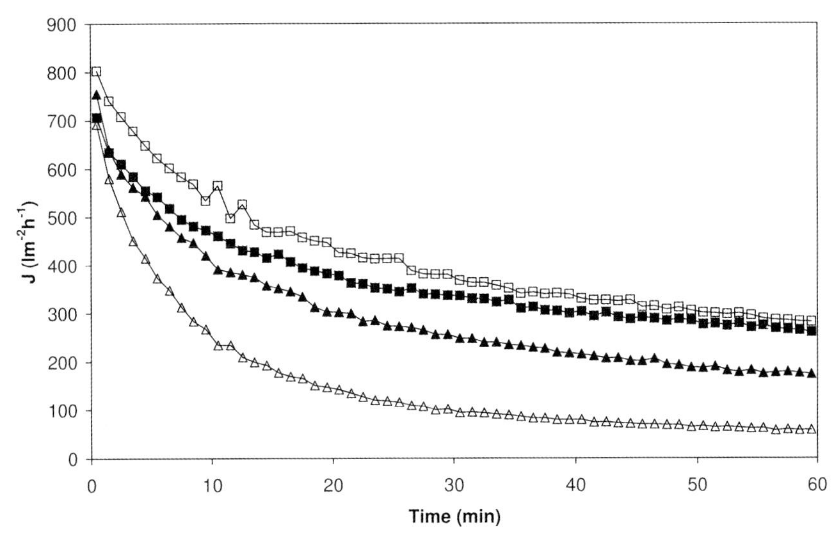


**Figure 8.4** The effect of fluorophore labeling and suspension washing of yeast suspensions. △ Unwashed fluorophore labeled suspension concentration 0.50 g L$^{-1}$, ▲ unwashed nonlabeled suspension concentration 0.50 g L$^{-1}$, □ washed fluorophore labeled suspension concentration 0.37 g L$^{-1}$, ■ washed nonlabeled suspension concentration 0.37 g L$^{-1}$. TMP 0.10 bar, CFV 0.20 m s$^{-1}$.

One could contemplate labeling only a portion of the cells by using a very small amount of fluorophore and assuming none remained in the supernatant, but interpretation of the resulting images would be difficult. The filtration behavior itself is discussed elsewhere [32].

### 8.5.3 Insights Into Cake Development

Having demonstrated that the filtration behavior for washed cells was not unduly affected by labeling, some insights into cake structure and development can be made.

Extensive in situ imaging of patchy monolayer yeast cell cake has been performed using the DOTM system [5] under constant permeate flux operation. The images obtained using MPM [32] were for low constant TMP with a low feed concentration. As the cake was thin, the structure was captured in a single top-down image. Although the membrane was not intentionally labeled, it manifested a tinge of green fluorescence, a result which could be explored more in future. This outcome was due to the fact that even after repeated washing of the yeast cells a minute concentration of acridine orange remained in the suspensions which was sufficient to label the membrane. The accidental marking was beneficial as it provided a pleasing contrast for the images (see [31, 32]).

Time lapse images of yeast cells depositing on the membrane can be found elsewhere [31, 32]. An image of the membrane taken after 15 min filtration is shown in Figure 8.5 [32]. The individual cells can clearly be seen and appear to be distributed randomly on the membrane; others [5, 9] have noticed a similar random distribution using DOTM type systems. The fact that cells with alkaline pH fluoresce green, whereas cells with acidic pH fluoresce red/purple, could possibly be exploited. A number of larger yeast cell aggregates are deposited on the membrane, one of which is marked with a triangle. The marked aggregate is estimated to contain five cells and as seen from a profile view (not shown) protrudes further into the flow than the deposited single cells.

The evolution of cake thickness, fractional area covered by washed yeast cells and normalized flux ($J/J_0$) with time is shown in Figure 8.6 for filtration of a dilute suspension under the same conditions as those for Figure 8.5. It can be seen in Figure 8.6 that, although there is clearly deposition of yeast cells on the membrane from the start of the filtration, the decline in the flux is slow due to the patchy nature of the deposit. Imaging of patchy cakes is often performed to determine the critical flux. Figure 8.6 together with the time lapsed images (not shown here) confirm that there may be deposition of cells on the membrane before the flux under constant TMP begins to fall significantly. This reinforces the contention of others (e.g. [5]) that imaging the membrane is the most sensitive approach for

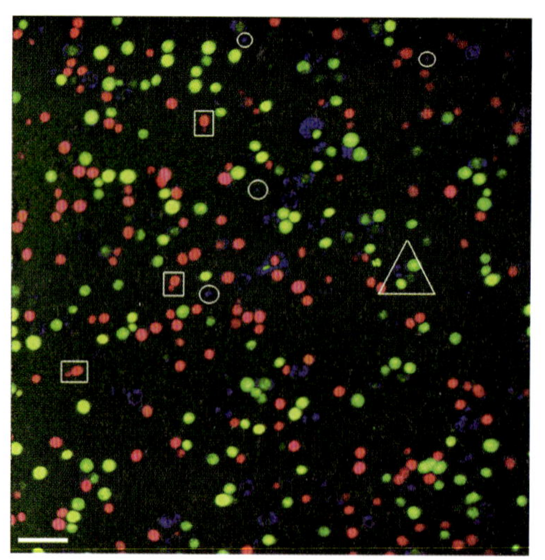

**Figure 8.5** Enlargement of a patchy yeast filtration cake formed after 15 min filtration. Concentration 7.4 mg $L^{-1}$, TMP 0.05 bar, CFV 0.08 m $s^{-1}$. Cell aggregates marked with a triangle. Budding cells marked with a rectangle. Deposited broken cell wall marked with a circle. Scale bar (left image) 20 μm.

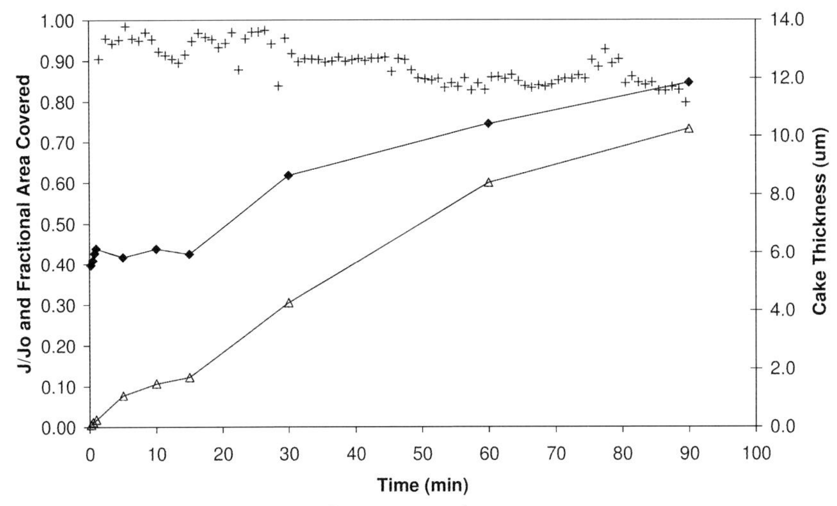


**Figure 8.6** Evolution of normalized flux ($J/J_0$, +), cake thickness (♦) and fractional area of the membrane covered (△) for constant TMP filtration of washed yeast suspension. Concentration 7.4 mg $L^{-1}$, TMP 0.05 bar, CFV 0.08 m $s^{-1}$.

detecting deposition and membrane fouling. The higher spatial resolution offered by multiphoton microscopy makes it possible to identify deposition of small species such as broken cell debris. This was not possible with the previous generation of optical techniques used for imaging deposition on the membrane such as DOTM as these were limited to the imaging of micron-sized particles. Thus MPM has the potential to be a more sensitive tool for the detection of the first point of deposition on the membrane and hence determination of the critical flux.

The fractional area of membrane covered and quoted cake thickness in Figure 8.6 were determined using the image analysis protocol outlined in Section 8.4.4. Each reported thickness is the average of the measured thickness at five random locations within the image. By 90 min the quoted cake thickness is 12 μm, approximately two cells and fractional coverage is over 70%.

Filtration with 0.37 g $L^{-1}$ washed yeast suspensions at TMP 0.10 bar and a cross-flow velocity (CFV) of 0.20 m $s^{-1}$ resulted in a much thicker cake and the evolution of cake thickness had to be estimated from drying and weighing of the cake as the MPM was found to be limited to about 40 μm. Incidentally after 60 min the estimated cake thicknesses from drying and weighing was 165 μm with a corresponding specific cake resistance of $5.48 \times 10^{11}$ m $kg^{-1}$, which is in excellent agreement with literature values [34, 35].

Although penetration of 200 μm was achieved through mouse skin [36], densely packed cakes of cells are less transparent that skin and most biological specimens, hence the observed limitation of around 40 μm. Notwithstanding this limitation, structural details of the top layers of the final cake and details as it develops can be obtained by MPM. It was found [32] that macrovoids (in the order of a cell) exist within a cake. Profile views of the developing cake were used to obtain this result but are not reproduced here because the images are slightly out of focus due to the unevenness of the membrane surface at the micron scale. Some images are given in [31, 32].

## 8.6 Imaging of Protein Fouling

### 8.6.1 Introduction

Apart from the work described below, the only imaging of protein fouling to date which used CLSM was imaging ex situ the fouling of fluorophore labeled BSA and ovalbumin on track-etched polycarbonate membranes [37]. Ex situ imaging allowed very high resolution and the images obtained where of excellent quality. In situ images using MPM have been obtained at Oxford [31, 38], using Isopore polycarbonate track-etched membranes fouled by BSA, ovalbumin and a mixture of the two. The use of in situ imaging allows for the development of the fouling to be characterized against time. Whilst the in situ imaging of the deposition of micron-sized particles is relatively common, no similar studies had been performed for proteins. Using MPM various questions can be addressed, such as:

- What role do protein aggregates play in the fouling process?
- For mixtures of proteins, what is the respective role of each component?
- Can the two-stage mechanism for protein fouling (that has been proposed in the literature as a result of filtration data analysis) be verified through multiphoton images?
- Likewise for cleaning; are the proposed mechanisms those observed in multiphoton images?

### 8.6.2 Effect of Fluorophore Labeling

The filtration performance of ovalbumin and the 50:50 mixture of BSA and ovalbumin is almost unaffected by fluorophore labeling of the proteins [38], whereas the BSA–fluorescein conjugate shows different behavior to native BSA, with a noticeable flux decline over the first 30 min, whereas the native BSA has an almost constant flux for 30 min, then begins to decline. The conjugation of fluorescein may change the surface properties of BSA increasing its tendency to aggregate or altering the protein–membrane interactions thus increasing the rate of fouling. Others [25] have noted that BSA exhibits some variability in filtration behavior, to the extent that lots of native BSA from the same supplier may perform differently. However the effect of labeling BSA deserves further investigation (see Section 4.3.1).

### 8.6.3 Using Images to Cross-check Fouling Models

From a top-down image it is impossible to ascertain whether the initial fouling occurs within the membrane pores, or at the openings of the pores. This can only be achieved using profile or side-on images. A side view of an internally fouled cylindrical pore membrane would consist of alternating vertical bands of fluorescence and darkness. The fluorescence bands are pores with protein deposited within them and the dark bands are nonporous areas, or pores that do not have any protein deposited within them. When a significant cake has formed on top of the membrane this is visible.

One difficulty when analyzing profile images is that the membrane was not intentionally made fluorescent and did not suffer any accidental labeling (cf. Section 8.5). Thus distinguishing the upper surface of the membrane is difficult. In future others might consider the use of acridine orange or an alternative to label the membrane itself. In Figure 8.7 an arrow has been added to mark the upper surface of the membrane; this should only be taken as a guide. The data shown are for a 0.8 μm membrane. Data for a 0.22 μm membrane is available elsewhere [38].

For BSA filtered through a 0.8 μm membrane the fouling rate increased with time for the entire experiment, indicating internal fouling (data available in [31]). From the profile images it can be seen that there is no significant cake deposition even after 30 min filtration. Thus the images correlate well with the fouling rate data. The fouling rate for ovalbumin decreased with time and showed no clear maximum. Ovalbumin fouling is rapid and given this the length of the initial internal fouling period may well be much shorter than that for BSA. Whilst the resolution of the resistance and fouling rate data gathered were too low to capture this event, the multiphoton profile images are sensitive enough to distinguish the internal fouling phase and confirm that ovalbumin does indeed foul via a

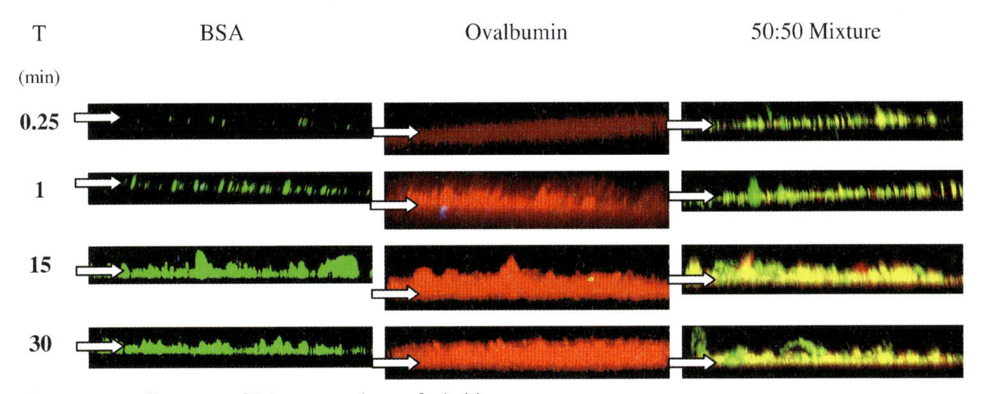


**Figure 8.7** Profile views of 0.8 μm membrane fouled by protein suspensions. Arrows indicate approximate position of membrane surface. TMP 0.10 bar, CFV 0.20 m $s^{-1}$, total concentration 50 mg $L^{-1}$.

two-stage fouling mechanism. The images support a two-stage fouling mechanism with a very short internal phase estimated to last less than 1 min.

It is important to note that even during the initial phase of the two-stage mechanism where internal fouling dominates, external fouling may also occur. Comparison of the top-down and side-on multiphoton images provided evidence of this. As the track-etched membranes have cylindrical non-interconnecting pores and a porosity of less than 20%, no more than a 20% coverage would be observed if fouling were purely internal in the initial phase. However, this was not observed to be the case [38]. Clearly the ability to acquire both top-down and profile views is powerful and should lead to better estimation of parameters.

### 8.6.4
### Deposition of Protein Mixtures

Multiphoton images are acquired and stored separately as red (R), green (G) and blue (B) channels. These are usually then merged to produce a single multicolored image. Figure 8.8 shows top-down projections of the R channel only, the G channel only and the merged RGB image of a membrane fouled with a 50:50 mixture at 5 min. The images were taken using a 4.0× digital zoom rather than the usual 1.5× zoom, giving a resolution in the $xy$ plane of 0.1 μm. The BSA–fluorescein

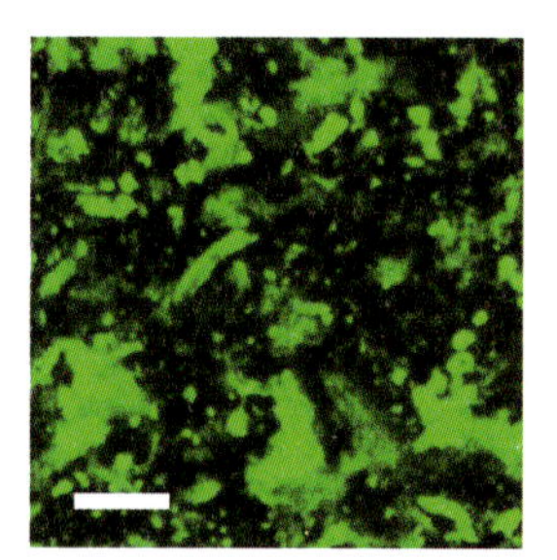

Green Channel – BSA Fouling

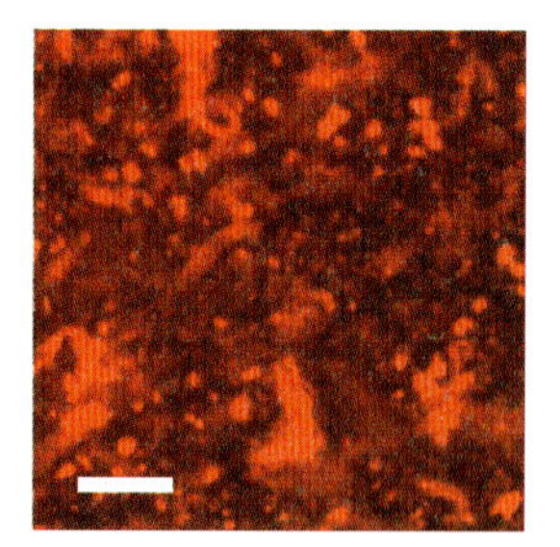

Red Channel – Ovalbumin Fouling

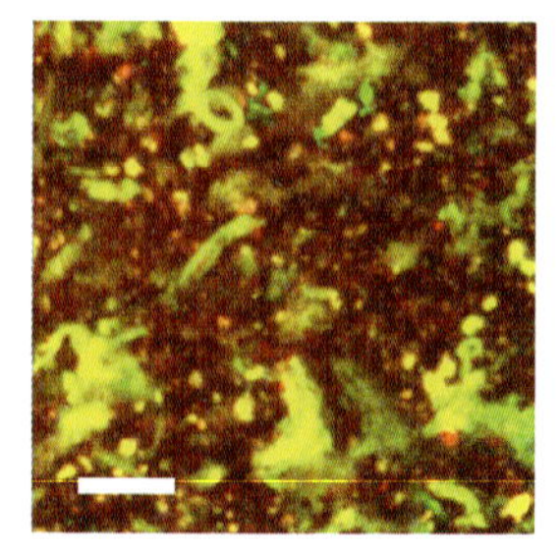

RGB Merge

**Figure 8.8** Images of the red channel only, green channel only and the RGB merge of a membrane fouled by a 50:50 mixture after 5 min. TMP 0.10 bar, CFV 0.20 m s$^{-1}$, total concentration 50 mg L$^{-1}$. Scale bar 10 μm.

conjugate and the ovalbumin Texas Red conjugate respectively produced signals only in the G and R channels. This was confirmed when filtering the single protein solutions. Thus the G channel showed the BSA fouling and the R channel showed the ovalbumin fouling. There was no signal in the B channel so it is not included in Figure 8.8. The image is representative of the fouling by the 50:50 mixture throughout the experiment. It is striking that most of the major identifiable structures in the RGB merged image can be seen in both the R and G channels. Thus the majority of the deposited aggregates contained both BSA and ovalbumin. This was confirmed by spectral analysis of the images.

A resolution of 0.1 μm does not allow the imaging of individual proteins molecules, but many structures can still be seen. These structures are protein aggregates of varying sizes. The smallest structures clearly identifiable structures are approximately 0.5 μm in diameter and may be protein aggregates that have fouled in the openings of pores. Much larger noncircular aggregates, with dimensions up to 10 μm can also be seen. These very large aggregates maybe caused by the peristaltic pump used in the filtration loop [39]. These images together with others provide visual confirmation of the important role played by protein aggregates in the fouling of microfiltration membranes.

An important question is where the initially deposited aggregates form, in solution or at the membrane. As large aggregates are present at very short filtration times [38], this suggests the tentative conclusion that the initial aggregates form in solution.

### 8.6.5
### Chemical Cleaning of Protein Fouled Membranes

Chemical cleaning of microfiltration membranes is a common practice and is intrinsically linked to the fouling of the membrane. The multiphoton system provides an ideal opportunity to monitor the cleaning process. Some preliminary work [40] has been undertaken in this area. The addition of free chlorine significantly reduces the BSA fluorescence and completely removes the ovalbumin fluorescence and so the monitoring of foulant removal cannot be monitored if the cleaning agent contained free chlorine as one is unable to distinguish between removal and bleaching. However the commercial cleaning agent Ultrasil and NaOH solutions do not act as bleaching agents, thus a reduction in fluorescence can be interpreted as protein removal. Before imaging, flushing with buffer was found to be necessary [40].

Figure 8.9 shows the results of chemical cleaning, given as the ratio of flux buffer (pure water) flux after chemical cleaning to the buffer flux taken at the start before protein filtration started. During the cleaning process this is analogous to recovery. The filtration of buffer immediately after protein filtration results in only marginal recovery of the flux. This is to be expected as protein-fouled membranes almost universally require chemical cleaning. The cleaning agents used in this study failed to achieve 100% flux recovery, or even the usually accepted standard of 95% recovery. The probable reason is the low temperature at which cleaning was

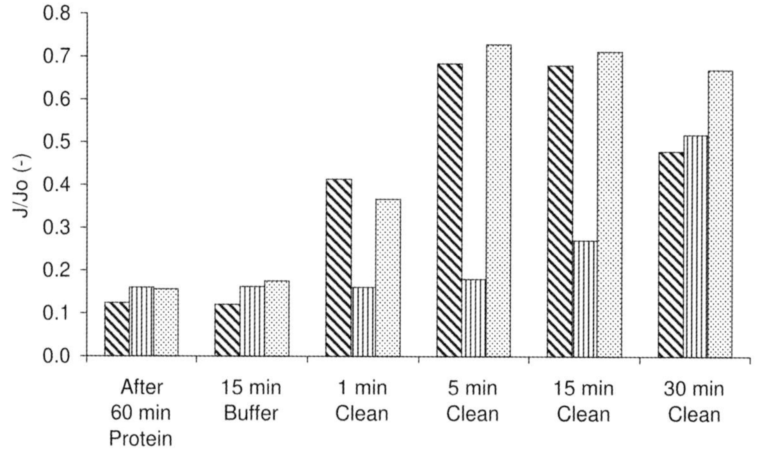


**Figure 8.9** Flux recoveries of chemical cleaned membranes fouled by a 50:50 protein mixture for 60 min. Total protein concentration 50 mg $L^{-1}$. Cleaning agents: Ultrasil 53 (diagonal stripes), sodium hydroxide (vertical stripes), free chlorine (dots). TMP 0.10 bar, CFV 0.20 m $s^{-1}$. Cleaning solution concentration 1 g $L^{-1}$.

undertaken. Here the cleaning was performed at room temperature, whereas it is normal to clean at an elevated temperature to increase the kinetics of the process [41]. It had been judged that the module design and concerns over the presence of high temperature fluids near the MPM made the investigation of the effect of temperature on the cleaning process unfeasible. An enhancement of procedures to enable a rapid and safe switch of a membrane module from a cleaning rig to the MPM will be required if more realistic data are to be gathered.

It is also interesting to note that the Ultrasil 53 rapidly removed aggregates from the membrane surface and achieved approximately 75% flux recovery within 15 min but recovery declined after 30 min cleaning due to re-fouling of the membrane by ovalbumin which had previously been removed from the membrane and was present in the cleaning solution. This was easily seen from the MPM images. With regard to the NaOH cleaning process, the MPM images showed that large aggregates remained on the membrane surface even after 1 h of cleaning.

## 8.7
## In Situ Characterization of Cell–Protein Fouling

### 8.7.1
### Introduction

Whilst numerous real fermentation broths have been studied [18–20], a popular model protein–cell system is the mixture of washed yeast cells and bovine serum albumin (BSA) [21, 22, 33]. The more complex mixture of washed yeast cells and a mixture of proteins has also been studied [23]. In all cases a cake of yeast cells was found to form on the membrane surface. A cake maybe considered a filter aid if the cake captures foulant species, stopping these species reaching the membrane, and the consequent reduction in fouling at the membrane outweighs any increase in specific resistance due to the blocking of the interstices in the cake by captured species [24].

In our work, the yeast was labeled with calcofluor white solution. BSA–fluorescein conjugate fluoresces green and ovalbumin–Texas red conjugate fluoresces red. Further details of the Oxford work can be found in [42].

### 8.7.2
### Filtering Protein Solutions Through Preformed Yeast Cakes

Having deposited a cake of yeast cells by filtering yeast only solutions ($0.37\,g\,L^{-1}$) for 1 h, the feed was switched to protein-only solutions ($50\,mg\,L^{-1}$). Maintaining a stable cake of yeast cells during the filtration of the protein-only solutions proved

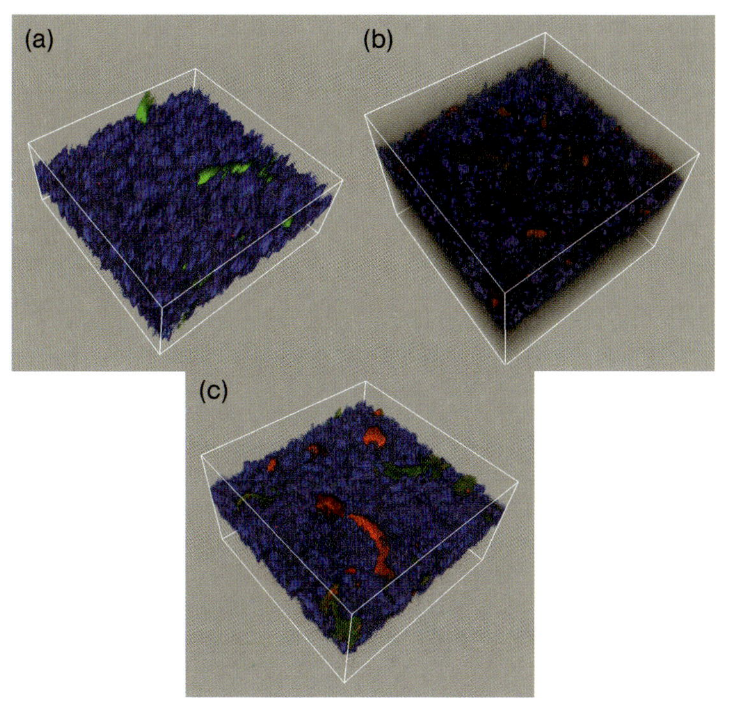


**Figure 8.10** Projections of 3D MPM images of protein aggregates deposited on the upper layers of a pre-deposited cake of yeast cells (blue) after 15 min filtration. The frames are 137 μm square. (a) BSA (green), (b) ovalbumin (red), (c) 50:50 mixture. TMP 0.10 bar, CFV $0.2\,ms^{-1}$.

difficult; and this resulted in variable filtration data. Perhaps a preformed cake consisting of a mixture of washed labeled yeast cells and unwashed and unlabeled cells would have been more effective. Although the filtration data itself was variable some images were obtained. In Figure 8.10, the upper layers of the cake (which remained on the membrane after 15 min of filtration of protein-only solution) were acquired, showing extensive capture of protein aggregates. Together with significant deposition on the surface of the cake some aggregates have penetrated and become lodged within the voids between cells in the cake. The number of ovalbumin aggregates was higher than the number of BSA aggregates which was not unexpected as ovalbumin shows a greater tendency to aggregate as evidenced previously when filtering protein-only solutions. The captured aggregates are large, several microns in maximum dimension. Some aggregates contained both green and red, indicating the presence of both BSA and ovalbumin, within a single aggregate.

## 8.8 Conclusions

The work detailed in this chapter shows that MPM can be used to provide in situ images of fouling by suspensions of washed yeast and fouling by proteins. Time-lapse 3D images have been obtained using the MPM and then projected from a number of angles. The multiphoton system is particularly well suited to the imaging of protein fouled membranes, perhaps better suited than it is to the imaging of particulate fouling where the limit to the depth that can be imaged is about the length of eight yeast cells. However when filtering proteins, the resultant external fouling layers are relatively thin and so do not present the same problems as those found with particulate suspensions. Interestingly, penetrating through the Isopore membranes to image within the pores was not an issue as the membranes themselves are thin and optically transparent.

A number of the important questions concerning protein fouling of microfiltration membranes can be addressed by MPM. Although native proteins are much smaller than the pores of microfiltration membranes, the protein-containing solutions cause severe fouling. The fouling is the result of protein aggregates. MPM has the ability to visualize protein deposition both on the surface and within the pores of the membrane. The use of multiphoton microscopy has provided unique insights into protein fouling of membranes. To use the system the proteins must be labeled with fluorophores. The filtration behavior of every fluorophore labeled protein solution should be compared with that of native protein solution prior to imaging. Some details on protein aggregate structure have been obtained and, in a general sense, co-localization observed. There is the potential to address this in far greater detail.

The chemical cleaning of protein fouled membranes was also investigated using the multiphoton microscope. There is potential for further work in this area.

## References


1 V. Chen, H. Li, A.G. Fane, *J. Membrane Sci.* **2004**, *241*, 23–44.
2 R. Chan, V. Chen, *J. Membrane Sci.* **2004**, *242*, 169–188.
3 M. Mackley, N. Sherman, *Chem. Eng. Sci.* **1992**, *47*, 3067–3084.
4 R. Wakeman, *Trans. IChemE Part A* **1994**, *72*, 530–540.
5 H. Li, A.G. Fane, H.G.L. Coster, S. Vigneswaran, *J. Membrane Sci.* **1998**, *149*, 83–97.
6 H. Li, A.G. Fane, H.G.L. Coster, S. Vigneswaran, *J. Membrane Sci.* **2000**, *172*, 135–147.
7 W.D. Mores, R.H. Davis, *J. Membrane Sci.* **2001**, *189*, 217–230.
8 W.D. Mores, R.H. Davis, *J. Membrane Sci.* **2002**, *208*, 389–404.
9 S.K. Kang, A. Subramani, E. Hoek, M. Deshusses, M. Matsumoto, *J. Membrane Sci.* **2004**, *244*, 151–165.
10 S. Chang, A.G. Fane, *J. Chem. Tech. Biotechnol.* **2000**, *75*, 533–540.
11 S. Chang, A.G. Fane, S. Vigneswaran, *Chem. Eng. J.* **2002**, *87*, 121–127.
12 W. Denk, J.H. Strickler, W.W. Webb, *Science* **1990**, *248*, 73–76.
13 A. Diaspro, C.J.R. Sheppard, *Two-Photon Microscopy: Basic Principles and Architectures*, in: A. Diaspro (ed.) *Confocal and Two-Photon Microscopy. Foundations, Applications and Advances*, John Wiley and Sons, New York, **2002**.
14 V.E. Centonze, J.G. White, *Biophys. J.* **1998**, *75*, 2015–2024.
15 W. Denk, K. Svoboda, *Neuron* **1997**, *18*, 351–357.
16 A. Periasamy, P. Skoglund, C. Noakes, R. Keller, *Microsc. Res. Tech.* **1999**, *47*, 172–181.
17 J.M. Girkin, D.L. Wokosin, *Practical Multiphoton Microscopy*, in: A. Diaspro (ed.) *Confocal and Two-Photon Microscopy. Foundations, Applications and Advances*, John Wiley and Sons, New York, **2002**.
18 S.L. Li, K.S. Chou, J.Y. Lin, H.W. Yen, I.M. Chu, *J. Membrane Sci.* **1996**, *110*, 203–210.
19 D.M. Krstic, S.L. Markov, M.N. Tekic, *Biochemical Eng. J.* **2001**, *9*, 103–109.
20 I. Hassairi, R. Ben Amar, M. Nonus, B.B. Gupta, *Bioresour. Tech.* **2001**, *79*, 47–51.
21 N. Arora, R.H. Davis, *J. Membrane Sci.* **1994**, *92*, 247–256.
22 Y. Ye, V. Chen, *J. Membrane Sci.* **2005**, *265*, 20–28.
23 C. Guell, P. Czekaj, R.H. Davis, *J. Membrane Sci.* **1999**, *155*, 113–122.
24 D. Hughes, R.W. Field, *J. Membrane Sci.* **2006**, *280*, 89–98.
25 S.T. Kelly, A.L. Zydney, *J. Membrane Sci.* **1995**, *107*, 115–127.
26 C. Guell, R.H. Davis, *J. Membrane Sci.* **1996**, *119*, 269–284.
27 E. Tracey, R.H. Davis, *J. Colloid Interface Sci.* **1994**, *167*, 104–116.
28 W.J. Mulholland, M. Kendall, *Opt. Des. Eng. Proc. SPIE* **2004**, *5249*, 557–566.
29 W.J. Mulholland, *Targeted Epidermal DNA Vaccine Delivery*, DPhil thesis, University of Oxford, **2006**.
30 W.J. Mulholland, M.A.F. Kendall, N. White, B.J. Bellhouse, *Phys. Med. Biol.* **2004**, *49*, 5043–5058.
31 D. Hughes, *Novel Tools for the Analysis of Fouling in Microfiltration*, DPhil thesis, University of Oxford, **2006**.
32 D.J. Hughes, U.K. Tirlapur, R.W. Field, Z.F. Cui, *J. Membrane Sci.* **2006**, *280*, 124–133.
33 V. Kuberkar, R.H. Davies, *J. Membrane Sci.* **2001**, *183*, 1–14.
34 T. Tanaka, R. Kamimura, R. Fujiwara, K. Nakanusha, *Biotechnol. Bioeng.* **1994**, *43*, 1094–1101.
35 M. Meireles, C. Molle, M.J. Clifton, P. Aimar, *Chem. Eng. Sci.* **2004**, *59*, 5819–5829.
36 U.K. Tirlapur, W.J. Mulholland, B.J. Bellhouse, M. Kendall, J.F. Cornhill, Z.F. Cui, *Microsc. Res. Tech.* **2006**, *69*, 767–775.
37 M. Ferrando, A. Rozek, M. Zator, F. Lopez, C. Guell, *J. Membrane Sci.* **2005**, *250*, 283–293.

**38** D.J. Hughes, Z. Cui, R.W. Field, U.K. Tirlapur, *Langmuir* **2006**, *22*, 6266–6272.

**39** S. Chandavarkar, *Dynamics of Fouling of Microporous Membranes by Proteins*, Ph.D. thesis, Massachusetts Institute of Technology, Cambridge, Mass., **1990**.

**40** R. Field, D. Hughes, Z. Cui, U. Tirlapur, *Desalination*, **2008**, in press.

**41** M. Cheryan, *Ultrafiltration and Microfiltration Handbook*, Technomic, Lancaster, Penn., **1998**.

**42** D.J. Hughes, Z. Cui, R.W. Field, U.K. Tirlapur, *Biotechnol. Bioeng* **2007**, *96*, 1083–1091.

# Part II
# Electrical, Laser and Acoustic Techniques for Membrane Process Characterization

# 9
# Electrical Characterization of Membranes*

*Juana Benavente*

## 9.1
## Introduction

Membranes are selective barriers placed between two fluid phases (usually liquid phases) which control the transport of particles (and/or heat) from one phase to the other. Different stages can be considered when the transport of molecules or ions through membranes is studied [1]: first, the solute should enter into the membrane phase from the feed/donor solution, then the solute is transported through the membrane, and finally, it desorbs at the permeate/receiving solution. Depending on their structure, symmetric polymeric membranes can be catalogued as porous and nonporous (or dense; swollen network) membranes, while asymmetric membranes formed by a thick porous sublayer and a dense "active" layer are successfully used for desalting applications (nanofiltration, reverse osmosis). Different interactions can be involved in the transport of matter through membranes (solute/membrane, solute/solvent, solute/solute), and in the case of charged membranes, electrical interactions can play an important role in both the rejection and the transport of charged particles across membranes [2], which might affect the effectiveness of a particular separation process. Membrane charge might be associated with existing charged groups or radicals in the membrane matrix (charged membranes) or it could be acquired after contact with a polar medium [3, 4].

Characterization of membranes used in filtration processes (microfiltration, ultrafiltration, nanofiltration, reverse osmosis) is usually carried out by hydrodynamic measurements (hydraulic permeability, retention) [5–10], but electrical measurements such as streaming potential (SP; or electroosmotic flow) and membrane potential (MP) are used for characterizing, respectively, the membrane/solution interface (zeta potential, surface charge density) and the effective membrane fixed charge and ion transport numbers in the membrane [11–27]. Moreover, great and rapid development of membranes for fuel cells

* A list of nomenclature is given at the end of this chapter.

*Monitoring and Visualizing Membrane-Based Processes*
Edited by Carme Güell, Montserrat Ferrando, and Francisco López

ISBN: 978-3-527-32006-6

application during the past ten years has clearly increased interest in electrical characterization, mainly the determination of membrane conductivity, which is a basic parameter for this kind of application and it is generally obtained from alternating current (a.c.) measurements, mainly by impedance spectroscopy (IS) analysis [28–32].

This chapter presents a general description of the three kinds of measurements (SP, MP, IS) used for electrical characterization of membrane in "working conditions", that is, with the membranes in contact with electrolyte solutions; and the information achieved from them is briefly indicated. The main attention focuses on the characterization of those membranes commonly used in traditional separation processes (from diverse materials and with different structures), and IS measurements are considered in more detail. Membrane and matrix material electrical parameters are obtained, but the measurements also provide thermodynamics and geometrical/structural information.

## 9.2
## Electrical Measurements

As previously indicated, the most common measurements for membrane electrical characterisation are:

1. Streaming potential (or electroosmotic flow), which allows the electrical characterization of the membrane/electrolyte solution and provides information on changes related to the adsorption/deposition of particles either on the pore wall (filtration or transmembrane streaming potential, FSP) or on the membrane surface (tangential streaming potential, TSP). Streaming potential also gives characteristic membrane material information (adsorption Gibbs function, number of adsorption sites, isoelectric point, etc.) when measurements are performed at different concentrations or pH.

2. Membrane potential (MP) gives information on the effective fixed charge associated with the bulk membrane phase, Donnan equilibrium between the membrane, and the adjacent solution, and the relative flow of ion through the membrane (ion transport number).

3. Impedance spectroscopy (IS) allows both electrical and geometrical characterization by analyzing the impedance plots and using equivalent circuits as models. If membrane (or active layer) thickness is known, conductivity can be determined from electrical resistance values. Moreover, from capacitance results the thickness of dense membranes (or active layers) can be estimated if the material dielectric constant is known.

### 9.2.1
### Streaming Potential

When a solid charged surface is in contact with an electrolyte, a distribution of charge in the liquid phase, named "electrical double layer" (e.d.l.), can be produced [33]. According to the Stern model for the e.d.l., the total charge at the solid/solution system consists of three components, as represented in Figure 9.1a [34]: (1) the fixed charges at the membrane surface ($\Sigma_o$), (2) the charges in the Stern layer ($\Sigma_s$), which are associated with the adsorption of ions, (3) the charges within the diffuse part of the electrical double layer ($\Sigma_d$). As electroneutrality is assumed, the total charge equals zero, this means: $\Sigma_o + \Sigma_s + \Sigma_d = 0$.

The Stern model considers that the adsorbed charges (ions) are strongly linked to the solid but those in the diffuse part of the e.d.l. can easily move. These ions should move along to the solid wall when an external pressure ($\Delta P$) is applied, and an electrical potential or "streaming potential" between the two ends of the system can be measured under stationary conditions [33]. That is, as a result of the charge accumulation at the low-pressure side, an electrical field is generated, as indicated in Figure 9.1b, which causes a backflow of ions until the steady state is reached (when the convective, $I_{cv}$, and conduction, $I_{cd}$, currents equilibrate one to each other: $I_{cv} = I_{cd} \rightarrow I_{total} = 0$). The streaming potential ($\Delta ø_{st}$) is the measurable electrical potential difference between the two ends of the capillary or solid/liquid system and it gives information about the electrostatic charge at the shear plane.

A simple relation between the streaming potential and the zeta potential ($\xi$) or electrical potential at the shear plane can be obtained by the Helmholtz–Smoluchowski equation [33]:

$$\zeta_{H-S} = (\lambda_e \eta \varepsilon_o \varepsilon)(\Delta ø_{st}/\Delta P) \tag{9.1}$$

where $\lambda_e$ and $\eta$ are the solution conductivity and viscosity, respectively, while $\varepsilon$ represents the membrane dielectric constant, and $\varepsilon_o$ is the vacuum permeability. Equation (9.1) is only valid for weakly charged walls and wide channels or pores, when different effects such as surface conductivity and double layer overlapping can be neglected. In other cases, surface conductivity ($\lambda_s$), which takes into account

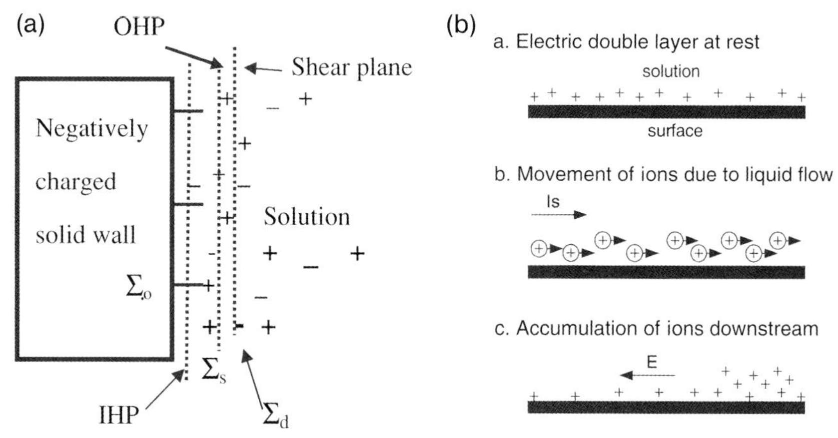


**Figure 9.1** (a) Membrane–solution interface: inner Helmholtz plane (IHP) and outer Helmholtz plane (OHP). (b) Streaming potential formation.

the contribution of the fixed charges into the shear plane, must be considered [33, 35]. Although $\lambda_s$ can not be determined experimentally, zeta potential can be obtained from measured streaming potential by using the Fairbother–Mastin approximation [36].

Surface charge density at the shear plane or electrokinetic surface charge density ($\sigma_e$) can be determined from zeta potential values by [33]:

$$\sigma_e = (2RT\varepsilon_o\varepsilon\,\kappa/z_iF)\sinh(z_iF\zeta/2RT) \tag{9.2}$$

where $z_i$ and $\kappa$ are the ion valence and the reciprocal Debye length (or double layer thickness) respectively, $R$ and F are the gas and Faraday constants, and T is the thermodynamic temperature of the system. Variation of electrokinetic charge density with pH or solution concentration allows the determination of characteristic membrane material characteristics such as the isoelectric point (pH at which $\zeta = 0$) or adsorption parameters if certain adsorption models are assumed [34].

## 9.2.2 Membrane Potential

The electrical potential difference at both sides of a membrane separating two solutions of the same electrolyte but different concentrations ($c_1$, $c_2$) is called "membrane potential" ($\Delta ø_m$). The Teorell–Meyer–Sievers (or TMS) theory [37, 38] assumes the membrane potential can be considered as the sum of three terms associated with two different contributions:

1. Two Donnan potentials, one for each membrane/solution interface, related to the exclusion of co-ions in the membrane:

$$\Delta ø_{Don} = (RT/F)\ln\left\{(wX_f/2c) + \left[(wX_f/2c)^2 + 1)^{1/2}\right]\right\} \tag{9.3}$$

where $w = +1$ or $-1$, for anionic or cationic membranes, and $X_f$ is the concentration of fixed charge in the membrane.

2. A diffusion potential due to the different mobility or transport number ($t_i$) of the ions in the membrane (for diluted solutions concentration was used instead of solution activity, $c_i \approx a_i$):

$$\Delta ø_{dif} = (RT/F)[(t_-/|z_-|) - (t_+/|z_+|)]\ln(c_1/c_2) \tag{9.4}$$

Combining these two expressions, the membrane potential can be expressed as [39]:

$$\begin{aligned}\Delta ø_m = (RT/wzF)(&\ln\{c_1[(1+4y_1)^{1/2}+1]/c_2[(1+4y_2)^{1/2}+1]\\ &+ wU\ln[(1+4y_1)^{1/2} - wU]/(1+4y_2)^{1/2} - wU\})\end{aligned} \tag{9.5}$$

where $y_j = (z_jk_sc_j/wX_f)^2$, $k_s$ is the salt partition coefficient, and U is a parameter related to the transport number of ions in the membrane ($U = [t_+/|z_+|] + [t_-/|z_-|]$).

Ion transport number ($t_+$ for cations, $t_-$ for anions) represents the fraction of the total current carried by each ion ($t_i = I_i/I_T$); and for single salts: $t_+ + t_- = 1$.

For highly porous and/or slightly charged membranes (or when the external salt concentration is much higher than the membrane fixed charge ($c \gg X_f$), the Donnan potential can be neglected, and the membrane potential can be considered as a diffusion potential [39]:

$$\begin{aligned} \Delta\text{ø}_m \approx \Delta\text{ø}_{dif} &= (RT/F)([t_+/|z_+|] + [t_-/|z_-|])\ln(c_1/c_2) \\ &= (RT/F)U\ln(c_1/c_2) \end{aligned} \quad (9.6)$$

According to Equation (9.6), a linear relationship should exist between membrane potential and $\ln(c_1/c_2)$ in these cases, which can be used to check the validity of the assumed conditions and to determine the ion transport numbers in the membrane.

### 9.2.3 Impedance Spectroscopy

Impedance Spectroscopy (IS) is an a.c. technique for electrical characterization of materials and interfaces based on impedance measurements carried out for a wide range of frequencies ($10^{-6} \leq f(Hz) \leq 10^9$), which can be used for the determination of the electrical properties of homogeneous (solids and liquids) or heterogeneous systems formed by a series array of layers with different electrical and/or structural properties (for example membrane/electrolyte systems), since it permits us a separate evaluation of the electrical contribution of each layer by using the impedance plots and equivalent circuits as models, where the different circuit elements are related to the structural/transport properties of the systems [40, 41].

When a linear system is perturbed by a small voltage (a sine wave input): $v(t) = V_o \sin \omega t$, the current intensity is also a sine wave:

$$i(t) = I_o \sin(\omega t + \phi) \quad (9.7)$$

where $V_o$ and $I_o$ are the maximum voltage and intensity, while $\omega = 2\pi f$ and $\phi$ represent the angular frequency and the phase angle, respectively. The impedance is defined as $Z(\omega) = v(t)/i(t)$, and due to the fact that both the amplitude and phase angle of the output may change with respect to the input values, the impedance is expressed as a complex number ($Z = Z_{real} + j\, Z_{img}$), which can be separated into real ($Z_{real}$) and imaginary ($Z_{img}$) parts by algebra rules. The simplest system is the electrochemical cell (that is: *electrode // solution (c) // electrode*) where transport of charge between the electrodes (electrical resistance) and charge adsorption on the electrodes (capacitance) are involved, and the equivalent circuit consists of a parallel resistance–capacitor (RC) circuit, then:

$$Z_{real} = \{R/[1 + (\omega RC)^2]\} \quad \text{and} \quad Z_{img} = -\{\omega R^2 C/[1 + (\omega RC)^2]\} \quad (9.8)$$

The analysis of the impedance data can be carried out by complex plane $Z^*(\omega)$ method by using the Nyquist plot ($-Z_{img}$ versus $Z_{real}$). The equation for a (RC) circuit gives rise to a semi-circle in the $Z^*(\omega)$ plane with intercepts on the $Z_{real}$ axis at $R_\infty$ ($\omega \to \infty$) and $R_o$ ($\omega \to 0$), being $(R_o - R_\infty) = R_s$ the resistance of the system

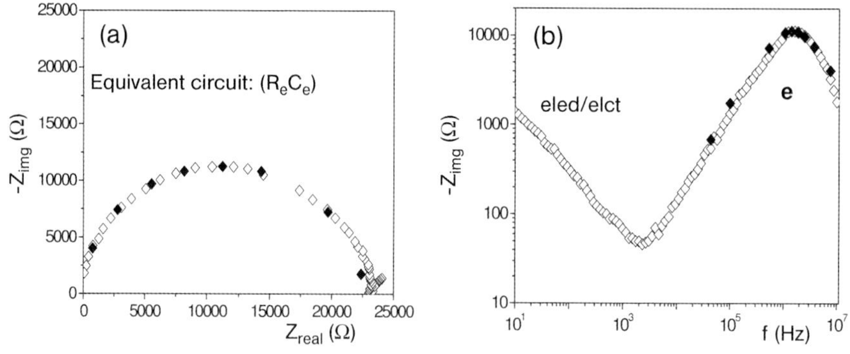


**Figure 9.2** Nyquist (a) and Bode (b) plots for the electrode/electrolyte solution (c) electrode system: (◇) experimental and (◆) calculated values ($R_e = 11\,800\,\Omega$, $C_e = 5 \times 10^{-12}$ F).

(see Figure 9.2a), while the maximum of the semi-circle equals $0.5(R_o - R_\infty)$ and occurs at such a frequency that $\omega RC = 1$, where $RC = \tau$ represents the relaxation time [42].

The Bode plot ($-Z_{img}$ versus f) is another common impedance plot which allows the determination of the interval of frequency associated with a particular relaxation process. Figure 9.2b shows the Bode plot for the electrolyte solution ($f > 2 \times 10^3$ Hz) with a maximum frequency between 1–2 MHz, and the electrode/electrolyte contribution for $f < 2 \times 10^3$ Hz. Electrical parameters ($R_i$, $C_i$) were determined by fitting the experimental data and calculated $Z_{real}$ and $Z_{img}$ values for certain frequencies are also drawn in Figure 9.2 (solid symbols). The good agreement existing between experimental and calculated values can be considered as a probe of the adequacy of the nonlinear program used for calculation [43].

However, complex systems usually present a distribution of relaxation times and the resulting plot is a depressed semi-circle, which is associated with a non-ideal capacitor or constant phase element (CPE), and its impedance is given by [42]:

$$Q(\omega)^* = Y_o(j\omega)^{-n} \tag{9.9}$$

where $Y_o = R_o \tau_o^{-n}$ is the admittance and $n$ is an empirical parameter ($0 \leq n \leq 1$). In these cases, an equivalent capacitance ($C^{eq}$) can be determined [44]:

$$C^{eq} = [(RY_o)^{(1/n)}]/R \tag{9.10}$$

When $n = 0.5$, the circuit element is called a Warburg impedance, W, which is associated with a diffusion process according to Fick's first law.

Membranes in contact with electrolyte solutions are heterogeneous systems (electrode/solution (c)/membrane/solution (c)/electrode) and generally two subsystems with different dielectric properties can be considered (the membrane and the electrolyte solution between the electrodes and the membrane surfaces), and consequently different contributions associated with relaxation processes having place in each subsystem should exist. Figure 9.3 shows examples of the impedance plots for different membrane/electrolyte systems, where significant differences related to membrane structure can be observed: (a) a dense aliphatic–aromatic membrane [45] (Figure 9.3a, b), (b) a polysulfone ultrafiltration membrane (Figure 9.3c, d), (c) a porous polycarbonate (PC) membrane of 10 μm thickness, and (d) dense sulfonated EPDM–polypropylene membrane (Figure 9.3e, f).

As can be observed, two semicircles were obtained for dense membrane/NaCl solution and the equivalent circuit is: $(R_eC_e)$–$(R_mC_m)$, that is, a series association of two resistance–capacitor subcircuits, one for the electrolyte solution and other for the membrane; while data for the ultrafiltration membrane corresponds to a depressed semicircle (associated with a CPE) and the equivalent circuit for the total membrane system is: $(R_eC_e)$–$(R_mQ_m)$. However, for the porous membrane/NaCl solution and charged membrane/KCl solution systems, a unique relaxation process (a semicircle) was obtained, which makes it impossible to evaluate the separate contributions associated with the membrane and the electrolyte solution; in both cases the equivalent circuit is given by a parallel association of resistance and capacitor representing the total membrane system $(R_{sm}C_{sm})$.

Nevertheless, these two latter systems (shown in Figure 9.3e, f) represent two completely different situations, as can be seem when the corresponding impedance plots for 0.002 M NaCl and 0.01 M KCl solutions are also considered (measurements carried out without membranes in the test cell):

1. The electrical resistance for the thin and porous PC membrane/electrolyte system $(R_{sm})$ is slightly higher than that determined for the electrolyte $(R_s)$; and membrane electrical resistance could be obtained by subtraction: $R_m = R_{sm} - R_s$.

2. The electrical resistance for the dense and charged DCM membrane/KCl solution system is clearly lower than that corresponding to the KCl solution, which indicates a higher charge in the membrane layer than in a 0.01 M KCl solution of the same membrane thickness.

On the other hand, when ideal capillary porous membranes are studied (that is, membranes with constant pore section and pore length equal to membrane thickness, $\Delta x_p = \Delta x_m$), the membrane resistance can be considered as that corresponding to the parallel association of the electrolyte filling the pores (assuming the resistance of the solid membrane matrix is much higher than the solution) and $R_m$ values might also give information on membrane porosity [46].

Moreover, when two-layer membranes (asymmetric or composite) are studied, the total system is: electrode/solution (c)/membrane (dense/porous layers)/solution (c)/electrode; and the impedance plots can then present three relaxation processes (two at the lowest frequencies which are associated with the membrane itself, plus the contribution of the electrolyte solution at high frequencies with $f_{max} \sim 10^6$ Hz), as can observed in Figure 9.4. Note, equivalent circuit: $(R_eC_e)$–$(R_1Q_1)$–$(R_2C_2)$.

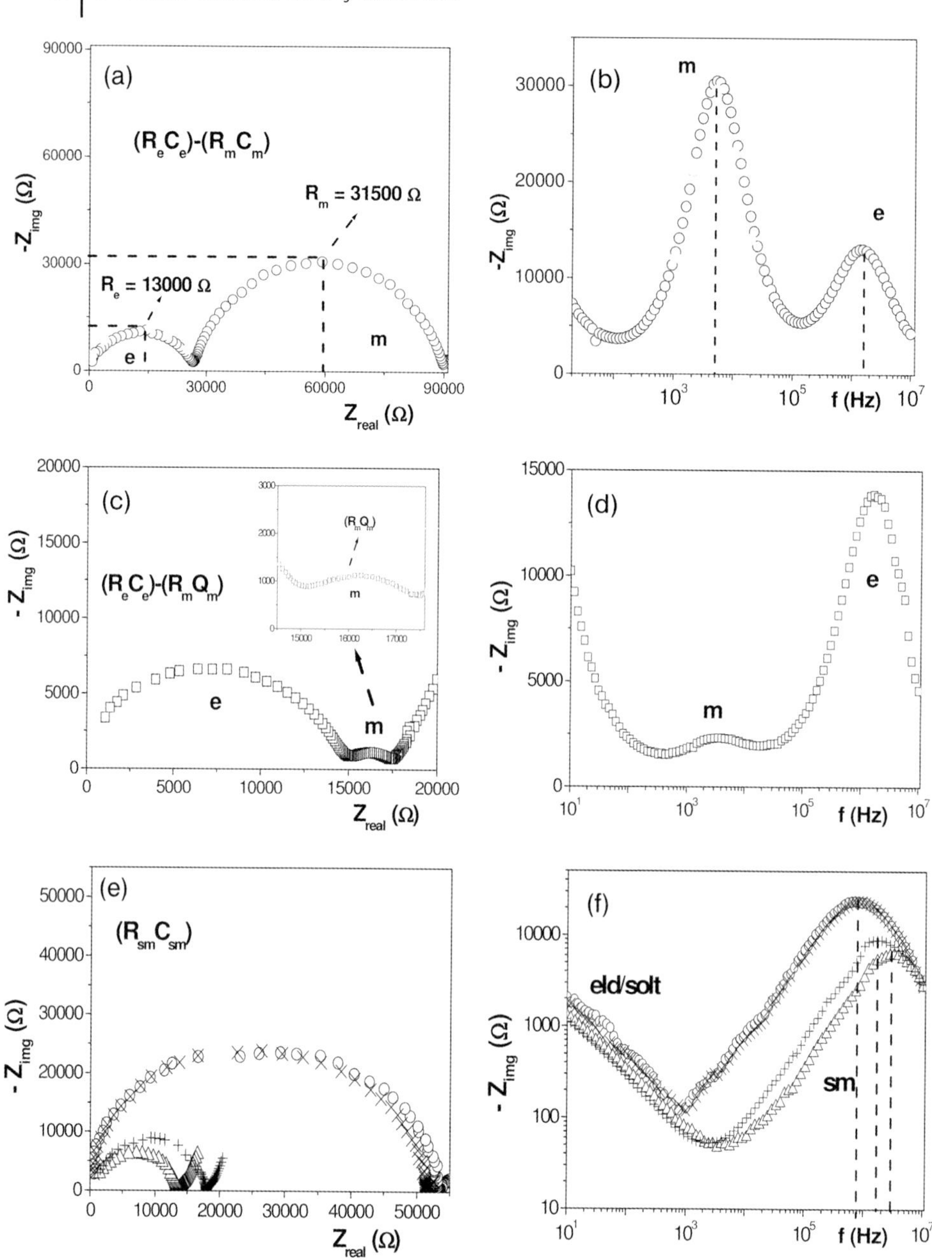


**Figure 9.3** Impedance plots for different membrane systems. (a, b) Dense polymeric membrane and 0.005 M NaCl. (c, d) Polysulfone ultrafiltration membrane and 0.002 M NaCl. (e, f) PC porous polycarbonate (Cyclopore) membrane in contact with 0.002 M NaCl solution (○) and 0.002 M NaCl solution (x); dense charged membrane (S-EPDM-PP) in contact with 0.01 M KCl solution (△) and 0.01 M KCl solution alone (+).

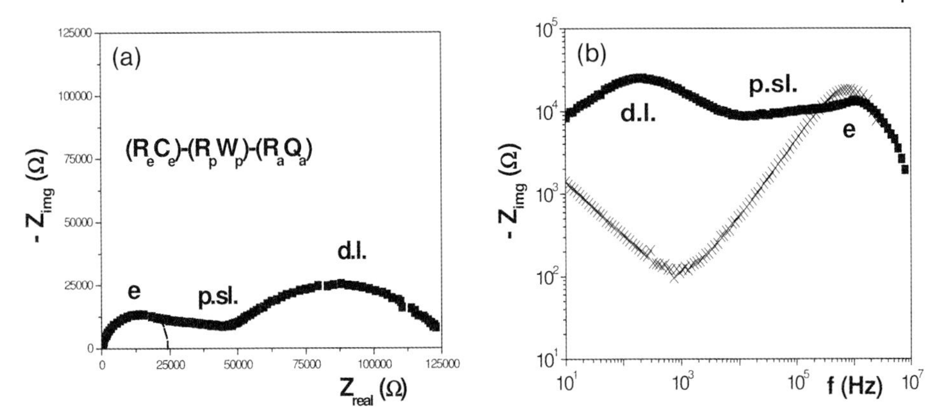


**Figure 9.4** Nyquist (a) and Bode (b) plots for: (■) electrode/electrolyte solution (c)/dense layer–porous sublayer/electrolyte solution (c)/electrode and (x) electrode/electrolyte solution (c)/electrode. e = electrolyte; p.sl. = porous sublayer; d.l. = dense layer.

In fact, the possibility of using IS measurements for separate determination of parameters associated with an individual layer in multilayer systems is one of the most important applications of this kind of electrical characterization, but as was indicated in this section, qualitative information on membrane structure can also be obtained from impedance plots.

All these examples show the interest of IS measurements, particularly when the membranes are in contact with electrolyte solutions, since qualitative information on membranes structure can be obtained. Other impedance representations (impedance modulus $|Z|$ and/or $\tan\phi = (Z_{img}/Z_{real})$ versus frequency) as well as different complex magnitudes such as dielectric constant and modulus or dielectric loss can also be determined from IS measurements and they are also commonly used in the literature [47, 48].

## 9.3 Experimental

### 9.3.1 Membranes and Solutions

Membranes from different materials and with diverse structures (porous, dense, asymmetric, composites) currently used in different filtrations processes (ultrafiltration, nanofiltration, reverse osmosis) and other separation applications as well as the effect of fouling on porous membranes were considered by studying the following samples:

1. A commercial composite polyamide/polysulfone membrane for reverse osmosis (HR95) and an ultrafiltration polysulfone membrane (PS-Uf) similar to the porous support of the composite membrane, both from DDSS (Denmark). Both membranes present similar thickness, $\Delta x_m = 165 \pm 5\,\mu m$ and the hydraulic permeability is $L_p^{PS\text{-}Uf} = 10.0 \times 10^{-10}$ m/s Pa and $L_p^{HR95} = 8.5 \times 10^{-12}$ m/s Pa, for the PS-Uf and the HR95, respectively, while the rejection coefficient for the reverse osmosis HR95 membrane is $\sigma^{HR95} = 99.5\%$ [7].

2. Fouling modifications was investigated by using the porous polysulfone PS-Uf sample after maintaining one of the membrane surfaces 72 h in contact with a solution containing: (i) 0.5 g/L of bovine seroalbumin (BSA), (ii) 0.5 g/L of diethylaminoethyl dextran (DEAE-dextran) supplied by Pharmacia Fine Chemical (Sweden). These fouled samples are called PS-Uf/BSA and PS-Uf/DEAE-dextran, respectively. At neutral pH the selected macromolecules present opposite charges: BSA has a negative charge while DEAE-dextran is a polycation due to $NH^+$ groups (3.2% approximately nitrogen groups) [49, 50].

3. An experimental asymmetric lignosulfonated modified polysulfone/polyamide membrane (sample PS/PA-LS20) obtained and kindly provided by Prof. R. García-Valls and Dr. X. Zhang. Membrane geometrical parameters are: total thickness of 117 μm and porosity of 42%, while the thickness of the lignosulfonated modified top layer is 2.7 μm [51]; values were determined from SEM pictures analyzed using IFME software [52].

4. A commercial composite polyamide/polysulfone membrane for nanofiltration (sample NF45) from Filmtec; membrane thickness, hydraulic permeability and salt rejection are: $\Delta x_m = 155 \pm 3\,\mu m$, $L_p^{NF45} = 10.8 \times 10^{-12}$ m/s Pa, $\sigma \approx 25\%$ ($c_{NaCl} = 0.01$ M).

5. A dense regenerated cellulose membrane (sample RgC) from Cellophane Española, S.A. (Burgos, Spain), with the following wet thickness and swelling degree: $\Delta x_m = 56 \pm 3\,\mu m$, $Sw = 50 \pm 4\%$.

Measurements were carried out with the membranes in contact with aqueous electrolyte solutions at different concentrations and room temperature ($25 \pm 3\,^\circ C$). Before the different measurements, the membranes were maintained in contact with a solution of the appropriate concentration during a certain time ($8\,h \leq t \leq 20\,h$), depending on the membrane structure.

### 9.3.2
### Streaming Potential Measurements

Streaming potential (SP) equipment basically consists of: (1) the measuring cell, (2) a mechanical drive unit to produce the pressure that drives the electrolyte solution from a reservoir into the measuring cell, and (3) two Ag/AgCl electrodes to measure the pressure-induced streaming potential. Two different kinds of SP measurements were performed: (a) filtration streaming potential (FSP), when the solution passes through the membrane, and (b) tangential streaming potential (TSP), which consists in the formation of an artificial channel by facing up two samples of the same membrane, or their "active" layer in the case of asymmetric/composite membranes. A more detailed description of the equipments and experimental procedures is given in [49, 53]. These measurements were carried out varying the concentration of the solutions at constant pH, but zeta potential–pH dependence for FSP measurements was also considered by changing the pH ($3 \leq pH \leq 9$) and keeping the concentration constant ($c = 0.001$ M).

In order to use the Fairbother–Mastin approximation for surface conductivity correction [36], measurements at a concentration high enough to neglect electrokinetic phenomena (0.1 M NaCl) were also made and zeta potential was determined by means of Equation (9.11) [53]:

$$\begin{aligned}\zeta &= \{[\lambda_e + (\lambda_S/h)]\eta\}/(\varepsilon_o\varepsilon_r)(\Delta\o_{st}/\Delta P) \\ &= (\eta/\varepsilon_o\varepsilon_r)(\lambda_o^a R_a/R_c)(\Delta\o_{st}/\Delta P)\end{aligned} \tag{9.11}$$

where $h$ is the half thickness of the solution channel (or pore radii), $\lambda_e^a$ is the conductivity of the solution at high concentration, $R_c$ represents the electrical resistance of the studied system, $R_a$ is the electrical resistance with high concentration solution (0.1 M), and the other parameters have already been indicated.

### 9.3.3 Membrane Potential and Impedance Spectroscopy Measurements

The test cell used for the electrochemical measurements was similar to that described elsewhere [54] and basically consisted of two glass half-cells of around 20 $cm^3$ volume each, with the membrane tightly clamped between them by using silicone rubber rings. A magnetic stirrer was placed at the bottom of each half-cell to minimize concentration polarization at the membrane surface, and measurements were carried out at a stirring rate of 525 rpm.

- The electromotive force ($\Delta E_f$) between both sides of the membranes caused by a concentration gradient was measured by two Ag/AgCl electrodes (reversible to chlorine solutions) connected to a digital voltmeter (Yokohama 7552; 1 G$\Omega$ input resistance). Measurements were carried out by keeping constant the concentration of the solution at one side of the membrane, $c_1$, and gradually changing the concentration of the solution at the other side, $c_2$, from $10^{-3}$ M to 0.1 M. Between two series of measurements, electrodes were maintained in a $10^{-3}$ M solution and the asymmetry potential was checked; if it exceeded 5 mV the AgCl layer of the electrodes was refreshed by electrolysis. Membrane potential, $\Delta\Phi_m$, was obtained by subtracting the electrode potential, $\Delta\Phi_e = -(RT/z\_F)\ln(c_1/c_2)$, from the corresponding $\Delta E_f$ measured value.
- Impedance spectroscopy (IS) measurements were carried out by using an impedance analyzer (Solartron 1260) controlled by a computer, which was connected to the solution in both half-cells via Ag/AgCl or platinum electrodes, which were larger that the membrane area to ensure the uniformity of the electric field (border effect can be neglected) and small holes to permit solution counter-diffusion and reduce concentration polarization near the membrane. The

experimental data were corrected by software as well as the influence of connecting cables and other parasite capacitances. The measurements were carried out using 100 frequency values in the range from 1 Hz to $10^7$ Hz at a maximum voltage of 0.01 V, with the solutions in both half-cells having the same concentration. IS measurements without membranes placed in the membrane holder were also performed to determine the impedance associated with each electrolyte solution.

## 9.4 Results of Electrical Measurements for Different Types of Membranes

### 9.4.1 Streaming Potential

Streaming potential as a function of the applied pressure for different membranes is shown in Figure 9.5. Trans-membrane or filtration streaming potential (FSP) for composite HR95 membrane, its porous support (membrane PS-Uf) and the DEAE–dextran fouled PS-Uf porous membrane is shown in Figure 9.5, where different effects of interest can be observed, depending on the membrane structure and solution:

- The different streaming potential sign when clean and DEAE–dextran fouled UF samples are compared, indicating the adsorption of positively charged DEAE–dextran macromolecule on the pores of the PS-Uf membrane (Figure 9.5a); this point clearly establishes the possibility of using streaming potential measurements as a tool to identify membrane fouling/modification, which is one of the most important and common applications of this technique.
- The zero value for PS-Uf porous membrane streaming potential at $\Delta P = 0$, while for composite membrane a value $\Delta\Phi_{st} \neq 0$ was obtained, which is attributed to the concentration profile across the dense active layer according to its high salt rejection (>96% for HR95) and the membrane potential associated with this concentration difference between both surfaces of the dense layer (Figure 9.5b) [7, 55]. This fact is one of the reasons for the use of tangential streaming potential measurements when high rejection membranes (RO or NF) are studied.

Figure 9.5c shows tangential streaming potential with dense RgC and nanofiltration NF45 membranes, which characterizes the external membrane surface (instead of "internal surface" or pore wall/solution interface), as represented in Figure 9.5d. From the slopes of the straight lines shown in Figure 9.5a, c, the streaming potential coefficient $\Phi_{st} = \Delta\varphi_{st}/\Delta P$ was determined and zeta ($\zeta$) potential was obtained from those values by using Equation (9.1).

Variation of zeta potential with salt concentration for some of the studied membranes is shown in Figure 9.6; a practically constant value for the whole range of concentration was obtained for PS-Uf and PS-Uf–dextran fouled samples (Figure 9.6a, FSP values), but a clear decrease for the absolute value of $\zeta$ potential

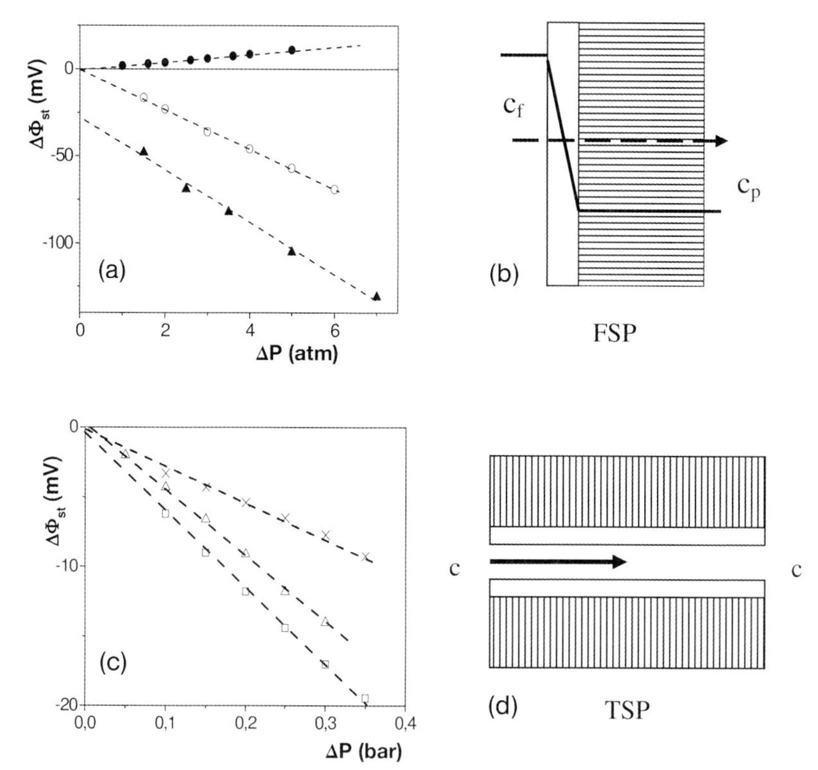


**Figure 9.5** (a) Transmembrane (filtration) streaming potential vs applied pressure: (○) PS-Uf membrane, (●) PS-Uf-DEAE-dextran fouled membrane, (▲) composite HR95 membrane. (b) Schematic representation of concentration profile in dense and porous layers for a composite RO membrane under filtration condition. (c) Tangential streaming potential for dense RgC membrane and NaCl solutions (□) 0.001 M and (x) 0.005 M, and composite nanofiltration NF45 membrane and 0.002 M KCl solution (△).

with the increase of solution concentration was obtained in the case of TSP measurements as can be seen in Figure 9.6b, where a comparison of results corresponding to membrane NF45 with KCl and NaCl solutions is made, these results show higher $\zeta$ values for NaCl than for KCl solutions, while differences hardly exist at the highest concentrations. Figure 9.6b also shows a comparison of $\zeta$ potential for NF45 membrane and KCl solutions determined from two series of measurements performed after maintaining the membrane in contact with aqueous solutions for three weeks. A significant reduction in $\zeta$ values was obtained which could be due to modification (or hydration) of the polyamide top layer, as was already determined by X-ray photoelectron spectroscopy (XPS) spectra analysis for other polyamide/polysulfone membranes [24], and this could be one of the reasons for reported differences in $\zeta$ potential values for similar (even the same) membranes [56]. It should be pointed out that FSP measurements allow the determination of both streaming potential and hydrodynamic permeability, which gives extra and important information mainly related to fouling effects on

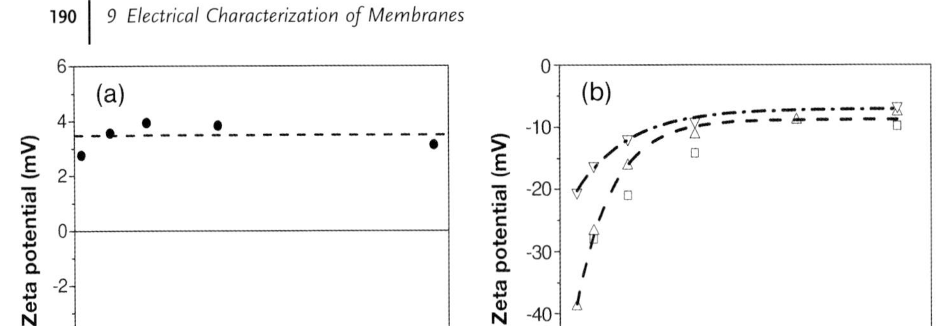


**Figure 9.6** (a) Zeta potential versus NaCl concentration for PS-Uf (○) and PS-Uf-DEAE–dextran fouled (●) membranes determined from transmembrane streaming potential. (b) Zeta potential versus salt concentration for composite NF45 membrane determined for tangential streaming potential: (△) KCl solutions and (□) NaCl solutions; (▽) KCl solution 2° series of measurements.

membranes; particularly, the DEAE–dextran fouling of PS-Uf membrane indicated in Figure 9.6 causes a reduction of approximately 65% in water permeability, attributed mainly to pore plugging according to streaming potential results [57], but this kind of information is not accessible from TSP measurements.

Electrokinetic surface charge density was obtained from $\zeta$ potential by Equation (9.2) and its variation with ion molar fraction, $\chi_i$, for the studied membranes is shown in Figure 9.7, while $\chi_i$ $(i = -, +)$ is given by:

$$\chi_i = (cv_i M_{H2O})/[\rho. - c(v_- M_- + v_+ M_+ (v_- + v_+) M_{H2O})] \approx (cv_i M_{H2O})/\rho. \tag{9.12}$$

These results seem to indicate adsorption of anions on the surfaces of PS-Uf and RgC membranes, but the reduction of the negative value of NF45 membrane might be associated with cation adsorption.

Normally, homogeneous or heterogeneous adsorptions of ions are considered by using Langmuir or Freundlich isotherms, respectively, in order to determine characteristic membrane materials parameters.

- In the case of homogeneous adsorption (Langmuir isotherm), free adsorption energy can be separated into two contributions, one related to zeta potential and the other to the chemical adsorption ($\Delta G^{ch}$), and the following expressions are obtained [34, 58]:

$$[A_1/(\sigma_e + \sigma_o)] - A_2 = \exp(-z_+ F\zeta/RT\chi_-) \tag{9.13}$$

$$N = A_1/A_2 \exp(-z_+ F\zeta/RT) \tag{9.14}$$

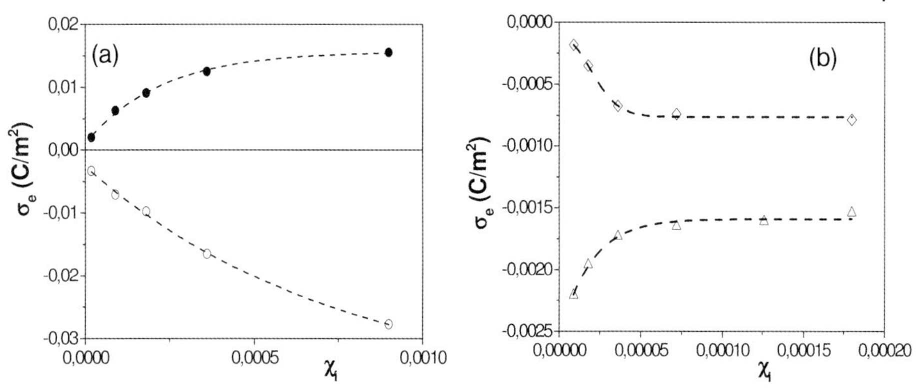


**Figure 9.7** Electrokinetic surface charge density versus ion molar fraction. (a) PS-Uf (○), PS-Uf-DEAE-dextran fouled (●) membranes. (b) composite NF45 membrane (KCl solution; △), dense RgC membrane (NaCl solution; ◇).

where $A_1$ and $A_2$ are two empirical parameters which allow the determination of the chemical adsorption free energy ($\Delta G^{ch} = -RT \ln A_2$) and the total number of adsorption, N.

- If heterogeneous adsorption is considered (Freundlich isotherm), the following dependence between $\sigma_e$ and $\chi_i$ is expected:

$$\sigma_e(\chi_i) = \sigma_o + A\chi_i^B \qquad (9.15)$$

where $\sigma_o$ represents the surface charge density on the solid wall, which can be zero for noncharged surfaces; the maximum number of available sites per membrane area, $N_{max}$, and the average value for the adsorption free energy, $<\Delta G_{ads}>$, can be determined by:

$$N_{max} = (A/z_{-}e)[\sin(\pi B)/(\pi B)] \qquad (9.16)$$

$$<\Delta G_{ads}> = -(RT/B) \qquad (9.17)$$

In both cases, the sign of the molar Gibbs free energy also provides information on chemical reactions or spontaneous adsorption of ions on the solid. Table 9.1 shows the estimated parameters for porous polysulfone (FSP measurements) and the polyamide top layer of NF45 (TSP measurements) assuming heterogeneous adsorption.Clear differences in the number of adsorption sites and mean free energy can be seen, depending on the membrane material, but practically similar values were obtained for NF45 membrane independently of the electrolyte studied.

As was indicated previously streaming/zeta potential is commonly used for estimation of membrane modification associated with fouling and to determine the main fouling mechanisms (basically deposition/adsorption of particles on the membrane surface or into the pore wall), in order to establish cleaning procedures.

**Table 9.1** Maximum number of adsorption sites per surface unit, $N_{max}$, and mean adsorption free energy, $<\Delta G_{ads}>$, determined from Equations (9.16) and (9.17).

| Membrane | $N_{max}$ ($m^{-2}$) | $<\Delta G_{ads}>$ (J/mol) |
|---|---|---|
| PS-Uf (NaCl)[a] | $1.6 \times 10^{18}$ | $-4.0 \times 10^{3}$ |
| NF45 (NaCl)[b] | $1.2 \times 10^{16}$ | $-25.4 \times 10^{3}$ |
| NF45 (KCl) | $3.6 \times 10^{16}$ | $-21.6 \times 10^{3}$ |

[a] From [34].
[b] From [58].

Due to the different electrical behavior of PS-Uf (negative) and DEAE–dextran (positive), pore plugging can be assumed for the results shown in Figure 9.7, but a more difficult assignation might exist when both membrane and foulant present the same kind of charge. In these cases, measurements of streaming potential at different pHs (for a given concentration) permits us to get useful information, as can be seen in Figure 9.8, which shows the variation of zeta potential with pH for clean PS-Uf and PS-Uf-BSA fouled membranes. A comparison of the isoelectric point obtained for PS-Uf ($pH \sim 3.2$) and PS-Uf-BSA fouled ($pH \sim 5.0$) samples indicates the presence of BSA within the membrane pores (BSA isoelectric point at $pH \approx 4.8$) and pore plugging can be considered as the main membrane fouling mechanism.

These results clearly show that streaming potential measurements allows the characterization of membrane surface/solution interface (external and pore walls or internal surfaces) giving information not only on the fouling particles/macromolecules but also on characteristic membrane material parameters.

### 9.4.2
### Membrane Potential

The measured electrical potential difference at both sides of a membrane ($\Delta E$) separating two solutions of the same electrolyte ($c_1$, $c_2$) but at different concentrations ($c_1 = 10^{-3}$ M NaCl; $10^{-4}\,M \leq c_2 \leq 0.3$ M) as a function of $\ln(c_1/c_2)$ is shown in Figure 9.9a for dense RgC, composite HR95, its porous support or sample PS-Uf, and the dextran–DEAE fouled PS-Uf membranes. As can be observed in all cases, almost linear $\Delta E - \ln(c_1/c_2)$ relationships were obtained. However, these values do not represent the membrane potential, since electrode potential [which linearly depends on $\ln(c_1/c_2)$] is also included. Figure 9.9b shows membrane potentials, $\Delta\Phi_m$ (obtained by subtracting the corresponding electrode potential to the $\Delta E$ measured values), for the same membranes and clear differences can be observed among the studied membranes, depending on its structure and when they are compared with $\Delta E$ measured values. For comparison, theoretical values for an ideal cation exchange membrane [$t_+ = 1$, then $\Delta\Phi = (-RT/F)$

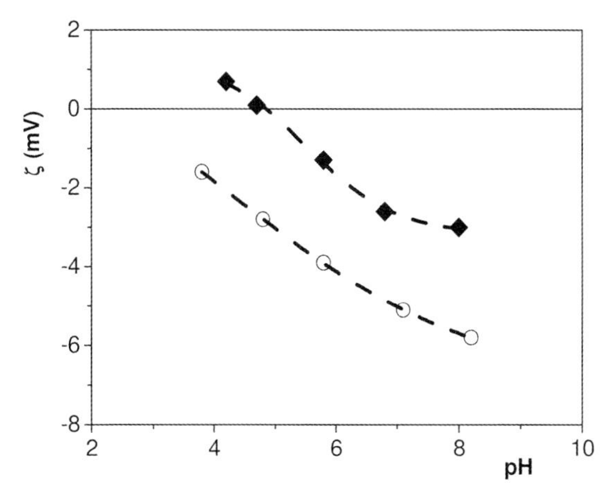


**Figure 9.8** Variation of zeta potential with pH for porous PS-Uf membrane (○) and PS-Uf-BSA fouled membrane (♦), c = 0.001 M NaCl.

$\ln(c_1/c_2)$] are also drawn in Figure 9.9b as a solid line. Two different behaviors are observed in this picture depending on the membrane and concentration ratio: dense RgC, composite HR95, and porous PS-Uf membranes (all of them with negative charge according to previous and literature streaming potential results [59]) present negative values for $c_1 < c_2$ and they follow the same tendency (although with different values) as the ideal cation exchange membrane but for $c_1 > c_2$ the opposite tendency was obtained. However, a linear decrease for

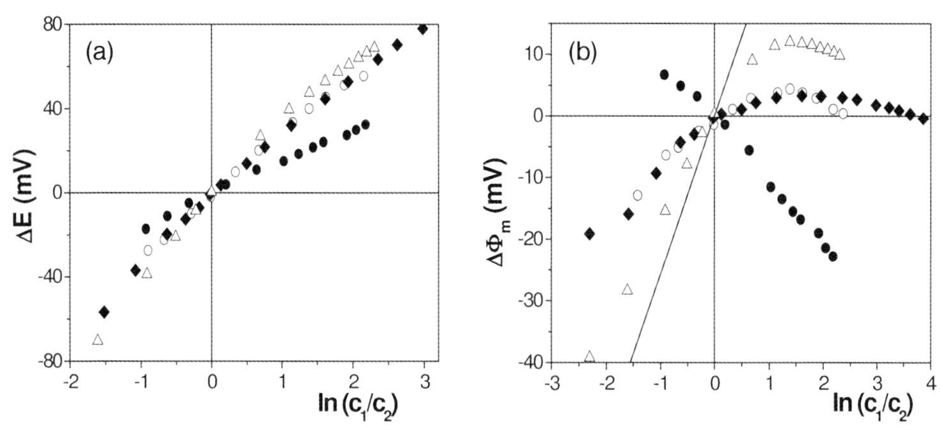


**Figure 9.9** (a) Electromotrice force $\Delta E_f$ versus $\ln(c_1/c_2)$ for different membranes in contact with NaCl solutions ($c_1 = 0.01$ M): PS-Uf (o), PS-Uf-DEAE-dextran fouled (●), HR95 (♦) and RgC (△). (b) Membrane potential, $\Delta\phi_m = \Delta E_f - \Delta\phi_{electrode}$, versus $\ln(c_1/c_2)$ for the same membranes.

membrane potential across the whole concentration range was obtained for the positively charged PS-Uf-DEAE–dextran fouled sample.

The linear dependence obtained at low $c_1$ values seems to be related to the membrane negative fixed charge ($X_f$) and the exclusion of anions (co-ions) from the membrane phase (Donnan exclusion), but at high concentrations ($c \gg X_f$) that effect can be neglected and the membrane potential is associated with a diffusion potential due to the different mobility of anions and cations through the membrane [39].

Membrane fixed charge concentration, $X_f$, can be obtained from $\Delta\Phi_m = f(c)$ curves shown in Figure 9.9b by Equation (9.18) [12, 60]:

$$X_f = 2\, z_i k_s U c_{ext}/(1 - U^2)^{1/2} \tag{9.18}$$

where $c_{ext}$ represents the concentration at the maximum of the curve (determined by considering the maximum (or extremal) condition, $(d\Phi_m/dc)_{ext} = 0$), $k_s$ is the salt partition coefficient in the membrane, and $U = [(t_-/|z_-|) - (t_+/|z_+|)]$. Ion transport numbers at high solution concentrations when $\Delta\text{ø}_m \sim \Delta\text{ø}_{dif}$ (that is, when $c \gg X_f$) can be obtained from the diffusion potential expression:

$$\begin{aligned} 1:1 \text{ electrolyte}, \Delta\text{ø}_{dif} &= -(RT/F)[t_+ - t_-]\ln(c_1/c_2) \\ &= (RT/F)[1 - 2t_+]\ln(c_1/c_2) \end{aligned} \tag{9.19}$$

$$\begin{aligned} 2:1 \text{ electrolyte}, \Delta\text{ø}_{dif} &= -(RT/F)[t_+ - (t_-/2)]\ln(c_1/c_2) \\ &= (RT/F)[1 - (3t_+/2)]\ln(c_1/c_2) \end{aligned} \tag{9.20}$$

Table 9.2 shows the fixed charge concentration and average cation transport number for the studied membranes. The observed $<t_+>$ for HR95 and PS-Uf membranes are slightly higher than solution transport number ($t_{Na^+}{}^{o} \approx 0.385$), indicating a week electronegative character. The PS-Uf-DEAE–dextran fouled membrane has electropositive character ($t_{Na^+}{}^{PS\text{-}Uf\text{-}BSA} < t_{Na^+}{}^{o}$), in agreement with streaming potential results, while the dense RgC membrane behaves as a slight cation exchanger membrane [59, 61].

In order to elucidate whether membrane potential might give information related to membrane electrical modification, different regenerated cellulose membranes were studied and their $\Delta\Phi_m - \ln(c_1/0.01)$ dependence is shown in Figure 9.10: (1) a porous sample (PRgC) from Sartorius, (2) the dense RgC membrane previously used after chemical treatments (immersion for 24 h in 1 M $H_3PO$ or NaOH, samples RgC($H_3PO_4$) and RgC(NaOH), respectively), (3) the dense RgC membrane and $BaCl_2$ solutions.

Clear differences depending on the membrane structure were obtained, as can be observed in Figure 9.10a, but chemical treatment hardly affects membrane potential values, which show the same concentration dependence as well as those measured with $BaCl_2$ solutions. For comparison calculated solution (NaCl) diffusion potential is also represented in Figure 9.10a; and its values hardly differ from those obtained for PRgC membrane, indicating that the transport of ions through the membrane pores do not differ from solution. Average cation transport number in the membranes, $<t_+>$, was obtained by the slopes of the straight

**Table 9.2** Average cation transport number, $<t_+>$, correlation coefficient ($r$) for Equation (9.19) or Equation (9.20), effective fixed charge concentration, $X_f$, and average cation permselectivity, $<S(+)>$, for composite HR95, polysulfone support PS-Uf, DEAE–dextran fouled PS-Uf and dense regenerated cellulose RgC membrane.

| Membrane | $<t_+>$ | r | $X_f$ (M) | $<S(+)>$ (%) |
|---|---|---|---|---|
| HR95 | $0.46 \pm 0.02$ | 0.993 | $-1.5 \times 10^{-3}$ | 12.2 |
| PS-Uf | $0.42 \pm 0.03$ | 0.990 | $-1.2 \times 10^{-3}$ | 5.7 |
| PS-Uf DEAE–dextran | $0.31 \pm 0.01$ | 0.996 | – | 19.5[a] |
| RgC | $0.52 \pm 0.03$ | 0.988 | $-1.4 \times 10^{-2}$ | 22.0 |
| RgC[b] | $0.51 \pm 0.02$ | 0.998 | $-8.6 \times 10^{-3}$ | 11.6 |
| RgC (NaOH) | $0.55 \pm 0.04$ | 0.986 | $-1.1 \times 10^{-2}$ | 26.8 |
| RgC ($H_2PO_4$) | $0.53 \pm 0.03$ | 0.991 | $-1.2 \times 10^{-2}$ | 23.6 |
| Porous RgC | $0.39 \pm 0.02$ | 0.993 | – | – |

[a] Anionic permselectivity $<S(-)>$.
[b] Measurements carried out with $BaCl_2$ solutions.

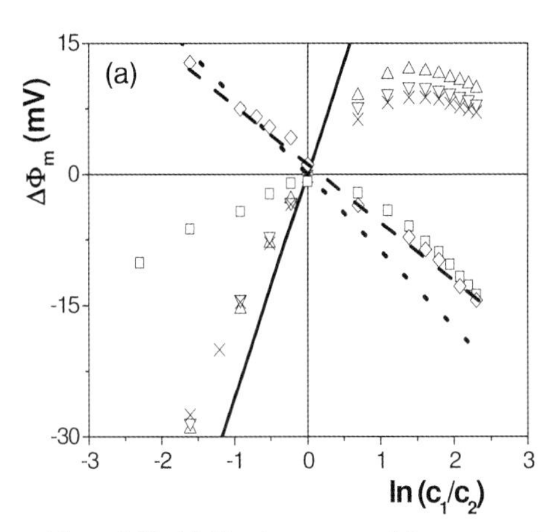


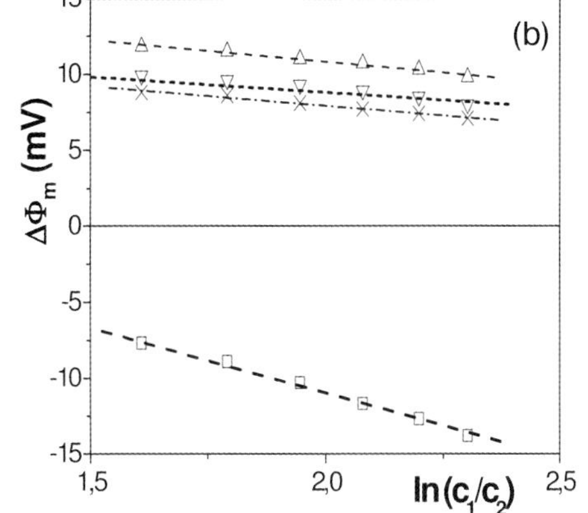


**Figure 9.10** (a) Membrane potential versus $\ln(c_1/c_2)$ for dense regenerated cellulose RgC ($\Delta$), RgC($H_3PO_4$) ($\nabla$) and RgC(NaOH) (x) with NaCl solutions, and RgC with $BaCl_2$ solutions ($\square$); porous regenerated cellulose PRgC sample ($\diamond$). Ideal cation exchanger (solid line), NaCl solution diffusion potential (dashed line), $BaCl_2$ solution diffusion potential (dotted line). (b) Linear dependence of membrane potential with $\ln(c_1/c_2)$ at high concentrations for the same dense RgC membrane and solutions that in (a).

lines shown in Figure 9.10b by means of Equation (9.19) or Equation (9.20) and they are indicated in Table 9.2.

It should be pointed out that, since water transport ($t_w$) was not considered in the previous discussion, then the ion transport numbers obtained refer to the solution and not to the membrane [62], and they are usually called apparent ion transport numbers ($t_i^{ap}$). Scatchard obtained the following relationship between apparent ($t_+^{ap}$) and true ($t_+^{m}$) ion transport number in a membrane [63]:

$$t_+^{ap} = t_i^m - 0.0018\, t_w(c_{avg}) \tag{9.21}$$

where $t_+^{ap}$ is the transport number obtained from Equation (9.19) or Equation (9.20) for a given pair of concentrations ($c_1$, $c_2$) and $c_{avg}$ is the average value of the concentrations at both sides of the membrane [$c_{avg} = (c_1 + c_2)/2$]. The following $t_+^{m}$ and $t_w$ values were obtained for dense RgC membrane in contact with NaCl and $BaCl_2$ solutions:

$$\begin{aligned} \text{NaCl}: (a)\ t_+^m &= (0.628 \pm 0.012), t_w = (145 \pm 12); \\ (b)\ t_+^m &= (0.693 \pm 0.006), t_w = (156 \pm 6) \end{aligned}$$

$$\text{BaCl}_2 : t_+^m = (0.527 \pm 0.009), t_w = (170 \pm 13)$$

where values in (a) and (b) for NaCl solutions correspond to measurements carried out by keeping concentration $c_1 = \text{constant} = 0.01$ M and changing concentration $c_2$ (*results from a*), or by keeping the ratio $c_2/c_1 = \text{constant} = 2$, but changing the concentration in both solutions (*results from b*). The slight differences for $t_+^{m}$ and $t_w$ values depending on the concentration gradients are related to some theoretical aspects, since experimental conditions for case (b) fulfils more adequately the thermodynamic approximations involved in the develop of the previous equations.

Another electrochemical parameter of interest determined from membrane potential results is the membrane permselectivity, S(i), which is a measure of the selectivity of counter ions over co-ions in a membrane [39]:

$$S(i) = (t_i - t_i^o)/(1 - t_i^o) \tag{9.22}$$

Permselectivity is useful when the transport of different electrolytes through a membrane is considered, since it is normalized to the value of the corresponding solution counter ion ($t_i^o$). The average cation permselectivity $<S(+)>$ for the studied membranes is also indicated in Table 9.2 (except for the positively charged PS-Uf–dextran fouled sample; then the anion permselectivity $<S(-)>$ is given). These results show that differences in membrane potential and ion transport numbers can be related to: (a) membrane structure, since rather similar values were obtained for porous polysulfone PS-Uf and regenerated cellulose PRgC membranes, which hardly differ from solution transport number, (b) membrane material, showing dense regenerated cellulose membrane with more electronegative behavior than the dense polyamide layer, and (c) the electrolyte used for membrane characterization, since the permselectivity of dense RgC membrane to $Na^+$ ions is 22% but is around 12% for $Ba^{++}$ ions.

### 9.4.3
### Impedance Spectroscopy

Impedance plots for HR95 and PS-Uf membranes are shown in Figure 9.11, where two different contributions can clearly be observed, associated with: the membrane (*m*) and the electrolyte solution between the electrodes and the membrane surfaces (*e*). For comparison, Figure 9.11 also indicates electrolyte impedance plots (without any membrane in the measuring cell) and the equivalent circuits associated with the different systems. As can be observed, a parallel (RC) circuit with only a relaxation process and maximum frequency around $10^6$ Hz (similar to that represented in Figure 9.2) was obtained for the electrolyte solution, while the circuit associated with the porous PS-Uf membrane (embedded in solution) consists of a parallel association of resistance with a CPE element ($R_mQ_m$); however, the most interesting result corresponds to the composite polyamide/polysulfone HR95 membrane, since two subcircuits (one for each layer) can be considered: (i) a parallel association of a resistance and a capacitor for the dense active layer ($R_dC_d$), (ii) a resistance in parallel with a CPE element for the porous sublayer ($R_pQ_p$), as was already obtained for the porous membrane.

Fitting the data shown in the impedance plots by using a nonlinear program [43] allows the different elements to be determined and their variation with NaCl concentration is shown in Figure 9.12; a decrease of electrical resistance can be observed when the salt concentration increases due to the concentration dependence of the electrolyte embedded in the membrane matrix or filling the porous support (Figure 9.12a). Significant differences in the electrical resistance values for PS-Uf membrane ($R_{PS}$) and the porous layer of HR95 composite membrane ($R_{pl}$) are attributed to the concentration profile in the skin layer and consequently in the

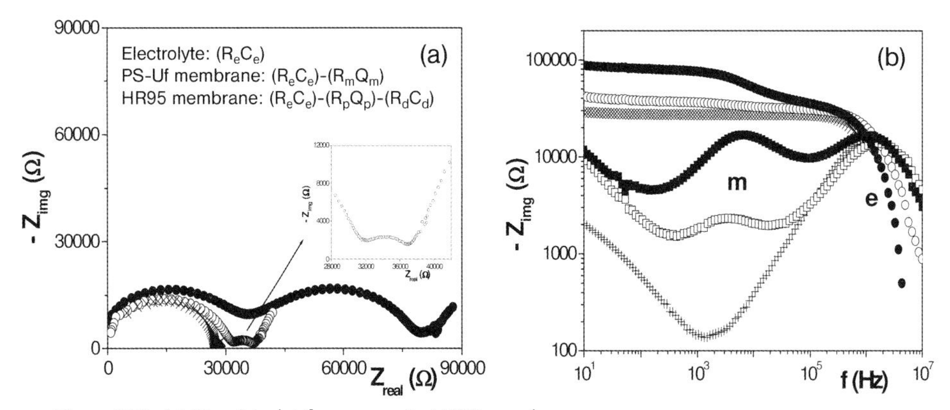


**Figure 9.11** (a) Nyquist plot for composite HR95 membrane (●), porous PS-Uf membrane (○) and NaCl solution (x). (b) Bode plots: HR95 membrane real part (●) and imaginary part (■), porous PS-Uf real part (○) and imaginary part (□), NaCl solution real part (x) and imaginary part (+).

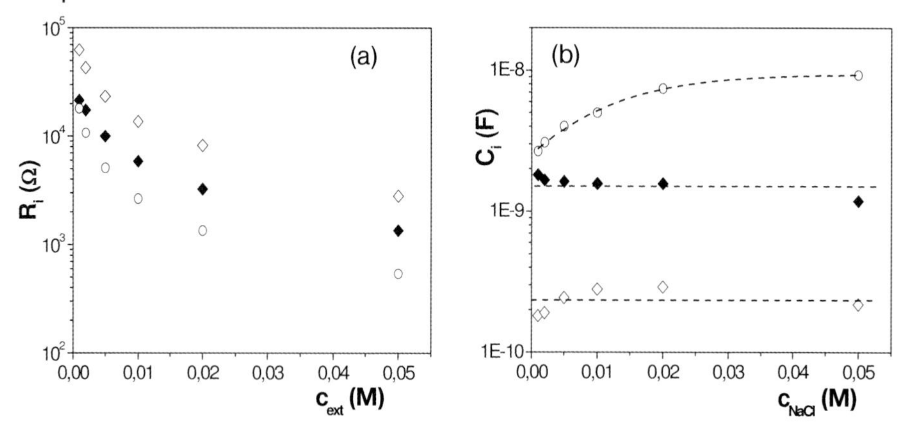


**Figure 9.12** Variation of membrane/layer electrical resistance (a) and capacitance (b) for dense layer of composite HR95 membrane (♦), porous sublayer of composite HR95 membrane (◇), porous PS-Uf membrane (○).

porous sublayer of the composite membrane, as a result of the high salt exclusion; then the average concentration of the solution filling the membrane interstices/pores can significantly differ from that corresponding to the external solution. Capacitance (or equivalent capacitance) for dense and porous layers hardly depends on salt concentration, as can be seen in Figure 9.12b, but $C^{eq}$ values for the porous PS-Uf membrane clearly increase when salt concentration increases.

It should be pointed out that, for simplicity, a two-layer model has been assumed for the HR95 reverse osmosis membrane but a more complex structure (including an "intermediate layer" with gradual changes in pore radii/porosity from one to another layer or three-layer model) could be more realistic [64, 65]. In this context, the partial inclusion of that intermediate layer in the thick porous sublayer could in some way affect the estimated values, but clearly it would modify those associated with a thin dense layer (the layer thickness, mainly).

As was previously indicated, IS measurements can also be used to determined membrane modifications; and Figure 9.13 shows Nyquist and Bode plots for PS-Uf and PS-Uf/BSA fouled membranes in contact with a NaCl solution. Here a significant increase in electrical resistance due to membrane fouling can be observed, but the electrolyte contribution hardly differs in both systems. In both cases, the equivalent circuit for the membrane–electrolyte system is given by: $(R_eC_e)$–$(R_mQ_m)$; that is, a series association of the electrolyte part, formed by a resistance in parallel with a capacitor and the membrane part, which consists of a parallel association of a resistance and a CPE or non-ideal capacitor $(R_mQ_m)$. Fitting the experimental data allows determination of the electrical parameters (resistance, capacitance) for the different NaCl solutions studied; and their variation with electrolyte concentration is shown in Figure 9.13c, d, respectively.

A decrease in the values of membrane electrical resistance with an increase in the NaCl concentration was obtained for both systems and the following resistance

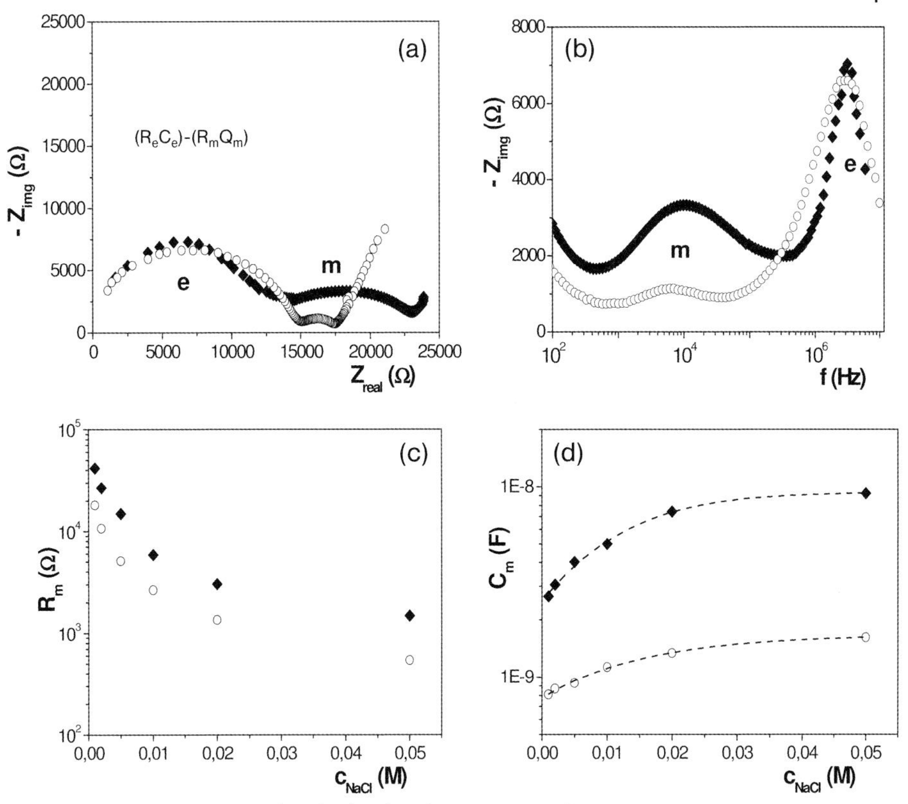


**Figure 9.13** Nyquist (a) and Bode (b) plots for porous PS-Uf (○) and PS-Uf-BSA fouled (♦) membranes in contact with $5 \times 10^{-3}$ M NaCl solution. Variation of membranes electrical resistance (c) and capacitance (d).

ratio was determined for the whole interval of concentrations: $R_m^{PS\text{-}Uf/BSA}/R_m^{PS} = (2.5 \pm 0.3)$. This value is very similar to the hydraulic resistance ratio ($R_H = 1/L_p$) for both membranes obtained from hydrodynamic measurements ($L_p = J_v/\Delta P$), that is, $R_H^{PS\text{-}Uf/BSA}/R_H^{PS} = (2.1 \pm 0.2)$, which seems to indicate that resistance modifications is directly related to the reduction of membrane pore radii/porosity. The increase of membrane capacitance also agrees with a reduction of the section for the transport of charge and, consequently, an increase for the adsorption surface.

A study of asymmetric lignosulfonated modified polysulfone–polyamide PS/PA-LS20 membrane was also carried out by IS measurements. Figure 9.14 shows the cross-section SEM micrographs of the PS/PA-LS10 membrane, which were analyzed by using IFME software [52]; these results indicate the membrane with a thickness of 92 μm presents an open central zone (porosity of 42%, mean pore radii of $2.95 \pm 0.13$ μm) and a dense top layer of 2.7 μm thickness and $(98.00 \pm 0.08)$ nm mean pore radii, where the lignosulfonate is mainly located [51].

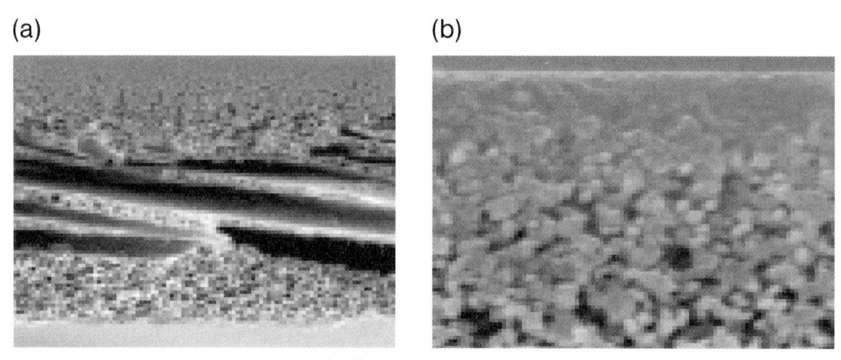


**Figure 9.14** Cross-section micrographs for asymmetric polysulfone/polyamide–lignosulfinate modified PS/PA-LS20 membrane (from [50]): (a) total membrane, (b) top layer (from [50]).

The asymmetric structure of this membrane could also be deduced from the Nyquist plot shown in Figure 9.15a, where a clear asymmetry in the electrical response can be observed, associated with the membrane contribution, which is in agreement with the membrane morphology given by SEM micrographs. The equivalent circuit for the total membrane system indicated in Figure 9.15a also considers its two-layer structure: $(R_eC_e)–(R_pW_p)–(R_dC_d)$.

Analysis of the impedance plots allows the determination of $R_p$, $W_p$, $R_d$, and $C_d$ for the different concentrations studied, and Figure 9.15b shows the decrease in the dense top layer and porous support electrical resistance with the increase of salt concentration. However, the dense top layer capacitance values hardly depend

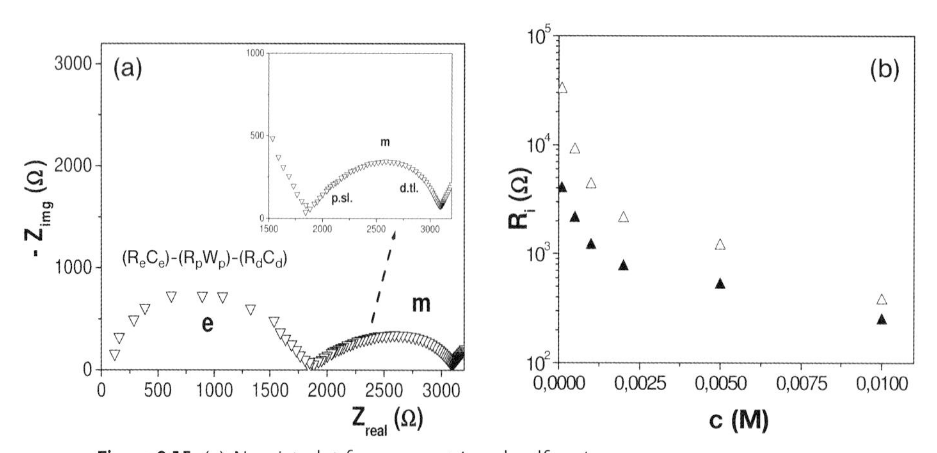


**Figure 9.15** (a) Nyquist plot for asymmetric polysulfone/polyamide–lignosulfinate modified PS/PA-LS20 membrane. (b) Variation with electrolyte concentration of electrical resistance for dense top layer (▲) and porous sublayer (△) of PS/PA-LS20 membrane.

on electrolyte concentration and the average value for the whole concentration range was considered for estimating the dense layer thickness ($\Delta x_{tl}$) by using the plane-parallel capacitor expression [41]:

$$C_d = (\varepsilon\varepsilon_o S_m / \Delta x_{tl}) \tag{9.23}$$

where $S_m$ is the membrane area and the other symbols have already been indicated. From this result a thickness of 3 μm was obtained (assuming $\varepsilon = 25$ [51]), which agrees very well with the value obtained from SEM micrographs, indicating the possibility of using IS measurements for determination of geometrical parameters for membranes or individual layers in the case of asymmetric/composite membranes.

Impedance plots for the other studied membranes (composite polyamide/polysulfone nanofiltration NF45 and RgC regenerated cellulose) are shown in Figures 9.16 and 9.17 respectively, and differences in the IS curves and equivalent circuits associated with both kinds of membranes are related to their different structures:

- The composite structure presented by NF45 membrane allows the assignation of two subcircuits for the membrane contribution, one for the dense layer (d.l.) and another for the porous sublayer (p.sl.), as indicated in Figure 9.16a. The comparison of the IS plots for the electrolyte solution alone (without any membrane) shows the adequacy of such assignation. From capacitance values the following average thickness for the dense layer was determined [66]: $\Delta x_{dl} = 0.23 \pm 0.05$ μm.
- Due to the high swelling degree of regenerated cellulose, a unique relaxation process for the RgC membrane/electrolyte system was obtained (already indicated) as can be seen in Figure 9.17, which shows the Nyquist and Bode plots for different NaCl concentrations. The results confirm the increase in membrane conductivity (decrease in electrical resistance) when the electrolyte concentration increases due to the higher number of charges embedded in the membrane matrix (and in the solution) and, as a result of this, a shift occurs to higher values of the maximum frequency (from 1 MHz at 0.001 M NaCl to 5 MHz at 0.01 M NaCl, represented by dashed lines in Figure 9.17a).

However, analysis of the IS diagrams only allows the determination of membrane system parameters ($R_{sm}$, $C_{sm}$), and the separate membrane contribution cannot be obtained in these cases, but subtraction of $R_s$ values (determined under similar conditions) might be considered as a way for $R_m$ determination.

Moreover, other transport parameters such as ionic permeability (or ionic diffusion coefficient) can be determined by combining some of the electrical characterization techniques indicated in this Chapter. Equations (9.24) and (9.25) present relationships among ionic permeabilities ($P_+/P_-$, cationic/anionic permeability respectively) with membrane potential ($\Delta\Phi_m$), concentration ratio at both membrane sides ($\gamma = c_1/c_2$), and membrane electrical resistance ($R_m$) at the average

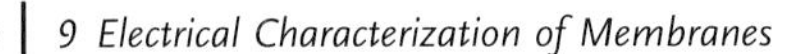


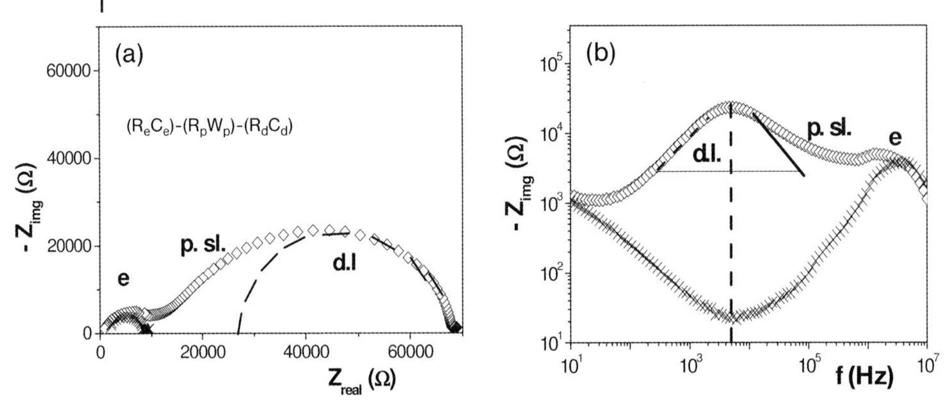


**Figure 9.16** Nyquist (a) and Bode (b) plots for asymmetric polysulfone/polyamide NF45 membrane in contact with a NaCl solution (◇) and the NaCl solution (×). Contributions of the dense layer (d.l.) and porous sublayer (p.sl.) are indicated.

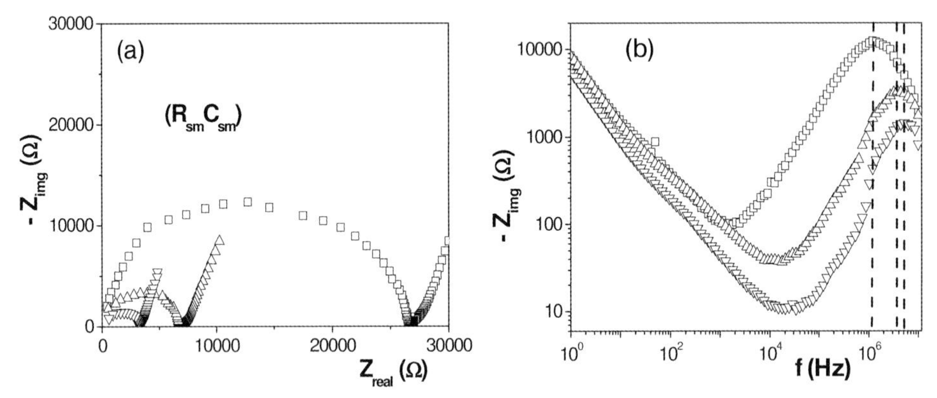


**Figure 9.17** Nyquist (a) and Bode (b) plots for dense and symmetric RgC membrane in contact with different NaCl solutions: 0.001 M (□), (△) 0.005 M, (▽) 0.01 M.

concentration $c^*$, while the other parameters have already been indicated [67]:

$$P_+/P_- = [\exp(F\Delta\Phi_m/RT) - \gamma]/[1 - \gamma\exp(F\Delta\Phi_m/RT)] \tag{9.24}$$

$$P_+ + P_- = (RT/F^2)(1/S_m R_m c^*) \tag{9.25}$$

Electrical resistance/NaCl concentration dependence for membrane PS-Uf was obtained by fitting the experimental data shown in Figure 9.12a, which allows the determination of $R_m$ at concentration $c^*$ ($c^* = (c_1 + c_2)/2$, where $c_1$ and $c_2$ are the solution highest values in membrane potentials shown in Figure 9.9b). Variation of anionic and cationic permeability with $c^*$ is shown in Figure 9.18, where the decrease in $P_+$ values with increasing NaCl concentration agrees with the screening of the small negative membrane charge (according to membrane potential and

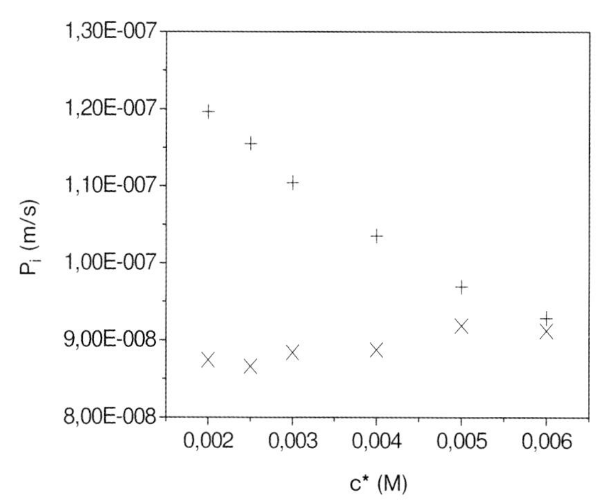


**Figure 9.18** Variation of ionic permeability with salt concentration: $P_+$ (+), $P_-$ (×).

streaming potential results) at high concentration, and practically similar ionic permeability values are reached at the highest concentrations.

Ionic permeabilities can be related to salt permeability by [67]: $P_s = 2\,P_+\,P_-/(P_+ + P_-)$. Then, by determining $P_+$ and $P_-$ average values for the concentration interval shown in Figure 9.18, the following average value for salt permeability was determined: $\langle P_s \rangle = (1.0 \pm 0.3) \times 10^{-7}$ m/s. This value is rather similar to that obtained from direct salt diffusion measurements [7] ($P_s = 3.6 \times 10^{-7}$ m/s) and shows the possibility of confirming values which have been estimated or determined from two different types of electrical measurements with those directly obtained from completely different (diffusion) experiments.

## 9.5 Conclusions

Theoretical and experimental descriptions of three different types of electrical measurements commonly used for membrane characterization have been presented. Streaming potential, membrane potential, and impedance spectroscopy are nondestructive techniques and they can be carried out with the membranes in contact with electrolyte solutions, which allows characterization of membranes in "working conditions", but they also permit us to established changes in the membrane characteristic parameters related to solution chemistry (ion size, charge, concentration, pH) and/or membrane structure.

Ion transport numbers across membranes (relative ion fluxes) and effective fixed charge can be obtained from membrane potential measurements, but a wide porous structure might mask the membrane material electrical characteristics (membrane/solution interactions). These can be determined from filtration streaming potential measurements or tangential streaming potential in the case of dense membranes, but important transport information associated with hydrodynamic permeation is missing in the latter cases. Other information related to different characteristic membrane material parameters (electrical/adsorption parameters) can also be obtained from these measurements.

Impedance spectroscopy allows the determination of electrical parameters related to the flow or adsorption of charge (electrical resistance or capacitance, respectively) for simple and heterogeneous systems consisting of a series array of layers with different electrical/structural characteristics (such as membrane/electrolyte systems), and it can be used for a separate characterization of membrane and electrolyte solution. Moreover, if composite membranes are studied, electrical parameters for each layer could also be determined as well as other structural/material information obtained from the estimated electrical parameters.

Membrane potential and impedance spectroscopy results can be combined in order to obtain other significant membrane transport parameters, such as ionic permeabilities (or ionic diffusion coefficients/mobilities if membrane geometry is known).

## Acknowledgments

Thanks are given Prof. G. Jonsson, head of the Membrane Laboratory at Technical University of Denmark (Kemiindustry Instituttet), where streaming potential and membrane potential measurements with membranes HR95 and PS-Uf (clean and fouled samples) were carried out, and thank Prof. R. García-Valls, Dr. X. Zhang, and Dr. C. Torras (Chemical Engineering Department, Universidad Rovira i Virgili, Tarragona, Spain) for submitting PA/PS-Lig membrane and SEM micrographs. Thanks are also given to the European Commission (Project IC15 CT 96-0826), CICYT (Spain; Project MAT2003-003328), and the Junta de Andalucía (Spain; Research Group FQM 258) for partial financial support.

## Nomenclature

| | |
|---|---|
| a.c. | Alternating current |
| $a_i$ | Solution activity |
| $c_i$ | Solution concentration |
| C | Capacitance |
| $C_I$ | Layer i capacitance |
| $C_d$ | Plane-parallel capacitance |
| $C^{eq}$ | Equivalent capacitance |
| e.d.l. | Electric double layer |
| E | Electric field |
| $E_f$ | Electromotive force |
| F | Faraday constant |
| $G_{ads}$ | Adsorption Gibbs free energy |
| $G^{ch}$ | Chemical part of adsorption Gibbs free energy |

$h$ Half thickness of the solution channel (or pore radii)
$I_{cv}$ Convection current
$I_{cd}$ Conduction current
$I_T$ Total current
$I_o$ Maximum a.c. current
IHP Inner Helmholtz plane
$k_s$ Membrane partition coefficient
$M_{H2O}$ Water molar mass
N Total number of adsorption sites per surface unit
$N_{max}$ Maximum number of adsorption sited per surface unit
OHP Outer Helmholtz plane
P Pressure
Q Non-ideal capacitor or constant phase element (CPE)
$R$ Gas constant
$R_I$ Electrical resistance for layer membrane, electrolyte or layer j ($i = m, e, j$)
S(i) Membrane permselectivity to ion i
$S_m$ Membrane surface
T Thermodynamic temperature
$t_i$ Apparent transport number in the membrane for ion i
$t_i^m$ True membrane transport number for ion i
$t_w$ Water transport number in the membrane
$V_o$ Maximum a.c. voltage
w Membrane charge sign
$X_f$ Membrane equivalent fixed charge concentration
$Y_o$ Admittance
Z Impedance
$Z_{real}$ Real part of impedance
$Z_{img}$ Imaginary part of impedance
$z_I$ Ion valency
$\varepsilon$ Dielectric constant
$\varepsilon_o$ Vacuum permeability
$\phi$ Phase angle between v(t) and i(t)
$\phi_{st}$ Streaming potential
$\Phi_m$ Membrane potential
$\lambda_e$ Solution conductivity
$\lambda_e^a$ Conductivity of the solution at high concentration
$\zeta$ Zeta or electrokinetic potential
$\eta$ Solution viscosity
$\kappa$ Reciprocal Debye length
$\nu_I$ Ion stoichiometric coefficient
$\omega$ Angular frequency
$\rho$ Solution density
$\Sigma_i$ Fixed charge at the solid surface ($i = o$), Stern layer ($i = s$), diffuse part of e.d.l. ($i = e$)
$\sigma_i$ Surface charge density ($\sigma_i = \Sigma_i/\text{area}$)

## References


1 M. Mulder, *Basic Principles of Membrane Technology*, Kluwer, Dordrecht, **1992**.
2 F.G. Helfferich, *Ion Exchange*, McGraw–Hill, New York, **1962**.
3 C.K. Lee, J. Hong, *J. Membrane Sci.* **1988**, *39*, 79–88.
4 R. Takagi, M. Nakagaki, *J. Membrane Sci.* **1990**, *53*, 19–35.
5 G. Jonsson, C.E. Boesen, *Desalination* **1977**, *21*, 1–10.
6 M. Cheryan, *Ultrafiltration Handbook*, Technomic, Pennsylvania, **1986**.
7 G. Jonsson, J. Benavente, *J. Membrane Sci.* **1992**, *69*, 29–42.
8 J.M.M. Peters, M.H.V. Mulder, *J. Membrane Sci.* **1998**, *145*, 199–209.
9 S. Bandini, D. Vezzani, *Chem. Eng. Sci.* **2003**, *58*, 3303–3326.
10 M. Ernst, A. Bismark, J. Springler, M. Jekel, *J. Membrane Sci.* **2000**, *165*, 251–259.
11 W. Pusch, *Charged Gels and Membranes I*, (ed. E. Sélégny), Reidel, Dordrecht, **1976**.
12 Y. Kimura, H.-J. Lim, T. Ijima, *J. Membrane Sci.* **1984**, *18*, 285–296.
13 J. Benavente, C. Fernández-Pineda, *J. Membrane Sci.* **1985**, *23*, 121–136.
14 H.-H. Swartz, V. Kudela, J. Lukás, J. Varík, V. Gröbe, *Collect. Czech. Chem. C.* **1986**, *51*, 539–544.
15 F. Yang, P.h. Déjardin, A. Schmitt, *J. Phys. Chem.* **1993**, *97*, 3824–3828.
16 K.J. Kim, A.G. Fane, M. Nyström, A. Pihlajamaki, W.R. Bowen, H. Mukhtar, *J. Membrane Sci.* **1996**, *116*, 149–159.
17 J. Hosch, E. Staude, *J. Membrane Sci.* **1996**, *121*, 71–82.
18 V. Compañ, T.S. Sorensen, A. Andrio, L. López, J. de Abajo, *J. Membrane Sci.* **1997**, *123*, 293–302.
19 T. Jimbo, A. Tanioka, N. Minoura, *Langmuir* **1998**, *14*, 7112–7118.
20 P. Fievet, B. Aoubiza, A. Szymczyk, J. Pagetti, *J. Membrane Sci.* **1999**, *160*, 267–275.
21 J.M.M. Peters, M.H.V. Mulder, H. Strathmann, *Colloids Surface A* **1999**, *150*, 247–259.
22 V.M. Barragán, C. Ruíz-Bauzá, *J. Membrane Sci.* **1999**, *154*, 261–272.
23 M. Pointié, *J. Membane. Sci.* **1999**, *154*, 213–220.
24 A. Cañas, M.J. Ariza, J. Benavente, *J. Membrane Sci.* **2001**, *183*, 135–146.
25 M.D. Afonso, G. Hagmeyer, R. Gimbel, *Sep. Purif. Technol.* **2001**, *22/23*, 529–541.
26 P. Fievet, M. Sbaï, A. Szymczyk, *J. Membrane Sci.* **2005**, *264*, 1–12.
27 R. de Lara, J. Benavente, *J. Colloid Interface Sci.* **2007**, *310*, 519–528.
28 G. Alberti, C. Bastolini, M. Casciola, F. Marmottini, G. Capannelli, S. Munari, *J. Membrane Sci.* **1983**, *16*, 121–135.
29 M. Oleinikova, M. Muñoz, J. Benavente, M. Valiente, *Langmuir* **2001**, *16*, 716–721.
30 D. Sangeetha, *Eur. Polym. J.* **2005**, *41*, 2644–2652.
31 P.M. Gomadam, J.W. Weidner, *Int. J. Energy Res.* **2005**, *29*, 1133–1151.
32 J.-S. Park, J.-H. Choi, J.-J. Woo, S.-H. Moon, *J. Colloid Interface Sci.* **2006**, *300*, 655–662.
33 J.R. Hunter, *Zeta Potential in Colloid Science*, Academic, London, **1988**.
34 J. Benavente, A. Hernández, G. Jonsson, *J. Membrane Sci.* **1993**, *80*, 285–296.
35 J. Liklema, *Fundamentals of Interface and Colloid Science*, Academic, London, **1993**.
36 F. Fairbrother, H. Mastin, *J. Chem. Soc. Trans.* **1924**, *125*, 2319–2330.
37 K.H. Meyer, J.F. Siever, *Helv. Chim. Acta* **1936**, *19*, 649–664.
38 T. Teorell, *Discuss. Faraday Soc.* **1956**, *21*, 9–16.
39 N. Lakshminarayanaiah, *Transport Phenomena in Membranes*, Academic Press, New York, **1969**.
40 J. Mijovíc, F. Bellucci, *Trends Polymer Sci.* **1996**, *4*, 74–81.
41 J. Benavente, J.M. García, R. Riley, A.E. Lozano, J. de Abajo, *J. Membrane Sci.* **2000**, *175*, 43–52.

42 J.R. Macdonals, *Impedance Spectroscopy*, Wiley, New York, **1987**.
43 B.A. Boukamp, *Solid State Ionics* **1986**, *18/19*, 136–140.
44 A.K. Jonscher, *Dielectric Relaxation in Solid*, Chelsea Dielectric, London, **1983**.
45 J. Benavente, J. de Abajo, J.G. de la Campa, J.M. García, *Sep. Sci. Technol.* **1997**, *32*, 2189–2199.
46 A. Cañas, J. Benavente, *Sep. Purif. Technol.* **2000**, *20*, 169–175.
47 H.-T. Kim, J.-K. Park, K.-H. Lee, *J. Membrane Sci.* **1996**, *115*, 207–215.
48 T. Sorensen, V. Compañ, *J. Chem. Soc. Faraday Trans.* **1995**, *91*, 4235–4250.
49 J. Benavente, G. Jonson, *Colloids Surface A* **1998**, *138*, 255–264.
50 J. Benavente, G. Jonsson, *Sep. Sci. Technol.* **1997**, *32*, 1699–1710.
51 C. Torras, X. Zhang, R. García-Valls, J. Benavente, *J. Membrane Sci.* **2007**, *297*, 130–140.
52 C. Torras, R. García-Valls, *J. Membrane Sci.* **2004**, *233*, 119–127.
53 M.J. Ariza, J. Benavente, *J. Membrane Sci.* **2001**, *190*, 119–132.
54 J. Benavente, A. Muñoz, A. Heredia, *J. Membrane Sci.* **1998**, *139*, 147–154.
55 J. Benavente, G. Jonson, *J. Membrane Sci.* **2000**, *172*, 189–197.
56 M.Ch. Wilbert, S. Delagah, J. Pellegrino, *J. Membrane Sci.* **1999**, *161*, 247–261.
57 G. Belfort, R.H. Davis, A.L. Zydney, *J. Membrane Sci.* **1994**, *96*, 1–58.
58 C. Molina, L. Victoria, A. Arenas, J.A. Ibañez, *J. Membrane Sci.* **1999**, *163*, 239–255.
59 C.J. van Oss, *Science* **1963**, *139*, 1123–1124.
60 H.-U. Demish, W. Pusch, *J. Colloid Interface Sci.* **1979**, *69*, 247–270.
61 K. Sakai, *J. Membrane Sci.* **1994**, *96*, 91–130.
62 A.J. Staverman, *Trans. Faraday Soc.* **1952**, *48*, 176–184.
63 G.J. Scartchard, *J. Amer. Chem. Soc.* **1953**, *75*, 2883–2887.
64 G. Jonsson, *Desalination* **1966**, *1*, 141–152.
65 R.R. Bhave, *Inorganic Membranes: Synthesis, Characteristics and Applications*, Van Nostrand Reinhold, New York, **1991**.
66 M.J. Ariza, A. Cañas, J. Benavente, *Surf. Interface Anal.* **2000**, *30*, 425–429.
67 S.G. Schultz, *Basic Principles of Membrane Transport*, Cambridge University Press, Cambridge, **1980**.

# 10
# X-ray Tomography Application to 3D Characterization of Membranes

*Jean-Christophe Remigy*

## 10.1
## Principles of X-ray Tomography

Tomography by X-ray is a nondestructive technique for the characterization of objects. Like all the tomographies, it is a question of obtaining 3D information on the internal and/or external structure of the observed object. This technique thus makes it possible to observe and to inspect objects (search for defects, fractures, etc.) or to measure characteristic dimensions (thickness, size of pores, porosity, etc). This chapter covers only the tomography of absorption, which is the simplest technique. More sophisticated methods exist such as phase contrast tomography [1, 2].

The object is exposed to X-rays and a detector records the distribution of the intensity (flux of photon) of X-rays having crossed the sample on a plane perpendicular to the axis of the rays (see Figure 10.1). An image is thus recorded, generally using a scintillator and a CCD camera, although other detectors are possible [1, 3]. A more complete description of the apparatus can be found in reference [4]. An image made up of various levels of gray is obtained and corresponds to the spatial distribution of X-rays having crossed the object.

The absorption of X-rays of wavelength $\lambda$ by a material is described by the Beer–Lambert law:

$$\frac{N}{N_0} = \exp\left[-\int_0^l \mu(x, y, z) \cdot dx\right] \qquad (10.1)$$

where $N$ and $N_0$ are, respectively, the number of photons having crossed the sample and the number of incident photons, $\mu(x, y, z)$ is the linear attenuation coefficient taken at the point $(x, y, z)$ and $l$ is the thickness crossed by the X-rays.

Equation (10.1) can be rewritten in the form of Equation (10.2). This equation shows that, by standardizing the intensity transmitted by the incident photons, it is possible to determine the integral of $\mu$ along the crossed thickness. As the object

*Monitoring and Visualizing Membrane-Based Processes*
Edited by Carme Güell, Montserrat Ferrando, and Francisco López

ISBN: 978-3-527-32006-6

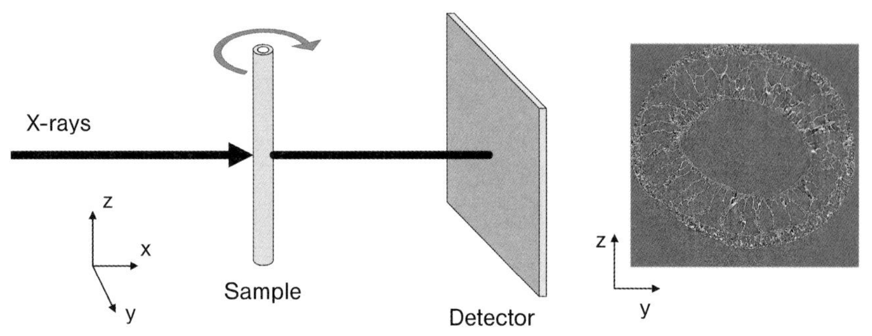


**Figure 10.1** Principle of X-ray tomography.

is supposed to be heterogeneous and composed of different materials, the recorded images correspond then to the spatial distribution of the $\mu$ integrated on the thicknesses of the different crossed materials.

$$\int_0^l \mu(x, y, z) \cdot dx = \ln\left(\frac{N_0}{N}\right) \tag{10.2}$$

The linear attenuation coefficient ($\mu$) is calculated, for a given wavelength, using Equation (10.3):

$$\mu = k.\rho.\frac{Z^4}{E^3} \tag{10.3}$$

where $k$ is a constant, ρ is the density of the sample, $Z$ is the atomic number and $E$ is the energy of the photon.

According to this equation, the linear attenuation coefficient $\mu$ is proportional to the density of the medium and the atomic composition of medium ($Z$ with the power 4) for a given energy $E$.

In the case of a heterogeneous medium, X-rays are absorbed differently according to the nature of crossed materials. The image obtained corresponds then to the projection of the internal structure of the object. The setting in rotation of the studied object on itself for angles ranging between 0° and 180° allows the frame grabbing of various projections of the structure of the object, as shown in Figure 10.2. It will be noticed that the images obtained between 180° and 360° are identical to those obtained between 0° and 180°. It is also important that the object does not move during the acquisition of the images: the images taken at 0° and 180° which are identical make it possible to check that the sample did not evolve or move during the experiment. Each image is thus made up of various pixels (elementary units of a 2D image, for example 1024 × 1024 pixels) in which each level of gray is proportional to $\mu$.

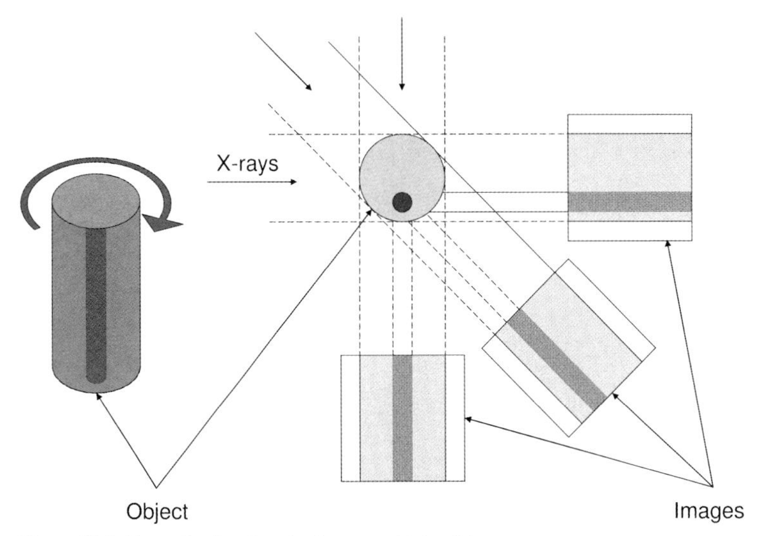


**Figure 10.2** Example showing the images obtained in tomography of the object drawn on the left of the diagram. The images are obtained for angles equal to 0°, 45° and 90°. The images are the results of projections of the internal structure of the object on the detector.

The use of techniques of reconstruction such as filtered back-projection [1] makes it possible to rebuild the initial structure in the form of a 3D image made up of voxels (elementary volume of the 3D image, for example $1024 \times 1024 \times 1024$ voxels). This technique thus aims at finding the value of $\mu$ for each voxel crossed by the X-rays. To obtain a faithful reconstruction, it is important to make sure in particular that the X-rays are not completely absorbed by the object and are absorbed (for all projections) only by the observed object. It is also necessary to observe the object at 180° to obtain a precise reconstruction although it is possible to find the structure, under certain conditions, when projections are truncated (local tomography) [1, 2].

The reconstruction gives a 3D image in which the level of gray of each voxel is proportional to the local $\mu$. This 3D image thus represents the 3D distribution of the local $\mu$. The local $\mu$ is the volume average of $\mu$ of each material. It is then possible to recognize each material composing the studied object. One thus obtains a numerical structure of the real structure of the observed object.

The use, or the development, of visualization software (commercial software includes: VGS, Amira; public domain software includes: imagej and so forth) and data processing allows the visualization of the required structure and to obtain the required information.

This simple presentation of the tomography does not show the complexity of measurement. Practically, in commercial apparatus (e.g. Phoenix X-ray, Xradia, Skyscan, X-Tech), the X-ray sources are not monochromatic, the rays are not parallel, the detectors are not ideal and so forth. Consequently, the images are more or less disturbed and artefacts appear (such as streaks, rings, aliasing at the corner distortion, etc.). These artefacts can be corrected during the visualization of data by the application of adequate algorithms (in the case of commercial scanners) or by the use of sources of monochromatic X-rays such as those delivered by synchrotrons. The uses of this type of radiation define synchrotron radiation microcomputed tomography (SRμCT): the beam is coherent, homogeneous, parallel, monochromatic and highly coherent. In these conditions, good quality images are obtained (high signal to noise ratio, high-contrast images), therefore less correction is necessary and the reconstruction is exact.

## 10.2
## Application of X-ray Tomography in the Membrane Field

The use of X-ray tomography is relatively new in the membrane field. The first experimental use of SRμCT was reported by Remigy et al. [5, 6], although Frank et al. used X-ray tomography in 2000 to observe a hemodialysis module [7]. They presented 3D reconstructed structures of UF and MF hollow fiber membranes. Yeo et al. published a paper in 2005 using X-ray microimaging (XMI) to observe the deposition of ferric hydroxide inside the fiber lumen [8] and later Chang et al. observed the flow characteristics in a hollow fiber lumen [9].

XMI allows the acquisition of only one projection of the internal structure (the sample does not turn around itself). A 2D radiography is then obtained by using a synchrotron radiation, whereas X-ray tomography allows 3D observations.

### 10.2.1
### X-ray Tomography and Membranes: the Limits and Design of Experiments

Several points have to be taken into account when setting up, such as contrast, acquisition time, X-ray energy and so forth. These points are based on our experience using SRμCT. The definition of an experiment using a commercial scanner is slightly more complicated and must take into account specificities of these apparatuses (polychromatic light, diverging beams, etc.).

1. The first is the thickness of the experiment or at least the thickness of the sample if the medium environment is the air. As described above, to recover information during the reconstruction step, the transmitted intensity should not be equal to zero (i.e. equal to the background). The thickness of the sample (or the experiment) should be small enough to meet this criterion.

   X-ray light can cross variable thicknesses of sample depending of the materials. These thicknesses must be taken into account for the experimental design. Thus, metals are crossed by X-rays less than plastics.

   The attenuation length, which corresponds to the thickness of material which reduces the intensity of the beams by a factor of two, gives a good idea of the thickness. It is noted that the wavelength of X-rays is an important factor in defining these lengths of absorption. Figure 10.3 presents attenuation length versus the X-ray energy for two materials representing two classes of membrane material: an organic membrane (polypropylene, Figure 10.3a) and an inorganic membrane (alumina, Figure 10.3b).

For an X-ray energy equal to 10 keV, the attenuation length is equal to $6 \times 10^3$ μm (6 mm) for the polymer, whereas it is equal to 250 μm for the alumina. It is clear that a greater thickness of membrane could be observed with the organic membrane (several millimetres) than with the inorganic membrane (<1 mm). For the same energy, the attenuation length of pure liquid water is slightly lower than $2 \times 10^3$ μm (2 mm), so the experiment could be conducted in

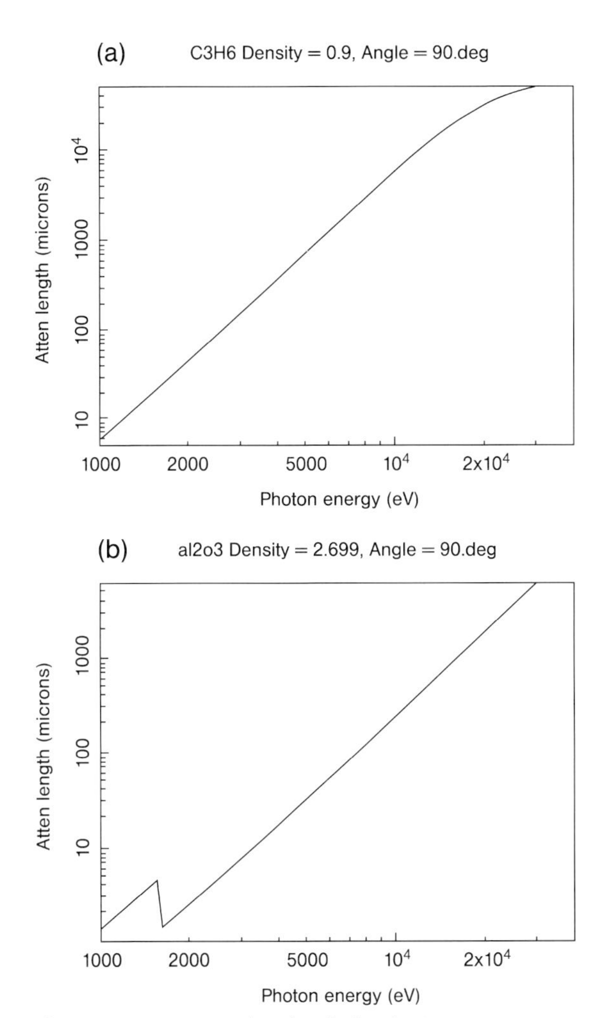


**Figure 10.3** Attenuation length calculated using http://henke.lbl.gov/optical_constants/atten2.html (10/01/07). (a) Polypropylene, (b) alumina.

the presence of water. Thus a wet membrane or module could be directly observed with a high energy source like the synchrotron sources. That also means that there is no need to prepare the studied sample if the sample is sufficiently thin.

2. The second point to take into account is the contrast of the observed objects (pore, membrane, particles, etc.) in terms of absorption of the rays.

   Contrast will determine the necessary exposition time and thereafter will fix the quality of the observations obtained. Thus, the observation of objects having close atomic composition ($Z$ power 4) will imply a weak contrast and will require, therefore, a longer duration to improve the noise to signal ratio. Real-time observation also necessitates a high contrast object.

   It is possible in certain cases to increase the contrast of a phase by the addition of atoms (addition of salts or molecules that are called phase contrast agents) that strongly absorb X-rays. The phase contrast agents are composed of electron-rich atoms. Thus, for example, it is possible to add sodium iodide to water to reinforce the contrast [7]. When adding a phase contrast agent, it is necessary to be sure that the phase contrast agent in the observed objects will take place only in one phase and will not be dispersed on the different phases of the object (i.e. the iodide must be only in the water phase, not in the membrane material). In the first case, the contrast will be reinforced, whereas in the second cases the contrast will be decreased.

   The design of an experiment must thus take into account this point: if it is possible to choose between different materials, it is better to use that which gives the largest contrast. Thus the fouling of polymeric membranes can be studied by using polymeric colloids (latex) or metal colloids. In the first case, contrast will be weak and the tomographic observations will be difficult. In the second case, the contrast will be high and the observations will be easier. This makes it possible to explain the use of the iron hydroxide during Yeo's experiment [8]. The observation of deposit formation during filtration is then largely improved by choosing particles containing metals when using organic membranes.

3. The third point to take into account is the required resolution defined by the size of the voxel in relation to the size of the observed sample.

   The voxel size defines the size of the smallest observable object or structure inside the sample. For example, objects of sizes superior to 0.7 μm could be considering if the size of the voxel is less than 0.7 μm.

   The image resolution depends essentially of the size of the sample and the camera pixel area. This is true for SRμCT and not exactly for commercial tomography, see references [9, 10]. The field of vision, which corresponds to the size of the observable object, can be coarsely calculated by multiplying the necessary size of the voxel by the image size (i.e. the number of pixels composing the image). For example, an experience conducted with a voxel size equal to 0.7 μm and a camera pixel area of 1024 × 1024 pixels will study a volume equal to about 700 × 700 × 700 μm (1024 × 0.7 = 716.8 μm): the sample will have to be smaller than this volume (in fact, the observed volume is a cylinder because of the axis of rotation).

The consequence is that the highest resolution (that is the smallest voxel size) could only be used to observe a small sample (membrane); a large sample (module) could only be observed at low resolution (i.e. high voxel size).

4. As the sample rotates during the experiment, it is preferable that the sample has an axis of rotation. Thus hollow fibers are particularly adapted to tomography, whereas flat sheet membranes are less adapted. In the same way, a more complex experiment, involving a filtration system for example, will have to be able to rotate without non-observed parts passing through the beam of X-rays.
5. The quality of the observations depends mainly on the immobility of the sample.

   The 3D reconstruction of the sample is carried out starting from a set of images of the sample. It is supposed that the observed object (and the internal structure or the phases of the object) does not move during the acquisition of these images. The immobility of the sample is checked after image acquisition since the captured images at 0° and 180° must be identical (i.e. the projections are the same). An obvious difference between these two images will be a sign of movement having taken place during the experiment. The 3D rebuilding will not then be precise.

   So the sample must be motionless during the time of the experiment. This depends on the resolution, the contrast and the number of images. For example, an exposure time of 1 s per image for an experiment requiring 1000 images taken around the sample leads to an experimental time equal to 1000 s (practically 20 min) during which the sample (its structure, its composition, etc.) should not evolve or move. Observation in real-time of the evolution of a structure (formation of a membrane, construction of a deposit of particles, etc.) is thus possible only if its evolution is much slower than the time of the experiment.

If the literature shows some experiments of tomography in real time, these experiments relate to phenomena and structures evolving slowly in time [11]. Until now, experiments examining membranes were made in static mode (i.e. the sample did not evolve during measurement). These choices were undoubtedly made because the time scale was short (a few seconds or minutes) for the experiments using membranes (formation of membrane, or deposits, etc.), whereas the acquisition time for evolving structures is in the range of minutes or hours.

### 10.2.2
### Reported Examples of the Application of Tomography on Membranes and Membrane Processes

This section presents a few examples of using tomography for the study of membranes or membrane processes. These examples are essentially based on our experiment performed when we used the ID 19 line at ESRF. Some of the observations on laboratory-made hollow fibers can be found in references [5, 6, 12]. The experimental conditions are described below. In the case of other examples from the literature, the experimental conditions are presented with the example.

The use of tomography in a static mode allows the recovery of the 3D structure of a membrane. In the membrane field, the size of the observed objects (for example pores or particles) is often in a range that covers from microns to nanometers. Taking into account that the lowest resolution (up to now) is 0.28 μm for an object size of 600 μm in the ID 19 line at ESRF [10], objects in the micron range could only be observed inside a small sample (one membrane). As rotation of the sample is needed, membranes with a rotation axis are preferred even if new techniques like synchrotron radiation computed laminography [13] allows the observation of flat samples. SRμCT enables complete visualization of a cross-section of the membrane without preparation. So a wetted membrane can be observed with the fluids that are used in the membrane process (water, bubble, etc.). In the case of hollow fibers, it is not necessary to cut the membrane because the axis of the fiber could be used as the axis of rotation. So X-ray tomography is a good tool to observe hollow fibers.

In their experiments, Remigy et al. [6, 13] used the ID 19 line [3] developed at ESRF. They used X-ray energy of 20.5 keV. This energy allows both a long attenuation length in water (about 10 mm) and high contrast. In this case, wet hollow fibers, 3–4 cm long, with an outer diameter around 1 mm were mounted on a sample holder. The observed volume was a cylinder 1 mm long (only 1 mm of the fiber was studied) and 1 mm in diameter for a pixel size equal to 0.7 μm. Regions of $750 \times 750$ μm were observed. We obtained 3D images ($2040 \times 2040 \times 1072$ voxels) with 256 gray levels (each gray level representing the absorption coefficient). Acquisition time was in the order of 30 min for 1200 projection images and around 100 reference images (images with no sample). Data acquisition for one sample typically represented 2–4 gigabytes and enabled the reconstruction of a $(2048)^3$ tomographic image.

#### 10.2.2.1 **Structural Characterization of Membranes**

Several articles or communications show the use of tomography for the structural characterization of membranes. Thus, studies were published relating to the mechanisms of membrane formation, the experimental acquisition of data such as porosity or tortuosity, the obtaining of realistic 3D models for the simulation of the flows in the membranes, or the detection of defects.

In 2007, we published an article in the *Journal of Membrane Science* [12] describing the tomography applied to hollow fibers. The work concerned the study of a polyvinydifluoride–copolyhexafluoropropylene hollow fiber manufactured using the phase inversion technique [14]. Figure 10.4 shows raw images of a section of the hollow fiber. The image from SRμCT (Figure 10.4a) is similar to the one that can be obtained by scanning electron microscopy. The difference between the two methods is that, when using SRμCT, the sample does not have to undergo preparation, whereas when using scanning electron microscopy, the sample is dried, frozen and broken. The SRμCT image is the image of the membrane structure in its wet state. The observed structure is thus the one that the membrane will have at the time of its use in an aqueous medium. Of course, it is possible to apply SRμCT to the study of membranes in the presence of other solvents in order to observe the impact of membrane swelling on the structure.

Figure 10.4b shows another interesting feature of SRμCT: the structure of the same sample can be observed in any plane within 3D space after carrying out only one experiment. The use of visualization software (like Amira® as used here) makes it possible to observe the structure from various viewpoints. Although the observation of individual different planes is possible with microscopic techniques, it is not possible to observe the whole sample at the same time.

The observation of the plane parallel to the surface of the membrane made it possible to show that the pores had a conical form and formed a honeycomb structure (see Figure 10.5). While progressing from surface towards the interior of the membrane, two close pores coalesce to form a larger pore. This fusion traces of the coalescence of the phases composing the membrane in formation. These observations imply questions about the phenomena which led to the formation of such structures. The authors proposed a process which suggests that these

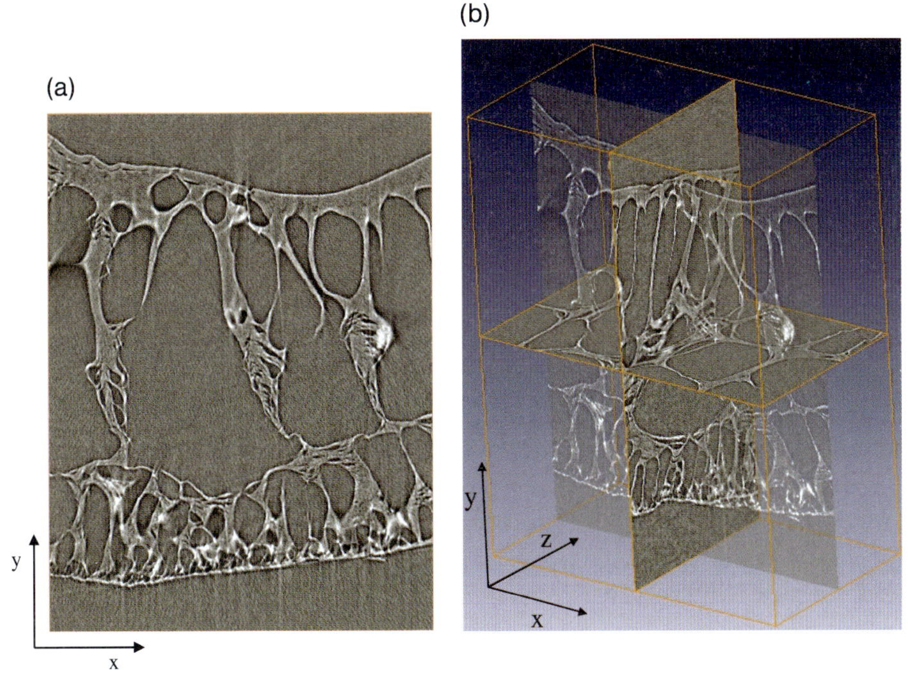


**Figure 10.4** Raw images of a microfiltration hollow fiber obtained by SRμCT. (a) Slice of hollow fiber, (b) *xy*, *yz* and *xz* planes of the same volume. Dimensions xyz of the yellow frame: 346 × 416 × 187 μm.

structures are formed by a coalescence induced by the presence, or not, of pores on the level of the skin.

Movies showing the 3D reconstructed structure of this membrane can be seen on Internet site in reference [12].

The processing of the data obtained by SRμCT makes it possible to assign each voxel to a material (water or membrane). Treatment of the data consists of segmentation, starting from the level of gray (related to chemical composition and density). Various types of segmentation are possible, manual or automatic. In first approaches, in the case of images with strong contrast, a simple thresholding allows the voxels of water to be separated from the voxels of membranes.

The volume of each material is then known and porosity can be calculated using Equation (10.4). By considering different volumes of the membrane, the authors can calculate the porosity of different slices of the membrane and find porosity within the range 70–85% for the most porous region and can estimate the smaller porosity to 24%.

$$\varepsilon = \frac{\text{number of voxels of liquid}}{\text{number of voxels of material}} \tag{10.4}$$

Another example of porosity determination is given by G.T. Vladisavljević et al. [15]. The purpose of this work was to obtain a structural characterization of the membrane which is used for membrane emulsification, but the article was more focused on emulsification. They investigated the microstructure of a Shirasu porous glass (SPG) membrane using high-resolution X-ray microtomography, using a desktop microscanner from Skyscan. From more than 600 images of the membrane cross-section perpendicular to the $z$-axis, they analyzed different cross-sections within the sample of SPG membrane. The investigated membrane thickness was 2.2 mm and the pixel resolution was 3–4 μm. They compared

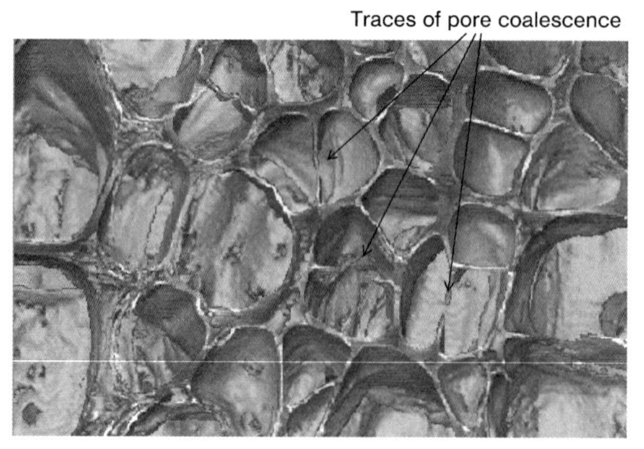


**Figure 10.5** Visualization of a plane parallel at the surface of the membrane: Description of the structure in honeycomb and the coalescence of the pores. Dimension 320 × 190 μm.

the data obtained from tomography with data previously obtained from Hg porosimetry. They obtained a good agreement between the different measures. The porosity was in the range 52.5–57.4% from tomography and in the range 50.4–58.1% from porosimetry. The pore diameter was in the range 18–21 μm from tomography, 0.4–20.3 μm from porosimetry and 9–24 μm from SEM.

Tomography data were obtained from different slices through the sample and indicated that the sample was homogeneous from the porosity point of view and pore size diameter.

The reader must note that the measured porosity (in both cases) takes account of pore sizes higher than the used resolution during the experiment (0.7 μm for Remigy et al., 3–4 μm for Vladisavljević et al.) [12, 15]. This explains why Vlasisavljević obtained a smaller pore diameter with Hg porosimetry than with tomography. The size of smaller pores (0.4 μm) is below the resolution (3–4 μm) used during the tomography experiment.

However, it is possible to measure the porosity of the membrane material (pore walls for instance) because the gray level of a voxel is proportional to the density (if the composition is known). Here in Figure 10.4a, the gray color corresponds to water (see the center of a pore) whereas the white color corresponds to the denser phase (i.e. membrane with a high polymer concentration). From Figure 10.4a, one can qualitatively deduce from the gray level of the pore wall that the pore walls are not homogeneous. By performing a calibration (not performed in this case), it would be possible to determine volumic local porosity in the walls of the pores of each voxel (0.7 $\mu m^3$) [1, 13, 16].

Tomography allows recovery of the 3D structure of the membrane for pore sizes higher than the chosen resolution (the best one is 0.28 μm at ESRF [10]) and finally a numerical structure is obtained. By using commercial software, some 3D characteristics could be measured, like pore size, pore density, porosity or tortuosity. The reader could also look at reference [17] where the study of polymer foams using SRμCT is presented.

However, the amount of information is incredibly high. Statistical methods are available when using commercial software but they are often not really suitable for membranologists who want to study mass transfer. For example, the porosity, calculated using Equation (10.4), takes into account all pores (with a pore size higher than the voxel size): both closed pores and open pores.

For mass transfer, only the communicating pores are interesting because the flux circulates through these pores and not the closed pores or the ended pores. New methods should be developed to find the correct pathway.

Sheppard applied a numerical algorithm on the 3D images in order to eliminate the closed and ended pores and to extract the communicating pore network; and an example of the method applied on sand can be found in reference [18]. Figure 10.6 provides an illustration of this. The blue tubes correspond to the network of pores where the water can flow. The largest pores are represented by the tubes with the largest diameter and so on.

From this extracted network, a theoretical permeability could be calculated over the whole membrane and compared with the experimental one. The advantage of

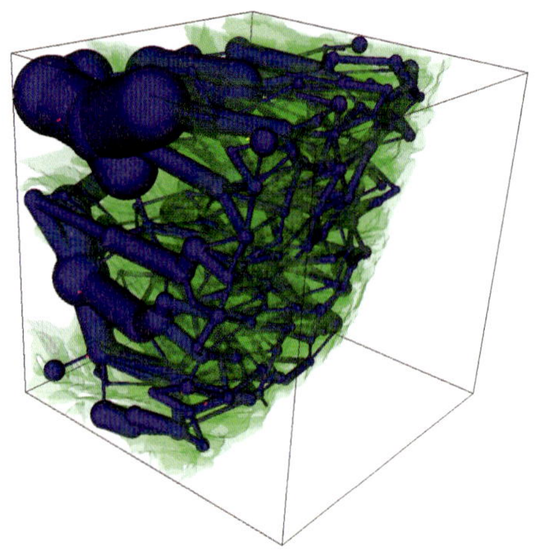

**Figure 10.6** Connected pore at the surface of the membrane presented in Figures 10.4 and 10.5 [21].

such a method, compared with a complete flow simulation applying the Navier Stocks equation and using the complete numerical structure, is to reduce the computational time and also extract the relevant information. It is, of course, obvious that one can generate a realistic grid, from the numerical structure, that can be used with finite element calculation and so on.

In 2004, a study directed by Kuo-Lun [19] was performed to study the connectivity of membrane using X-ray microtomography. Unfortunately, no publication has been written to date and although a Internet link gives a report on this work, it is not possible to take note of it. The main findings of the work were that X-ray microtomography was applied successfully to reconstruct the 3D pore structure of polycarbonate (PC) track-etched membranes and polyvinylidene fluoride (PVDF) membranes. The 3D structures provided an easy way to analyze pore interconnectivity. Results showed that the void ratio in two dimensions of PC and PVDF were circa zero and one, respectively, which implied that the structure of the PC membrane is a single straight-through while the connectivity of PVDF membrane is isotropic.

A visualization of the connectivity of a porous structure was presented by Remigy and Meireles [6]. By applying SRμCT on a PSU UF hollow fiber membrane having a sponge-like structure, as shown in Figure 10.7, we first recover the 3D structure (see Figure 10.8) and then apply a segmentation of the data that automatically selects all voxels having a gray level within a given range (0–120) and which are directly connected to a manually selected voxel. With this method, we can choose and select a pore within the sponge-like structure (i.e. select a voxel for the pore) and show (using segmentation) that the sponge structure forms a bicontinuous structure (in any direction) over the studied volume (i.e. the length of fiber is 280 μm) with a mean pore diameter equal to 2.7 μm. The results are presented in Figure 10.9, where the inner and outer skins appear together with the

**Figure 10.7** FESEM image of a PSU UF hollow fiber, showing a sponge-like structure.

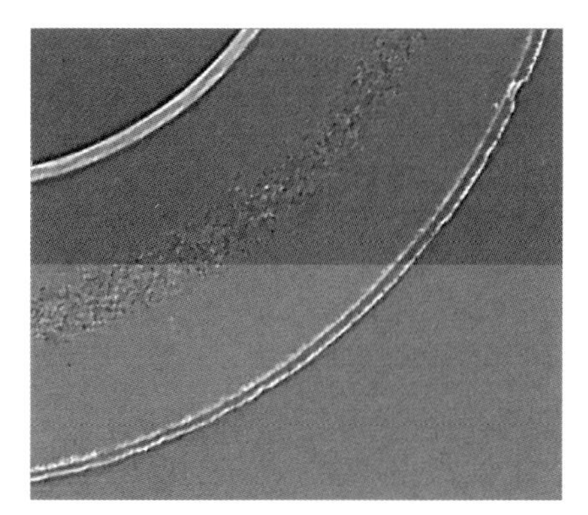

**Figure 10.8** SRμCT raw image of the PSU UF hollow fiber in Figure 10.7.

bicontinuous structure in orange. It should be noted that the porous structure having a pore size below the resolution (0.7 μm) does not appear here and is represented as a dense structure (white or gray according to porosity) on the raw tomographic image in Figure 10.8. The inner skin appears dense (white color) and does not show any pores whereas the outer skin is not well defined with large pores and a dense part, as shown in Figures 10.8 and 10.9.

Segmentation of the data allows a search for defects whose size is within a minimum equal to the resolution used. Surface sizes of the explored membranes correspond to sizes of several hundred microns up to 1 mm. By studying a polysulfone hollow fiber for ultrafiltration (presented in Figure 10.10), we sought the presence of defects on the membrane surface [6].

The membrane has a foam structure, in which the pores have a size less than 0.7 μm, with a large macrovoid. The two skins appear dense (white color) and thick. Some macrovoids are connected to the exterior. This is clearly visible in

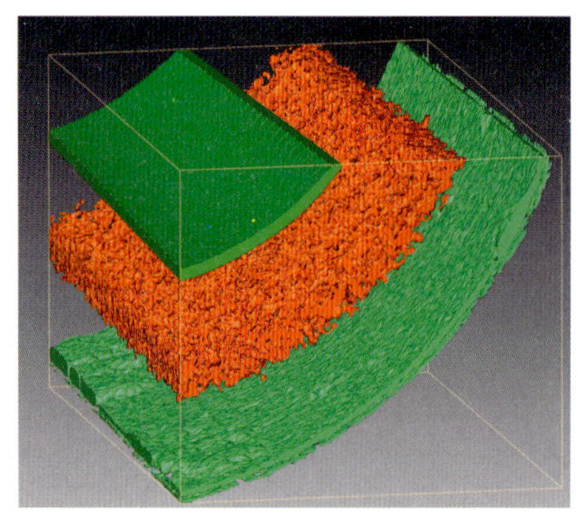

**Figure 10.9** Segmented structure in Figure 10.8, showing a bicontinuous structure of the sponge-like part of the membrane in orange (inner and outer skin in green); dimension of the orange box: $299 \times 253 \times 504\,\mu m$.

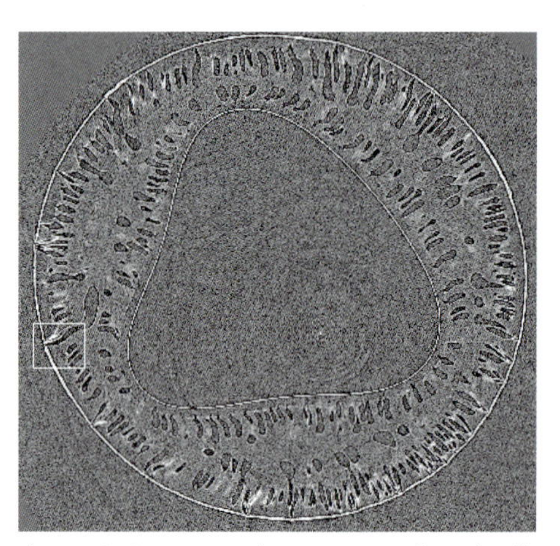

**Figure 10.10** Tomography raw image of an ultrafiltration polysulfone hollow fiber – outer diameter: 1 mm [6].

Figure 10.11 which shows different successive slices of the white square shown in Figure 10.10. Two successive slices are separated by $0.7\,\mu m$. In Figure 10.11, the external skin appears as a white layer. From the $Z=0$ image, one can see a black line across the skin linking the exterior to the macrovoid. This black line is not present on the $Z=-2$ and $Z=+2$ images and is visible on the $Z=-1$ and $Z=+1$ images. This line is the image of a pore which is considered a defect: its size is within the micrometer range whereas the MWCO of the membrane (21 kDa, PEG) is low (i.e. pore sizes in the nanometer range).

Visualization of the 3D reconstructed membrane is presented in Figure 10.12. In this figure, only the interfaces between water and the membrane are

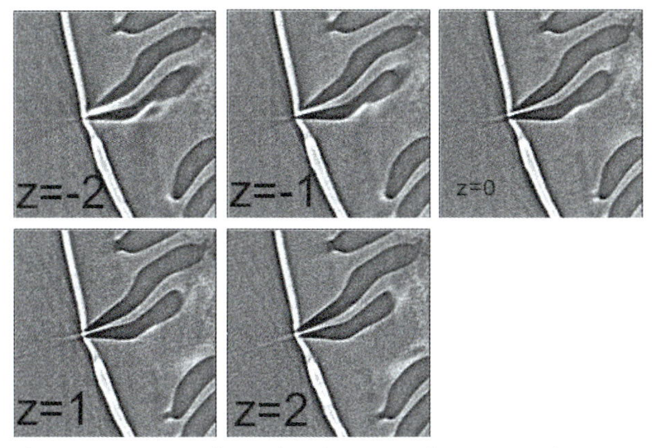


**Figure 10.11** X-ray radiography of PSU HF of successive slices of the membrane, z = −2, −1, 0, 1, +2 μm. See studied area in Figure 10.10, indicated by the white square [6].

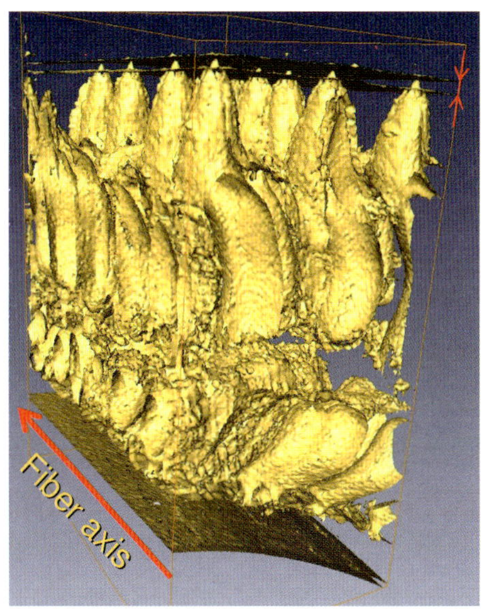


**Figure 10.12** Visualization of 3D reconstructed PSU HF of Figure 10.10; description of the distribution of the macrovoids.

represented. Thus, the external skin is presented as two interfaces (edges of the skin) separated by vacuum. The elimination of material makes it possible to better comprehend the structures. The distribution of macrovoids in space is obvious: they are aligned along the axis of the fiber and the top of each macrovoid is connected to the exterior through the external skin. The macrovoids thus form conical pores which emerge on the level of external surface: the defects.

If one can visually find such defects on raw images using a long and tedious process, a better way is to do it numerically. We have applied the same segmentation of images as above (i.e. selection of voxels that are directly connected to a selected voxel) but we progressively change the gray range corresponding to water. This range, for our images, varies from 0 to 120. As contrast (when using SRμCT) is reinforced when the edges are abrupt [1], the edges of the pores appear as a low level of gray (between 0 and 65 for the largest). Then, the more the pore size decreases and approaches the size of a voxel, the more the level of gray increases and tends towards the white color (i.e. 255). The selection of voxels by variation of the levels of gray makes it possible to select the largest pores first. Identification of defects in the space is thus possible.

Figure 10.13 shows the result of such a segmentation applied to the membrane in Figure 10.10. The internal channel is represented in blue; the internal skin (clear green) is well defined and appears dense. It corresponds to a MWCO of 8 kDa (PEG). The internal foam structure is not represented because it would hide the interesting structures. The external skin appears clear green.

It is remarkable that voxels, having the same level of gray as the external skin (and thus the same density), are located inside the foam structure. The superposition of the selected voxels and the reconstructed structure shows that these voxels belong to the wall of some macrovoid, a wall located at the entry into the macrovoid. The voxels represented in red correspond to water directly in contact with the membrane (beside the abrupt edges). The selection of these voxels makes it possible to identify the largest pores since the wall of the pores is abrupt. Thus, in Figure 10.13, four larger pores are detected. They are directly connected to a macrovoid.

In this experiment, the only detected pores have a size larger than 0.7 μm (resolution used). This detection can be made easily on large surfaces (corresponding to 1 mm of hollow fiber): it is then possible to detect one pore on 1 mm of fiber and

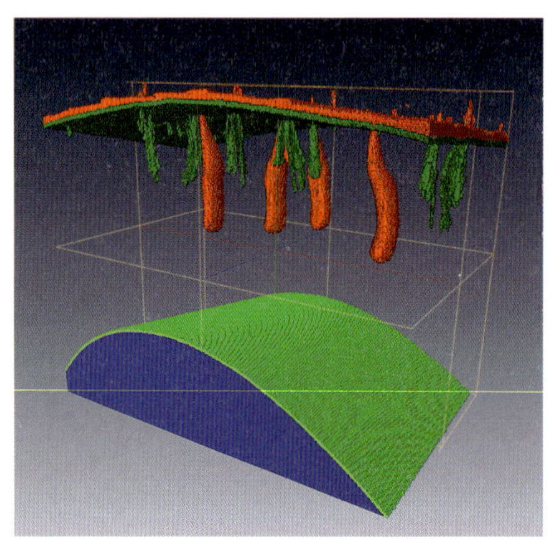

**Figure 10.13** Defect detection at the outer skin of the hollow fiber in Figure 10.10.

around the fiber in one experiment. That corresponds to a minimum density of approximately 35 000 pores $m^{-2}$ for pores of size larger than 0.7 μm. In comparison, the detection of the same defect using electronic microscopy makes it possible to detect $10^8$ pores $m^{-2}$ (one pore on a square surface of 100 × 100 μm). The observation by microscopy of $10^8$ defects $m^{-2}$ has a low statistical meaning: the observation of one defect on a square surface of 100 × 100 μm is a chance fact or, in contrast, is the fact that there are defects regularly distributed on the fiber surface. It is then necessary to multiply the observations, whereas one experiment is enough using X-ray tomography.

#### 10.2.2.2 Concentration, Fouling or Deposit Characterization

To date and according to the limits of our knowledge, there exists only one study making use of tomography for the study of fouling [7]. Other authors have used the absorption of X-rays to follow the formation of deposits [8]. We will summarily describe them by showing the limits of tomography with respect to the study of deposits and their formation.

Frank et al., in 2000, published a paper dealing with the visualization of concentration field in hemodialyzers [7]. Using a medical X-ray scanner, they observe a hemodialysis module (1.4 $m^2$) containing about 10 000 fibers (outside diameter 255 μm). To enhance the contrast, they add some sodium iodide (NaI, 0.1 M) to the water. The voxel size is approximately 5 mm in thickness and possesses a cross-section of 0.1875 × 0.1875 mm (note that they do not use the SRμCT but a commercial scanner). That means they cannot directly observe the particle, the fiber or the fouling, all of which are smaller in size.

For each voxel, the scanner obtains a level of gray which is the volumetric average of the levels of gray of each element included in this voxel. Using a mathematical procedure, Frank et al. finally retrieve the volume average concentration of the water (containing NaI), the air (inside the fiber) and the fiber. Taking a tomographic image every 5 min, they also obtain a pseudo real-time concentration in the hemodializer (i.e. the phenomena, faster than the time of acquisition, are averaged). From this visualization, they found that the concentrations are heterogeneous in the module, indicating that the fibers and flow repartition are not optimized.

The example of Frank et al. [7] is interesting because it shows some of the limits of tomography:

- Observation of a module is not possible, other than with a large voxel size. As the voxel size is roughly equal to the size of the area observed divided by the resolution of the camera, any increase in the observed area (object) results in an increase in the size of the voxel. Thus, the observation of a small thickness of deposit is not possible
- Observation in real time is only possible for slowly evolving phenomena. Frank et al. [7] acquired images every 5 min, whereas a deposit can take less time to formed.
- The observation of particles or the 3D structure of a deposit is not possible, other than in static mode. The observation of particles and the structure of a deposit requires having particles and, especially, a deposit layer larger than the size of a

voxel. Supposing that a reasonable size of deposit is obtained (let us say 10 times the resolution):

- If the particles are rather large (i.e. about the size of a voxel), it will be possible to have an idea of the structure of a deposit.
- If the particles are smaller than the size of a voxel, it will also be possible to obtain a 3D chart giving the local average densities of the material and, knowing the compositions, the local medium porosity.
- In both cases, it should be remembered that the smallest resolution obtained to date with an ESRF of 0.28 μm for an area is approximately 600 $\mu m^2$. The thickness of the deposit would be approximately 300 μm. Taking these dimensions into account, it will be possible to observe a single fiber having a strong fouling. A question comes then to mind: do these conditions make it possible to have interesting information with respect to the phenomena concerned during fouling?

To circumvent these difficulties and to obtain shorter acquisition times, it is possible to take only one simple radiograph. Yeo et al. [8] applied this principle in 2006 by using a technique called X-ray microimaging (XMI). They observed the deposition of particles of iron hydroxide (sizes from 0.1 μm to 10 μm) during a dead-end filtration into the lumen of a PAN fiber (nominal size pore 0.5 μm, outer diameter 0.8 mm). The images they obtained, using a pixel size equal to 1 μm, were the projections of the fiber structure (i.e. membrane plus pores filled with water) and the deposition of iron hydroxide. The acquisition time was short (1–10 s) and images were recorded every 3 min.

Finally, they concluded that the technique is able to observe the deposition of particles as a cake inside the lumen of a hollow fiber membrane and also to detect deposition and fouling within the membrane structure; and they suggested that it can also be applied to the evaluation of critical flux in cross-flow operation by observing the flux at which deposition begins.

This technique can indeed be used to detect the formation of a deposit and can be used to locate the deposits. However, the localization remains limited since the obtained image is the projection of a 3D structure onto a 2D plane. There is thus a superposition of the deposits located at the surface of the membrane onto those located within the pores. Finally, taking into account the resolution used (1 μm); the observation of a deposit thickness less than 1 μm remains difficult and questionable. Nevertheless, this technique can provide information inaccessible by other techniques and it thus deserves to be developed.

## 10.3 Conclusions

Tomography is a characterization technique giving access to 3D information which it is not possible to reach with other techniques. Applied to the field of membranes, it permits the observation of membranes in a state very close to their actual use, since this technique does not require preparation. It is necessary nevertheless to take into account the constraints inherent in the absorption of X-rays, the acquisition of images and the implementation of the technique.

This technique is new in the world of membranes, whereas it is old in medical applications. The main reason comes from the fact that resolution was too weak, until recently. Development of the SRμCT made it possible to reach resolutions (less than one micrometer) which are relevant for certain membrane applications. One can quote, for example, the study of 3D structures to understand the formation of microfiltration membranes or ultrafiltration membranes. Tomography is then used like a simple characterization technique.

The structures obtained for microfiltration membranes can be used like realistic numerical structures for a simulation of flow or mass transfer and they allow a relevant comparison between the model and the experiment. It is then a question of obtaining realistic structural models by comparing them with models created by man.

It is obvious that, in many cases, the resolution is too weak to obtain particularly relevant information. Thus, the user or the manufacturer of an ultrafiltration membrane is often more interested by the structure of the skin (selective part of the membrane) that by the subjacent part. The range of sizes studied is then nanometric (1–100 nm), a range which is not yet covered by the SRμCT, although efforts are being made in this direction. It should be noted that research groups aim at combining tomography with other techniques of microscopic observation in order to reach the 3D nanometric range: resolutions of 30 nm in 3D have been announced [20].

Progress in tomography should be followed by membranologists because research goes in the direction of the required observations.

Another field, whose progress is to be followed or developed, is the data analysis of raw data obtained by tomography. The treatment and analysis of the images, using techniques developed initially for the medical field for example, are to be adapted to membranes and, in many cases, it will be necessary to develop specific algorithms.

To conclude, tomography using synchrotron radiation is a technique with a future for the field of membranes. It offers a very important potential but this it is necessary to develop.

## References


1 J. Baruchel, J.-Y. Buffière, E. Maire, P. Merle, G. Peix (Eds.) *X-ray Tomography in Material Science*, Hermes Science, Paris, **2000**.

2 http://www.esrf.eu/files/Highlights/HL2006.pdf.

3 http://www.esrf.eu/UsersAndScience/Experiments/Imaging/ID19/BeamlineDescription/Detectors (December 2007).

4 http://www.esrf.eu/UsersAndScience/Experiments/Imaging/ID19/BeamlineDescription (December 2007).

5 J.-C. Remigy, M. Meireles, J.C. Rouch, L. Quesada, in: *ICOM*, Seoul, **2005**.

6 J.-C. Remigy, M. Meireles, *Desalination* **2006**, 199, *1/3*, 501–503.

7 A. Frank, G.G. Lipscomba, M. Dennis, *J. Membrane Sci.* **2000**, *175*, 239–251.

8 A. Yeo, P. Yang, A.G. Fane, T. White, H.O. Moser, *J. Membrane Sci.* **2005**, *250*, 189–193.

9 S. Chang, A. Yeo, A. Fane, M. Cholewa, Y. Ping, *J. Membrane Sci.* **2007**, *304*, 181–189.

10 http://www.esrf.eu/UsersAndScience/Experiments/Imaging/ID19/Beamline_description/Detectors/Frelon_intro/Frelon_optics (December 2007).

11 http://www.esrf.eu/events/conferences/past-conferences-and-workshops/X-ray-imaging-school/Presentations/01_Baruchel (December 2007).

12 J.C. Remigy, M. Meireles, X. Thibault, *J. Membrane Sci.* **2007**, *305*, 27–35.

13 http://www.esrf.eu/news/spotlight/spotlight37laminography/index_html/ (December 2007).

14 P. Menut, J.C. Rouch, J.C. Remigy, *Proc. 9 Coll. PROSETIA*, Toulouse, **2004**.

15 G.T. Vladisavljevi'c, I. Kobayashi, M. Nakajima, R.A. Williams, M. Shimizu, T. Nakashima, *J. Membrane Sci.* **2007**, *302*, 243–253.

16 H. Taud, R. Martinez-Angeles, J.F. Parrot, L. Hernandez-Escobedo, *J. Pet. Sci. Eng.* **2005**, *47*, 209–217.

17 A. Elmoutaouakkill, G. Fuchs, P. Bergounhon, R. Peres, F. Peyrin, *J. Phys. D Appl. Phys.* **2003**, *36*, A37–A43.

18 W.M. Mahmud, J.Y. Arns, A. Sheppard, M.A. Knackstedt, W.V. Pinczewski, *Transp. Porous Med.* **2007**, *66*, 481–493.

19 http://thesis.lib.cycu.edu.tw/ETD-db/ETD-search/view_etd?URN=etd-0901105-095650 (December 2007).

20 D. Attwood, *Nature* **2006**, *442*, 642–643.

21 A. Sheppard, unpublished data.

# 11
# Optical and Acoustic Methods for in situ Characterization of Membrane Fouling*

*J. Mendret, C. Guigui, P. Schmitz, and C. Cabassud*

## 11.1
## Introduction

Nowadays, ultrafiltration (UF) or microfiltration (MF) membrane processes are widely used because of their ability to remove particles, colloidal species and microorganisms from different liquids feeds. However a limitation inherent in the process is membrane fouling due to the deposition of suspended matter during filtration. Therefore the understanding of formation and transport properties of particle deposits responsible for membrane fouling is a necessary step to optimize membrane processes. Thus it is necessary to obtain local information in order to analyze and model the basic mechanisms involved in deposit formation and then to further predict the process operation. Besides, it is also useful to control the deposit formation and to plan preventive or curative actions with a controlled efficiency. Nonetheless, local parameters such as cake thickness and porosity are hardly reachable with conventional techniques.

The only way to obtain local information about the deposit structure at a given time during filtration is to use in situ and real-time measurement with a non-invasive method, that is to say a method which does not disturb the process and cake build-up during filtration or removal during hydraulic cleanings.

Over the past years, a number of non-invasive techniques have been used to observe in situ membrane processes but few are adapted to confined geometry like inside/out hollow fibers [1, 2]. Most of the time, one characterization method only provides one kind of information about the deposit: thickness or images or a structural parameter. This chapter focuses on the development and validation of two characterization methods to obtain complementary information (thickness and porosity) about the local structure of the deposit.

* A list of nomenclature is given at the end of this chapter.

*Monitoring and Visualizing Membrane-Based Processes*
Edited by Carme Güell, Montserrat Ferrando, and Francisco López

ISBN: 978-3-527-32006-6

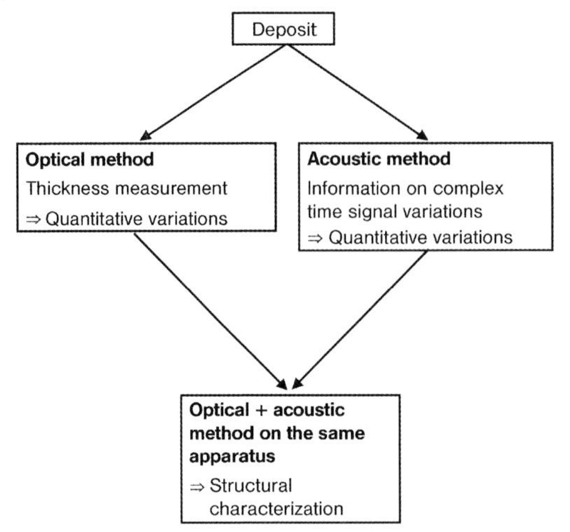


**Figure 11.1** Global approach adopted for the acoustic and optical methods development.

## 11.2
## Approach

Two characterization methods have been developed. The approach is to use those two methods in a complementary way to obtain a complete characterization of filtration cakes.

The first method is an optical method based on the use of a laser sheet at grazing incidence. With this method, it is possible to follow the deposit thickness and kinetics in situ and real-time. The second method is an acoustic method. The analysis of the acoustic wave of the deposit with time enables its thickness to be followed. Moreover, variations of the amplitude of the acoustic wave are linked to possible changes in the structural properties of the deposit. Nonetheless, only qualitative variations are available. The use of both methods on the same apparatus will enable a complete characterization to be obtained (Figure 11.1).

This chapter presents the principle of the methods and an example of results for each method in dead-end UF of clay suspension. Moreover, a discussion about the common use of those two methods for a better characterization of filtration deposits will be given.

## 11.3
## In Situ Deposit Characterization with an Optical Method Using a Laser Sheet at Grazing Incidence

### 11.3.1
### Optical Methods Using a Laser for Fouling Characterization

This paragraph is a brief review of optical methods using a laser which have been applied to fouling characterization.

#### 11.3.1.1 Optical Laser Sensor

With an optical laser sensor it is possible to measure a deposit thickness during filtration. The method is based on the use of a laser beam which is focused tangentially to the surface of a tubular membrane. The image of the focused point is then captured by a photomultiplier which measures the signal intensity. At beginning of filtration the energy collected by the photomultiplier is maximal and then it decreases as the deposit grows and absorbs the signal. Thus the intensity variation is directly related to the deposit thickness. A calibration curve enables the instantaneous signal magnitude to be associated to the deposit thickness. When all the signal intensity is absorbed, no more light is detected by the photomultiplier, which determines the maximum cake thickness that can be measured. Hamachi and Mietton-Peuchot [3] performed a dynamic measurement of deposit thickness using an optical He–Ne laser sensor ($\lambda = 543.5$ nm, 0.2 mW) during microfiltration of clay suspensions (particle diameter $dp = 2.45\,\mu$m) with an outside–in tubular membrane. They used a module with two glass windows. Thus, measurements of deposit thickness were performed at a single position, between the inlet and the outlet of the tubular membrane, assuming that the point is representative of the global deposit growth. The maximum cake thickness they could measure was 30 μm but they overcame this limitation by moving the sensor 30 μm away with a micrometric screw device. This displacement only takes a few seconds, so that recording was considered as continuous. However, with this technique, there is a maximum possible concentration for the suspension (375 mg $L^{-1}$ for [3]) above which the laser beam intensity was totally absorbed by suspended particles.

#### 11.3.1.2 Laser Triangulometry

In contrast to the optical laser sensor method developed by Hamachi and Mietton-Peuchot, laser triangulometry is an optical method based on the reflection of a laser beam from a surface whose position varies, like a developing particle layer. A CCD camera is used to capture the reflected beam. As the deposit grows, the reflected beam shifts and the measurement of this shift allows the calculation of the deposit thickness. Altmann and Ripperger [4] studied particle deposition and layer formation during the crossflow microfiltration of diatomaceous earth and silica particles with a commercial laser triangulometer. The cake layer height was measured in situ through a transparent window on the membrane module which was a 400 mm long channel. The width of the flow channel was 60 mm, the height 3 mm. This method enables deposit thickness to be measured in real time and offers good resolution (5 μm), depending on the camera and on objective resolution. For these characterization methods, the laser sensor was a beam and the surface of the characterization area was the same as the beam circle.

At Laboratoire d'Ingénierie des Systèmes Biologiques et des Procédés at INSA Toulouse (LISBP), an original method was developed [5] which is the main object of this chapter. The originality of the method presented is to use a laser sheet rather than a focused point: this allows a larger surface to be analyzed and the possibility to obtain a deposit topography (2D information).

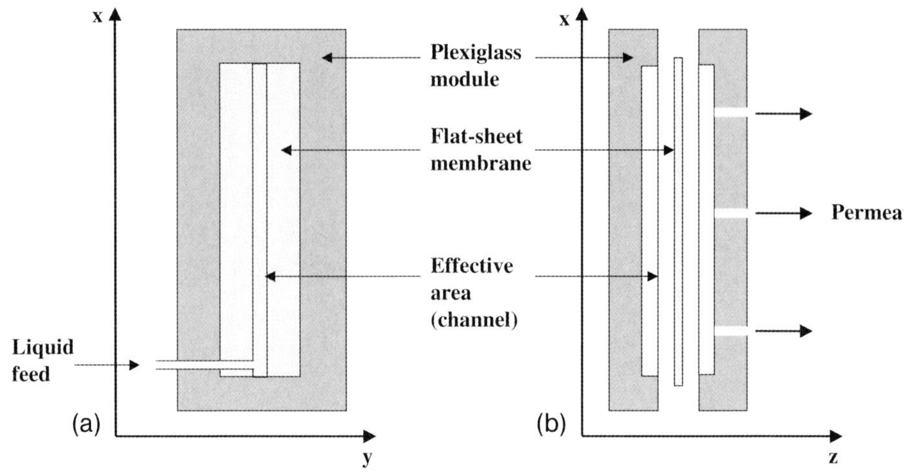


**Figure 11.2** Typical filtration module: (a) front view (if the feed flows from the bottom of the channel), (b) side view (cutaway in the channel).

### 11.3.2
### Filtration Set-up

A specific filtration set-up is used to measure in situ cake characteristics during filtration in a confined geometry like a rectangular channel. The filtration cell has to be transparent. Figure 11.2 gives an example of a typical filtration cell suitable for the developed method. Each of the two parallel plates of a flat transparent Plexiglas chamber is machined with a rectangular channel (example of the channel dimension: 2–4 mm large, 2 mm depth, whatever the height). A flat-sheet membrane is installed between the two plates; thus filtration is operated only on the channel surface. If filtration is operated at constant pressure, an electronic balance connected to a personal computer enables the permeate mass versus time to be registered in order to measure permeate flux time variations.

### 11.3.3
### Principle of the Optical Method

The principle of the developed experimental technique to measure the deposit thickness, laser sheet at grazing incidence (LSGI), is similar to that of laser triangulometry. A laser sheet is focused on the support, which is a membrane, at grazing incidence with a small angle $\theta$. When a deposit is growing on the support, the laser reflection is shifted from its original position and measurement of this shift, $\Delta$, allows calculation of the deposit thickness. The intersection between the laser sheet and the support is a line corresponding to the surface that is analyzed.

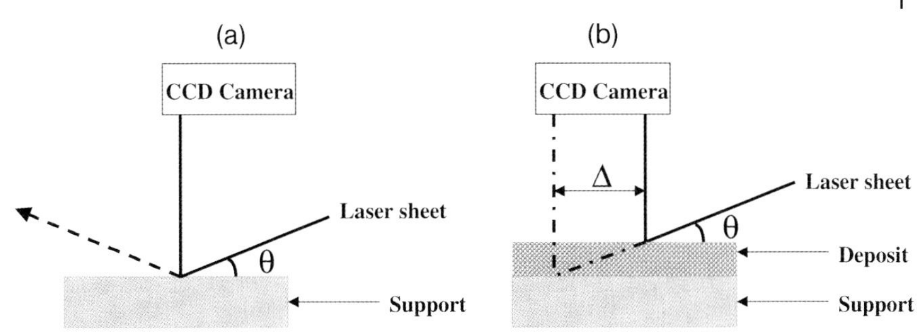


**Figure 11.3** General principle: (a) clean support, (b) fouled support.

A CCD camera perpendicular to the support plan captures the position of this intersection line (Figure 11.3). Knowing $\Delta$ and $\theta$ enables the deposit thickness, $e_d$, to be calculated thanks to the following relationship:

$$e_d = \Delta \tan \theta \tag{11.1}$$

### 11.3.4
### Measurement of Cake Thickness Growth

A laser sheet is positioned at grazing incidence relative to the filtration module (Figure 11.4a) and is focused on the membrane. The laser sheet is obtained using a Lasiris SNF-501L laser diode (power 3.5 mW, wavelength $635 \pm 5$ nm) and a specific lens positioned at the laser diode output. The impact line is perpendicular to the $x$ direction and is larger than the channel width. A CCD camera (cooled digital 12 bit Sensicam QE camera, with sensor size $1376 \times 1040$ pixels) records the laser light reflected by the membrane in the direction perpendicular to the membrane.

The part of the laser sheet which is reaching the membrane outside of the channel undergoes a single refraction along its path before reaching the membrane, at the interface between air and Plexiglas (Figure 11.4b). The part of the laser sheet which is reaching the membrane inside the channel undergoes two refractions: (1) at the interface between air and Plexiglas and (2) at the interface between Plexiglas and water (Figure 11.4b). Consequently, there is a shift between the positions of the laser line outside and inside the channel due to the refractive index difference. When a deposit is growing during a filtration experiment, the laser light is reflected at its surface and consequently the measured shift, $\Delta X$, decreases with time (Figure 11.4b). The shift variation as a function of time, $\Delta X(t)$, is measured on the video pictures.

The deposit thickness at time $t$, $e_d(t)$, is given by the following relationship:

$$e_d(t) = \frac{\Delta X(t) - \Delta X(t=0)}{\tan \theta_3} \tag{11.2}$$

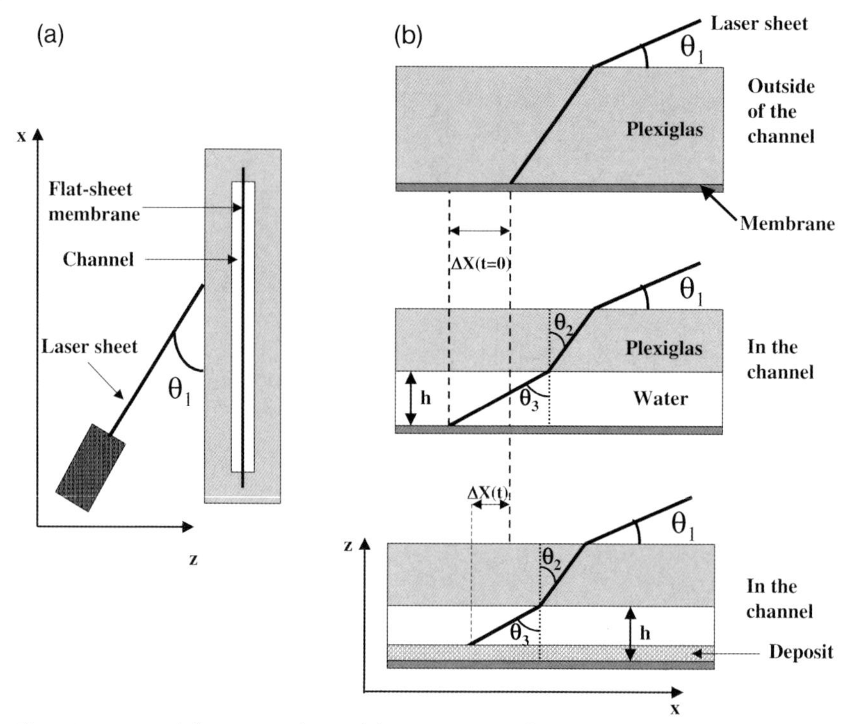


**Figure 11.4** Laser deflections in the module: (a) position of the laser sheet, (b) deflection of the laser inside the channel (top figure) and outside of the channel (bottom figure).

The shift at $t = 0$, when there is no deposit, is related to the channel height ($h$) and to the angles $\theta_2$ and $\theta_3$ by the relationship:

$$\Delta X(t = 0) = h \cdot (\tan\theta_3 - \tan\theta_2) \tag{11.3}$$

where $h$ is the channel height. The relation between $\theta_2$ and $\theta_3$ is given by Snell–Decartes's law:

$$n_2 \cdot \sin\theta_2 = n_3 \cdot \sin\theta_3 \tag{11.4}$$

where $n_2$ and $n_3$ are respectively the refractive indexes in Plexiglas and in water with values of 1.5 and 1.33, respectively. Then, Equation (11.3) can be written:

$$\Delta X(t = 0) = h \cdot \left[\tan\theta_3 - \tan\left[\arcsin\left(\frac{n_3}{n_2} \cdot \sin\theta_3\right)\right]\right] \tag{11.5}$$

Resolution of Equation (11.5) allows $\theta_3$ to be obtained from the measurement of $\Delta X(t = 0)$.

During our experiments, typical values of $\Delta X(t = 0)$, $\theta_2$ and $\theta_3$ are respectively 0.56 mm, 45° and 61°.

## 11.3.5
## Image Analysis

A typical image obtained by the CCD camera is presented in Figure 11.5. Images can be processed using a software like Matlab. For example, various regions of interest (ROI) can be selected on a given image. One can then obtain averaged intensity profiles of the reflected light for the various ROI, the averaging being performed over the width ($y$ direction) of each ROI. Such light intensity plots present a clear maximum (Figure 11.5), the existence of which is related to the fact that the incident light intensity profile is Gaussian over the width of the laser sheet. The position of the maximum, in pixel, is obtained the following way: the intensity profiles are first fitted with a sum of the quartic B-spline function, over 10 subintervals. Then, the $x$ position of the maximum of this analytical function is determined (Figure 11.5). This fitting step is necessary to smooth the experimentally found intensity profiles, which are slightly noisy despite the transverse averaging. For the given optical set-up and image acquisition system, the minimum width of the ROI allowing such a fit to find a clear maximum was 10 pixels, which gives the transverse resolution that can be achieved by this technique.

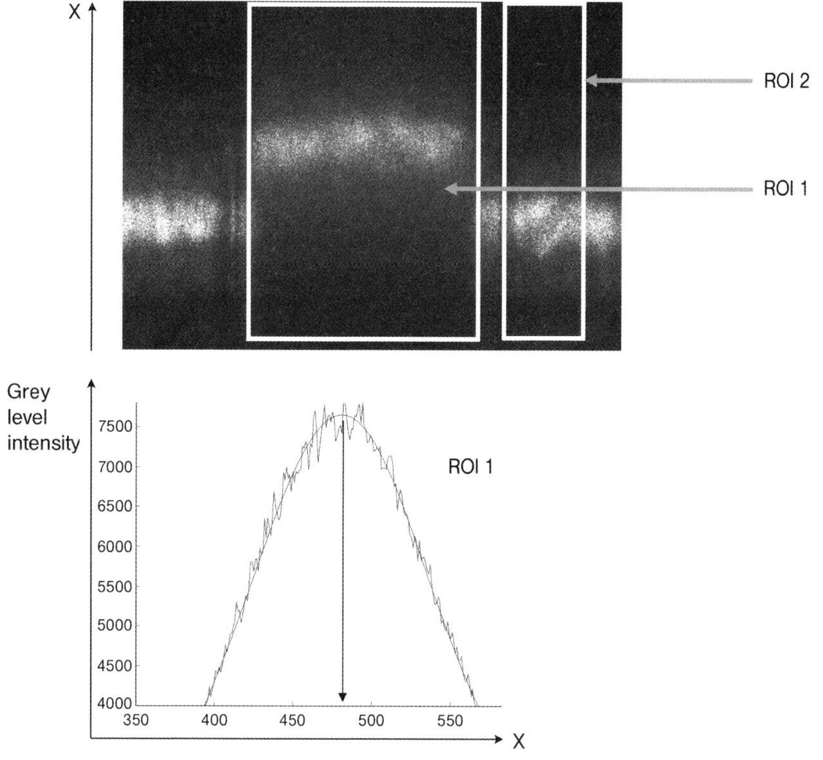


**Figure 11.5** Determination of the position of the maximum of gray level intensity for a region of interest.

The shift $\Delta X$ is found by measuring the difference between the position of the maximum intensity of the reflected light on a given ROI in the channel (for instance, ROI 1 in Figure 11.5) and the position of the maximum intensity of the reflected light in a reference area outside of the channel (ROI 2 in Figure 11.5).

The switch from pixels to length is done using an image of the channel without the laser sheet, using the fact that the channel width is precisely known. Measuring $\Delta X$ and $\Delta X(t=0)$ enables the cake thickness $e_d(t)$ to be determined using Equation (11.1).

## 11.3.6
## Capabilities and Limitations of the LSGI Technique

Compared to the laser triangulometry technique, the originality of this technique lies in the use of a thin laser sheet instead of a crude laser spot. Consequently, measurements of deposit thickness profiles in the direction transverse to the channel can be obtained, by dividing the channel in several ROIs as explained above. The lateral resolution of this technique is defined by the minimum lateral size of the ROIs that can be used, which is 10 pixels (30 μm with the magnification used in our application). However, it must be noted that it is not possible to obtain an accurate thickness value near the channel wall because of the signal being too noisy, probably due to light diffusion from the wall.

One can also consider to move the set laser camera in the $x$ direction to measure the deposit thickness at different $x$ positions along the channel height. As the laser sheet thickness is small, one could expect a good spatial resolution (about 300 μm) in the $x$ direction. Finally, this would allow to get a deposit thickness "map" as a function of $x$ and $y$.

The accuracy in thickness measurement is determined by several uncertainties. First, when determining the position of the maximum on the gray level intensity plot, the accuracy is estimated to be $\pm 1$ pixel. So, the accuracy of the determination of the difference $\Delta X(t) - \Delta X(t=0)$ is $\pm 2$ pixels. Then, another uncertainty happens when converting from pixel to meters as the measurement of the exact channel width is difficult to estimate. Indeed, this measurement is tricky due to parallax effects and light diffusion from the lateral walls. The latter effect is due to wall surface roughness: it cannot be avoided but may be reduced by polishing the channel walls. We estimate the accuracy on the channel width determination to be $\pm 4$ pixels. It could be lowered by using a larger CCD array and larger magnification for the optics. As can be seen (Equation (11.4)), the error on $\theta_3$ is difficult to estimate. The calculation of the averaged deposit thickness on the channel width was computed with and without the different uncertainties. The resulting global uncertainty on the cake thickness measurement is estimated to be $\pm 2.5$ μm with the magnification used in our application.

The minimum cake thickness that can be measured with this technique is determined by the conversion rate between pixels and length and, in our application, it was estimated to be 3 μm pixel$^{-1}$. This measurement resolution is limited by the pixel resolution of the CCD sensor and the zoom lens capability. It could be made more accurate by using a smaller angle $\theta_3$ ($\theta_3$ was 51 in our application). Anyway, a 3 μm resolution is good in comparison with those obtained with other characterization methods.

This experimental technique also allows the deposit growth to be characterized versus time by applying the method presented above for successive images and thus the cake growth kinetics to be obtained. The image acquisition rate is chosen by the operator (down to a few images per second) so that this measurement technique could be used for various membrane fouling kinetics.

Like many optical methods, the method presented in this chapter is limited by the concentration of the suspension. Indeed, at large concentration, light diffusion by the suspended particles is going to "blur" the image of the laser sheet. Thus, for each application there is a maximum concentration that can be used.

### 11.3.7
### Typical Results

It was seen in the previous section that the developed method enables deposit thickness for several locations on the channel length in conjunction with the permeate flow rate to be measured simultaneously. For a fixed position on the $x$ axis, it is possible in the same time:

- to determine the deposit kinetics (cake growth thickness with time variation),
- to obtain the deposit cross-section profile in the $z$ direction at time $t$,
- to measure permeate flux or pressure.

First experiments were performed for dead-end ultrafiltration of clay suspensions at $1\,g\,L^{-1}$. A module like the one presented in Figure 11.2 was used; it was disposed vertically. The channel geometry enables confinement phenomena occurring in a hollow fiber to be taken into account and their influence on the deposit to be observed. The channel had 28.2 cm height and 2 mm width and depth. The membrane was a polysulfone membrane (molecular weight cut off (MWCO) = 100 kDa) and the effective filtration area was $56.4 \times 10^{-5}\,m^2$. Filtration was operated in dead-end at constant TMP (80 kPa) for 4 h duration. The clay particles can be considered as small flat plates with the following dimensions $8 \times 6 \times 1$ μm (length × width × thickness). Feed suspension flowed from the bottom to the top of the channel and measurements were taken at 70% of the channel height from the bottom.

Figure 11.6 shows an example of the deposit thickness (mean value on the channel width) and permeate flux variations during filtration as a function of time.

The initial permeate flux is $204\,L\,h^{-1}\,m^{-2}$. It decreased rapidly during the first time period (from 0 s to 1500 s) and then more slowly to $20\,L\,h^{-1}\,m^{-2}$. At the same time, it was possible to measure the cake thickness growth. The final thickness was 55 μm for a total deposited mass of $80\,g\,m^{-2}$ and a relative flux decrease of 87%. Analysis of the variations of the flux decrease and of thickness growth with time enabled several layers with different structures to be distinguished. The

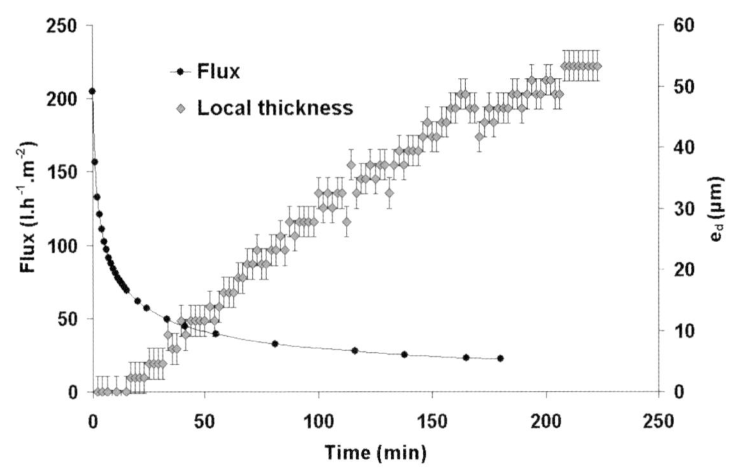


**Figure 11.6** Deposit thickness and permeate flux variations for dead-end ultrafiltration of bentonite (TMP = 80 kPa, concentration = 1 g $L^{-1}$, initial permeability ($L_{p0}$) = 1.8 L $h^{-1}$ $m^{-2}$ $kPa^{-1}$).

results were discussed in terms of cake build-up mechanisms and this allowed, thanks to the experimental approach, to validate the fact that modification of cake structure can appear with time and that the first-build cake layers are the more resistive even if they are quite thin (Figure 11.7).

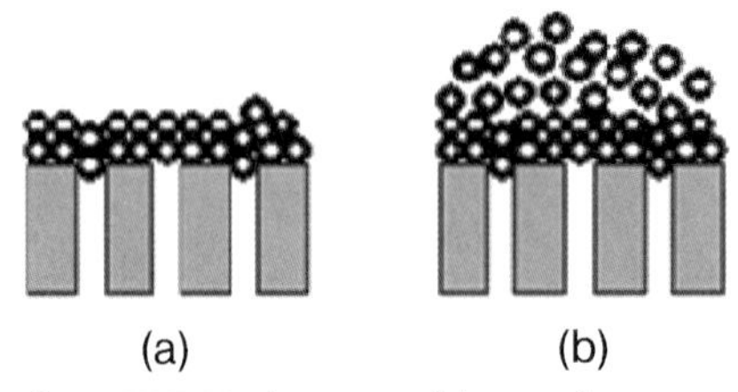


**Figure 11.7** Mechanisms of deposit formation: (a) first period, (b) second period; modification of cake structure.

Figure 11.8 gives an example of a deposit profile measurement at $t = 162$ min calculated by dividing the channel into ten areas (each area has a width of 70 pixels).

The deposit is thicker at the extremities, close to the walls. This could be attributed to wall effects leading to a local increased velocity and subsequently to a higher cake erosion, in the middle of the channel. The thickness profile measurement was compared to the mean thickness measurement. Thus, the integral of the profile curve on the channel width was estimated. A value of 0.162 $\mu m^2$ was found which is close to the value of the integral on the channel width of the mean thickness (0.158 $\mu m^2$).

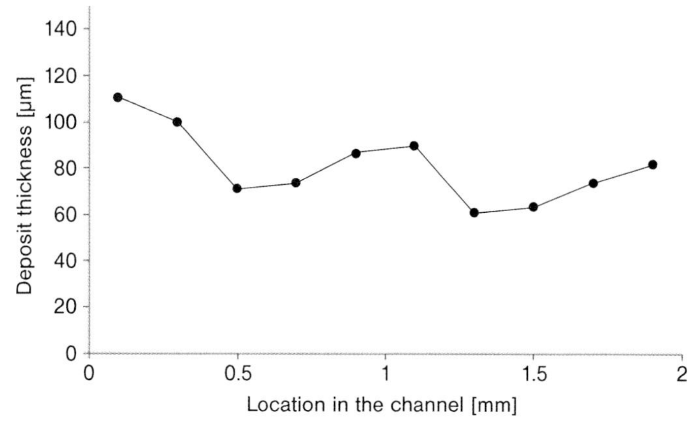


**Figure 11.8** Deposit profile for the data from Figure 11.6 at $t = 162$ min (TMP = 80 kPa, concentration = 1 g L$^{-1}$, $L_{p0} = 2.5$ L h$^{-1}$ m$^{-2}$ kPa$^{-1}$).

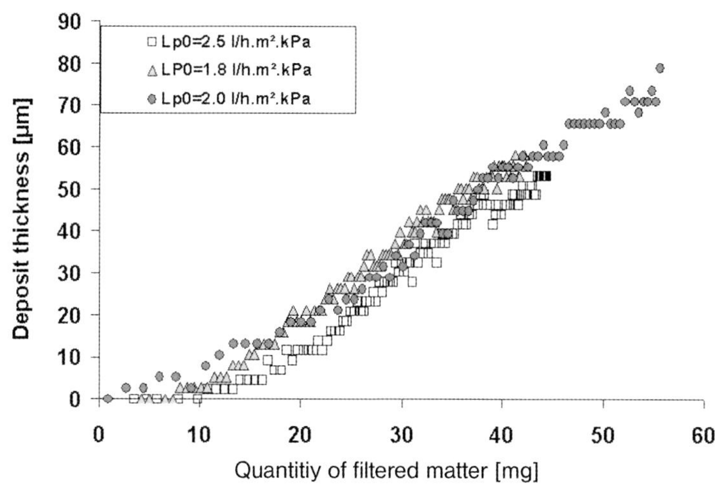


**Figure 11.9** Deposit thickness versus the quantity of filtrated matter (TMP = 80 kPa, clay concentration = 1 g L$^{-1}$).

Those results point out some capabilities of the method. First it is possible to follow the changes in the deposit structure with time and to distinguish several kinds of fouling layers. It is also possible to obtain the cross-section of the deposit and to detect wall effects on particle deposition.

Concentrations of 1 g L$^{-1}$ and 2 g L$^{-1}$ were tested for clay suspensions. In both cases, image processing allowed the determination of a sharp maximum in the light intensity profile and thus the deposit thickness measurement was possible. Higher concentrations will be tested but it currently appears that the LSGI method is well suited to a range of concentrations larger than for other optical techniques.

In order to access the results reproducibility, three experiments were performed under the same operating conditions: TMP = 80 kPa, clay concentration = 1 g L$^{-1}$ and feed suspension flowing from the bottom of the channel. The measurement was carried out at the same location for the three experiments (70% of the channel height from the bottom). Figure 11.9 shows the mean deposit thickness versus the quantity of filtrated matter for the three tests.

For each experiment, a new sample of the same membrane was used and the initial permeability was measured with pure water. Results reveal a good reproducibility concerning the mean deposit thickness value and the deposit growth dynamics with regard to the quantity of filtrated matter. The maximum percentage deviation of the data for the three replication runs from the average value is 8%, which is reasonable.

### 11.3.8 Verification of the Order of Magnitude

In order to compare the order of magnitude of the measured values of the deposit thickness, average thickness values were estimated using the cake filtration modelling at constant pressure considering clay particles as small flat plates with an equivalent diameter (measured with a granulometer) of 7 μm. Mean thickness along the channel was computed using data from the experimental filtration curve. The slope of the curve $t/V$ versus $V$ is linked to the global cake specific resistance:

$$\frac{t}{V} = \frac{1}{Q_0} + \frac{\alpha_{\text{global}} \mu_l c}{2 T_{MP} S^2} V(t) \tag{11.6}$$

where $\alpha_{\text{global}}$ is the global cake specific resistance, $\mu_l$ the liquid viscosity, $c$ the deposited mass per filtered unit volume, $S$ the filtration area and $Q_0$ the initial filtration flow rate.

In dead-end UF it is assumed that the deposited mass at the membrane surface per unit of volume is:

$$c = c_{\text{feed}} V \tag{11.7}$$

where $c_{\text{feed}}$ is the feed concentration. The deposit permeability ($L_p$) is related to the cake specific resistance ($\alpha_{\text{global}}$) through the deposit properties like the porosity ($\varepsilon_{\text{global}}$) and the particle density ($\rho_p$):

$$L_p = \frac{1}{\alpha_{\text{global}} \rho_p (1 - \varepsilon_{\text{global}})} \tag{11.8}$$

The deposit permeability can also be calculated using the Carman–Kozeny equation, when particle shape and deposit porosity are known:

$$L_p = \frac{1}{h} \frac{\varepsilon_{\text{global}}{}^3}{(1 - \varepsilon_{\text{global}})} \frac{1}{A_s^2} \tag{11.9}$$

Where $h$ is the Carman–Kozeny constant which is 4.17 for spherical particles. For nonspherical particles, a shape factor has to be taken into account for the particle-specific area calculation. This specific area, $A_s$, is defined as the particle area/particle volume ratio and can then be calculated from:

$$A_s = \frac{3f_h}{a} \tag{11.10}$$

where $a$ is the radius of a sphere with the same volume as the parallelepiped particle ($a = e_p lL$).

From Equations (11.8) and (11.9) the following expression is obtained and enables to estimation of the global porosity:

$$\alpha_{\text{global}} = \frac{4.17(1 - \varepsilon_{\text{global}})A_s^2}{\varepsilon_{\text{global}}{}^3 \rho_p} \tag{11.11}$$

The global porosity enables estimation of the deposit thickness:

$$e_{\text{global}} = \frac{V \times c}{S\rho_p(1 - \varepsilon_{\text{global}})} \tag{11.12}$$

Figure 11.10 shows the comparison between the global thickness ($e_{\text{global}}$, calculated from Equation (11.12)) and any local thickness ($e_{\text{local}}$, directly measured with the optical method) for the UF of a clay suspension at 1 g L$^{-1}$ and with a TMP of 80 kPa.

The comparison between measured and calculated values shows good agreement. The measured order of magnitude and cake growth dynamics seem to be correct. Nonetheless, the evolution of the measured thickness with the deposited mass is nonlinear. This is due to the modification of the deposit structure with time which is not taken into account in the model. The limitation of the model is

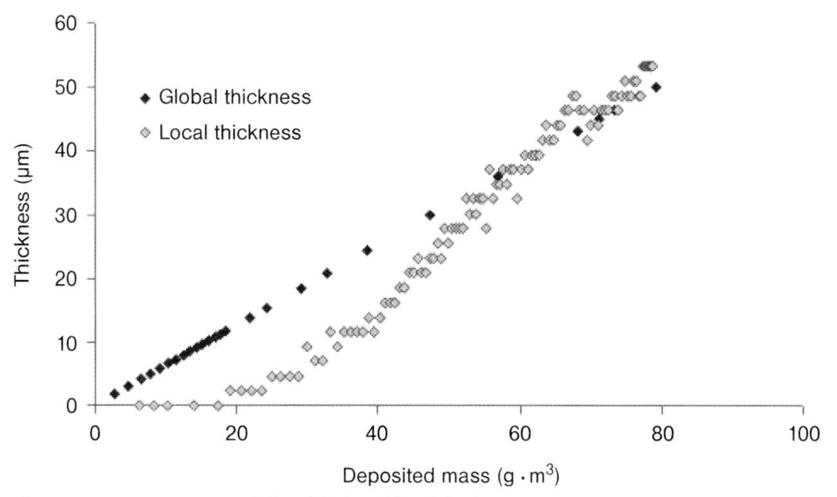


**Figure 11.10** Comparison of the global and local thickness (TMP = 80 kPa, concentration = 1 g L$^{-1}$, $L_{p0}$ = 1.8 L h$^{-1}$ m$^{-2}$ kPa$^{-1}$).

to assume constant deposit properties with time and space, whereas actual restructuration effects occur during filtration. This shift could also be due to a deposit thickness distribution along the channel length, whereas calculated values assume a uniform deposit distribution. In consequence, this method could be used to study the non-uniformity of the deposit on the membrane.

### 11.3.9 Conclusion

An optical method was developed and validated on the base of experimental results. With this method, it is still possible to measure with a good sensitivity the deposit thickness in a confined channel, versus time and for several locations on the fiber length in conjunction with flux measurements. This method could be used in order to better understand the formation mechanisms of deposits in association with operating conditions and to develop new models taking into account the time and space variations in the deposit properties.

However, this method presents two main limitations. First it is only applicable on a transparent module and second it is not possible to directly measure the deposit porosity. Thus, in parallel another method which would overcome those drawbacks and bring complementary information was investigated.

## 11.4 In Situ Deposit Characterization with an Acoustic Method

Ultrasonic measurements use sound waves to measure the location of a moving or a stationary interface. They are based on the introduction of high-frequency sound waves into an object to obtain information about it without altering it. Two basic quantities can be measured: the time of flight, or the time taken for the sound to travel through the sample, and the amplitude of the received signal which varies with cake structure parameters (porosity, material, etc.).

The general principle of acoustic measurements for layer characterization is to follow the ultrasonic echo reflected on the top of a layer and the one reflected on the bottom of the layer. One can write:

$$\Delta t = \frac{2e_d}{V_L} \tag{11.13}$$

In this expression, $V_L$ is the ultrasonic longitudinal velocity in the layer, $e_d$ is the layer thickness and $\Delta t$ is the time of flight between the two echoes. This time of flight is usually calculated with an intercorrelation method or using the FFT phase slope. The measurement of the time of flight allows the deposit thickness to be calculated.

The next section gives a review about studies using acoustic waves for the monitoring of membrane fouling.

### 11.4.1 Acoustic Methods for Fouling Characterization

Mairal et al. [6] were the first to use ultrasonic reflectometry for the observation of membrane fouling. They monitored in situ fouling of inorganics ($CaSO_4$) in a crossflow reverse osmosis system with a flat-sheet module. The relative amplitude of the ultrasonic amplitude was followed in conjunction with flux decline. Ultrasonic signal measurement provided a sensitivity to the dynamics of the fouling layer growth that was comparable to that observed from the flux decline behavior. Moreover, morphological characterization via optical microscopy and SEM was performed. During the experiments, several mechanisms of layer growth were observed depending on the orientation of the crystals. It was possible to measure a corresponding ultrasonic response to the changing morphology. Ultrasonic amplitude was also monitored during the cleaning phase in conjunction with the permeate rate. Both permeate rate and ultrasonic response increased rapidly and then stabilized. At this moment, morphological analysis of the membrane conducted after the cleaning phase confirmed that membrane was completely clean and free of any fouling deposits. This suggested that the ultrasonic technique might be useful for the monitoring of fouling removal in real time. However, in this study ultrasonic measurement can only provide relative information about the fouling layer (qualitative modification of the acoustic signal in comparison with the acoustic signal before fouling occurs). Moreover, an obvious modification of the signal appears after a long filtration time (about 70 h).

Li et al. [7] used ultrasonic reflectometry for the study of fouling on flat-sheet nylon membranes during crossflow microfiltration at 100 kPa. The feed solution was an effluent from a wastewater treatment plant. The acoustic signal had a frequency of 10 MHz. Results obtained from this study suggest that the combination of flux determination and ultrasonic measurement can provide a much clearer view of the fouling behavior of a membrane than flux decline alone that is also sensitive to charge accumulation and membrane compaction. Once fouling was initiated, the acoustic impedance difference and topographical characteristics at the feed solution/membrane interface will change resulting in a new echo. With this new echo, it was possible to quantify the fouling layer by calculating its thickness. Besides, the fouling echo amplitude is an indication on the state of the fouling layer. The more dense the fouling layer is, the better is the reflection (and thus the larger is the amplitude that is seen). Once again, sequential modes of fouling layer were observed by morphological analysis. The ultrasonic technique was capable of distinguishing individual modes of growth. So, in this study it was possible to obtain quantitative information but the fouling echo appears only after 240 min of filtration.

The method was also applied to an inside/out tubular membrane [8] during UF of 0.08 $g\,L^{-1}$ bovine serum albumine (BSA) solution. The tubular membrane used was a polyethersulfone (PES) membrane with a MWCO = 40 kDa, and with inside and outside diameters of 14 mm and 21 mm, respectively. It was possible to detect and distinguish the acoustic response signal from the various interfaces. The differential signal due to the fouling layer appeared after 10 min of fouling operation. They found a linear relationship between the amplitude of the differential signal and the fouling resistance.

The monitoring of membrane fouling with acoustic waves is one of the few non-invasive methods potentially applicable to commercial-scale modules. Nonetheless, amplitude measurements can only provide qualitative characterization (relative amplitude variations) of the cake formation process because of the lack of knowledge about the propagation mode of an acoustic wave in a deformable deposit.

Most of the studies about the use of acoustic waves for fouling characterization gave qualitative information about the cake layer (relative amplitude variations) and concerned flat-sheet membranes. The originality of the method presented here [9] is to give the possibility of extracting qualitative information by using another characterization method in parallel.

### 11.4.2
### Development of an Acoustic Method for Fouling Characterization

The acoustic method can be applied on various materials which do not need to be transparent. There is a no constraint for the filtration module expected that the filtration surface has to be larger than the acoustic sensor (generally a circle, diameter 4–10 mm).

A step in the development of the acoustic technique for the monitoring of membrane processes is to choose a frequency adapted to the membrane material. Therefore, several frequencies have to be tested (usually in the range 1–10 MHz). The adapted frequency must result in a workable signal, with large amplitude and well separated waves. Some frequencies can be "cut" by the membrane material, that is to say totally absorbed. Usually, each material has its own adapted frequency.

For the chosen frequency, deposit detection has then to be tested. Indeed, the frequency adapted to the membrane material must also detect particle deposits for the application which is studied. Generally, the signal on a clean membrane is modified by the formation of a deposit on the membrane surface. Nonetheless, it is very difficult to obtain information about the deposit structure without a signal treatment. Indeed, at the beginning of the filtration (little deposit), the echo due to the deposit is embedded with the one of the membrane (Figure 11.11). The deposit

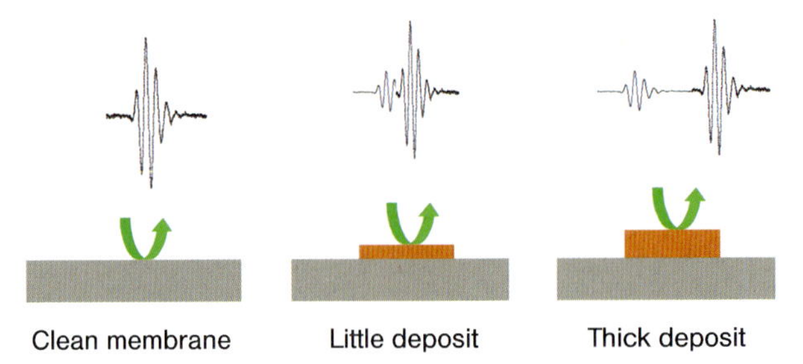


**Figure 11.11** Schematic view of the acoustic signal variation with the deposit thickness.

has to be thick enough (about hundreds of micrometers) to generate a well separated echo. An example of signal treatment is presented in the next section which concerns an application of the method and some typical results that can be obtained.

### 11.4.3 Example of the Development of an Acoustic Method for the Characterization of Dead-end UF Deposits

In order to obtain complementary information about the deposit structure by using both acoustic and optical methods, we developed the acoustic method for a particular application, which is the characterization of dead-end UF deposit [9].

As explained above, the first objective was to choose an adapted frequency for the polysulfone membrane. It was decided to work with 2.25 MHz (Figure 11.12). This frequency enable a clear and workable acoustic signal to be obtained. For higher frequencies, the acoustic signal was totally absorbed by the membrane and thus it was not possible to extract any kind of information.

The next objective was to see whether it was possible to detect a deposit on those membranes with an acoustic signal and to follow its kinetics. First tests were

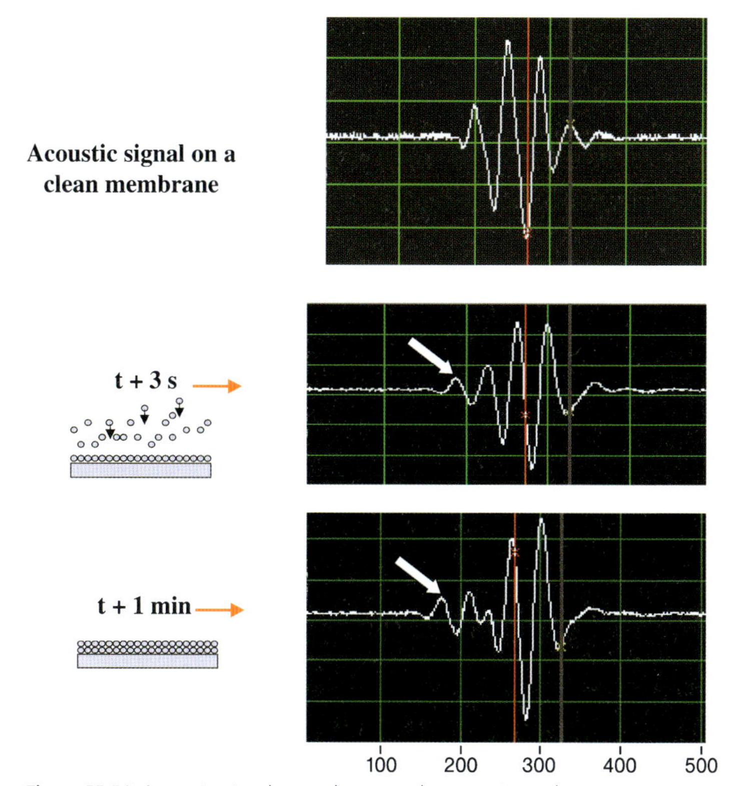


**Figure 11.12** Acoustic signal on a clean membrane at 3 s and at 1 min after the introduction of glass particles.

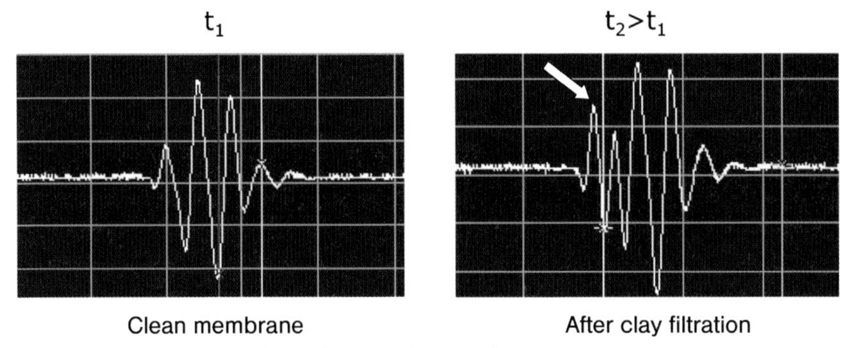


**Figure 11.13** Acoustic signal on a clean membrane and on the same membrane after filtration of a clay suspension.

performed with glass particles (mean diameter 22 μm) settling free on clean membranes. Figure 11.12 shows the acoustic signals at 3 s and 1 min after the introduction of glass particles into a cell with a polysulfone membrane (without transmembrane pressure). The final thickness of the deposit was about 200 μm. In comparison with the signal on a clean membrane, it is possible to see an additional echo due to the deposit of glass particles (shown by an arrow in Figure 11.12). This echo moves as the deposit grows. These experiments indicate that it is possible to detect a deposit on a polysulfone membrane and to follow its build-up kinetics.

Then, preliminary tests were performed with particle deposits. First, deposits were formed by filtration of clay suspensions (same particles as in Section 11.3.7) in a reference batch cell (Amicon, 100 $cm^3$) in dead-end mode. Acoustic experiments were performed ex situ, after filtration and formation of a 500 μm particle deposit. The acoustic signal was measured on a clean membrane and on the same membrane after filtration. As it can be seen in Figure 11.13, there was a modification of the signal and a new echo due to the clay deposit.

This was the case of a thick deposit which was not representative of a fouling deposit during the first filtrations (under 100 μm). Nonetheless, as mentioned above, a main difficulty is to generate a separated echo due to the deposit for slight thicknesses. A special signal treatment has to be used in order to decorrelate the signal due to the membrane from the signal due to the deposit. Thus, software was developed in order to treat the signal and to separate the deposit signal from the measured signal.

Figure 11.14 gives an example of signal treatment with the specific software. The experimental time (duration of the experiment) is plotted versus the time of flight of the echo (echo time). It is thus possible to separate the two echoes. The time of flight of the membrane echo is constant because the membrane surface has a fixed position. As expected, the time of flight of the deposit echo decreases as the deposit grows. Indeed, as the deposit grows, its surface becomes closer and closer to the sensor. Thus, a decreasing amount of time is taken to go from the sensor to the growing deposit surface and return to the sensor.

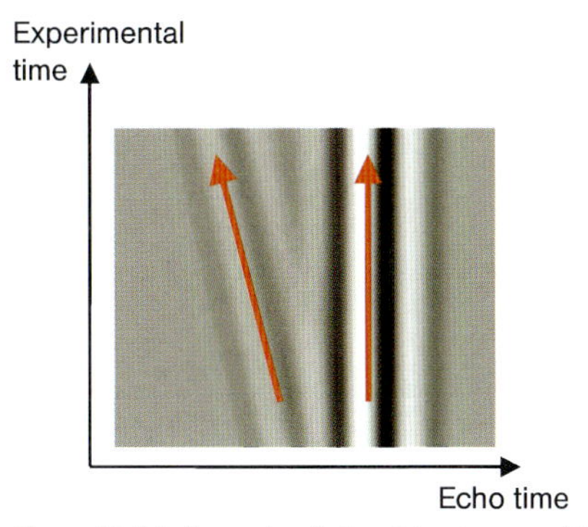


**Figure 11.14** Example of signal treatment with the developed software: deposit echo and membrane echo are separated from the beginning of the filtration.

## 11.4.4 In Situ Application of the Method

Those promising results convincingly demonstrated the interest of adapting the method in situ. So, the acoustic method was adapted to the same filtration module and membrane as in Section 11.3.7. The only difference concerns the channel width, which must suit the acoustic sensor. For this filtration module, the channel width is 4 mm.

As we do not know the acoustic velocity in particle deposit, the general principle of acoustic method for thickness measurement has to be modified. It was proposed to first measure the signal on the clean membrane, at $t=0$, then the deposit echo on a fouled membrane (Figure 11.15).

Then Equation (11.13) becomes:

$$t_1 - t_2 = \frac{2e_d}{V_w} \tag{11.14}$$

where $t_1$ and $t_2$ are respectively the time of flight of the membrane echo (echo 1 in Figure 11.15) and the deposit echo (echo 2 in Figure 11.15) and $V_w$ is the acoustic velocity in the suspension.

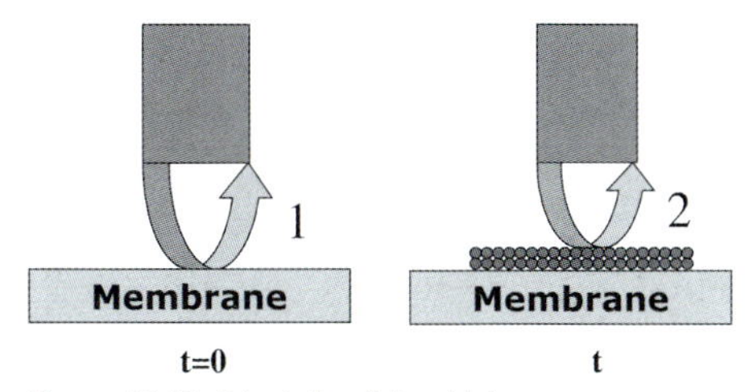


**Figure 11.15** Principle of the thickness measurement with the acoustic method presented in this chapter.

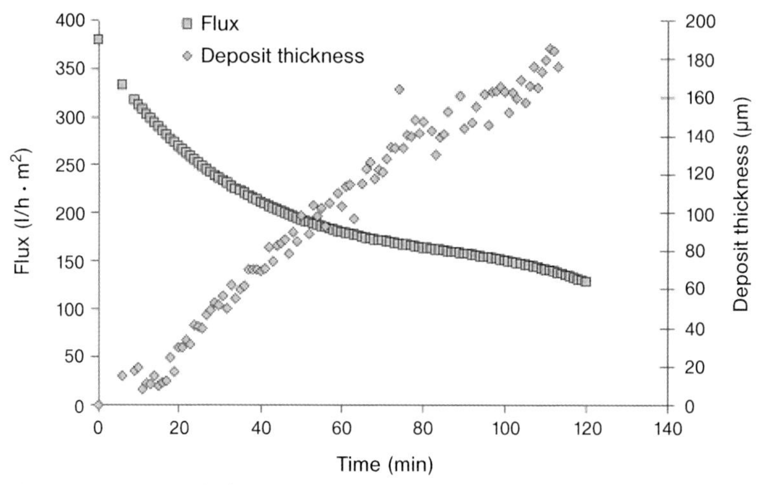


**Figure 11.16** Deposit thickness measured with the acoustic method and flux versus the quantity of deposited matter (TMP = 80 kPa, concentration = 1.5 g $L^{-1}$, $L_{p0}$ = 2 L $h^{-1}$ $m^{-2}$ $kPa^{-1}$).

First experiments were conducted with clay suspensions at 1 g $L^{-1}$ in dead-end mode. The TMP was constant and equal to 80 kPa. Figure 11.16 gives an example of results obtained with the acoustic method applied on the filtration module. The measured thickness and the flux are plotted versus time. The flux curve is a typical result encountered in dead-end filtration: a rapid decrease at the beginning of the experiment and then more slowly until a stabilized value (which was not reached at the end of this experiment). The final thickness is 180 μm for a total deposited mass of 120 g $m^{-2}$ and a relative flux decrease of 66%. Reproducibility of the measurement was tested and gave good results.

### 11.4.5
### Combined Use of the Two Methods: Approach

The goal of the study was to obtain several kinds of information (thickness, porosity) about a deposit. Therefore, two characterization methods were developed. First, the acoustic method gives time variations from a complex signal, which are qualitative variations. Second, the optical method gives thickness measurements, which are quantitative variations. Adaptation of the two methods on the same apparatus, a confined channel, enables a complete deposit characterization to be obtained.

First experiments were conducted using both methods on the same module. Figure 11.17 represents the ratio of the deposit thickness by the final thickness, named relative thickness, as a function of time for the two methods.

It appears that the two methods measure the same relative thickness variations. Nonetheless, there is a shift between the absolute thickness values. This shift

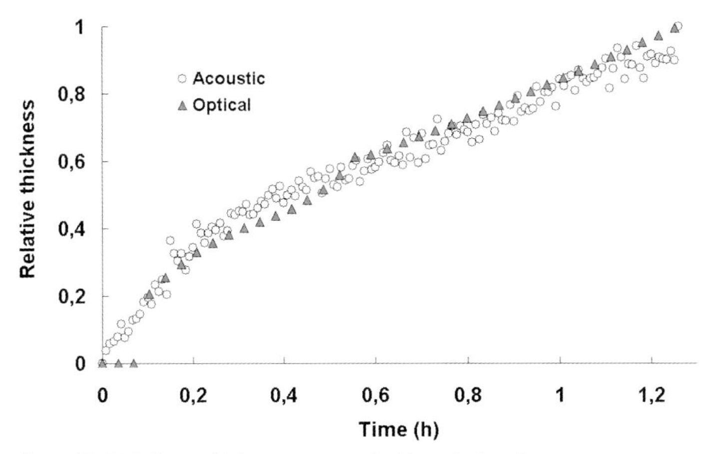


**Figure 11.17** Relative thickness measured with optical and acoustic methods versus time (TMP = 80 kPa, concentration = 1.5 g L$^{-1}$, $L_{p0}$ = 1.9 L h$^{-1}$ m$^{-2}$ kPa$^{-1}$).

could be due to a lack of knowledge about acoustic propagation and reflection in porous media. The next step will be to use the optical method in order to calibrate the thickness measurements with the acoustic method. Then porosity will be investigated by studying the variations of the signal amplitude.

## 11.5 Conclusion

In situ local characterization of cake formation is a key step towards a better understanding of filtration processes. Some non-invasive characterization methods have shown their efficiency for cake investigation on flat-sheet membranes but they give only one kind of information. Few studies concern cake characterization in a confined geometry like an inside-out hollow fiber membrane.

This chapter presents two original characterization methods that were developed in our laboratory (LISPB). Both allow the characterization of in situ fouling on organic membranes during filtration. Their common use on the same filtration apparatus could give complete characterization of filtration deposits.

The first innovative method (LSGI) is based on the use of laser sheet at grazing incidence. It enables the deposit thickness at several locations in the vertical direction to be measured and the deposit thickness profile in a channel cross-section ($z$) to be obtained. Moreover, with this method it is possible to follow the deposit kinetics in relation to flow rate variations. First results were obtained with clay suspensions and showed good reproducibility. Local measurements in conjunction with global measurements (flux or pressure) enable the deposit structure to be studied. It is now a tool that is used in our laboratory to study the suspension of suspension properties or of operating conditions on cake properties and filtration performances.

The second method uses acoustic waves. It gives information about the deposit porosity and can be used with a nontransparent module. With this method, it is also possible to follow the deposit kinetics. This method has to be calibrated because of the lack of knowledge about the theory of acoustic waves propagation in porous media. Thus, the common use of both methods on the same module will enable complementary information about the deposit to be obtained.

The developed methods are powerful characterization tools on a local scale and will enable a better understanding of fouling phenomena in filtration, leading to a better process operation control.

## Nomenclature

| | |
|---|---|
| $a$ | Radius of a sphere of volume $(= e_p L l)$, m |
| $A_s$ | Particle specific area, $m^{-1}$ |
| c | Deposit mass per filtered unit volume, $kg\,m^{-3}$ |
| $c_{feed}$ | Feed concentration, $g\,L^{-1}$ |
| $d_{eq}$ | Equivalent particle diameter, m |
| $e_d$ | Deposit thickness, m |
| $e_p$ | Particle thickness, m |
| $f_h$ | Heywood form factor |
| $h$ | Channel height, m |
| $J$ | Permeation flux, $m\,s^{-1}$ |
| $l$ | Particle width, m |
| $L$ | Particle length, m |
| $L_p$ | Permeability, $L\,h^{-1}\,m^{-2}\,Pa^{-1}$ |
| $L_{p0}$ | Initial permeability, $L\,h^{-1}\,m^{-2}\,Pa^{-1}$ |
| $n$ | Refraction index |
| $Q_p$ | Filtration flow rate, $m^3\,s^{-1}$ |
| $Q_0$ | Initial filtration flow rate, $m^3\,s^{-1}$ |
| $S$ | Filtration area, $m^2$ |
| $t$ | Time, s |
| $TMP$ | Transmembrane pressure, Pa |
| $V$ | Filtrated volume, $m^3$ |
| $V_L$ | Longitudinal acoustic velocity, $m\,s^{-1}$ |
| $V_w$ | Acoustic velocity in the suspension, $m\,s^{-1}$ |
| $\Delta X$ | Shift between the position of the laser sheet outside and inside of the channel, pixels |
| $\alpha$ | Cake specific resistance, $m\,kg^{-1}$ |
| $\varepsilon$ | Cake porosity |
| $\mu_l$ | Liquid viscosity, $Pa\,s^{-1}$ |
| $\rho_p$ | Particle density, $kg\,m^{-3}$ |
| $\theta$ | Inclination angle of the laser sheet |

## References


1. V. Chen, H. Li, A.G. Fane, *Journal of Membrane Science* **2004**, *241*, 23–44.
2. J.-C. Chen, Q. Li, M. Elimelech, *Advances in Colloid and Interface Science* **2004**, *107*, 83–108.
3. M. Hamachi, M. Mietton-Peuchot, *Chemical Engineering Science* **1999**, *54*, 4023–4030.
4. J. Altmann, S. Ripperger, *Journal of Membrane Science* **1997**, *124*, 119–128.
5. J. Mendret, C. Guigui, P. Schmitz, C. Cabassud, P. Duru, *AIChE Journal* **2007**, *53*, 2265–2274.
6. P. Mairal, A.R. Greenberg, W.B. Krantz, *Desalination* **2000**, *130*, 45–60.
7. J. Li, R.D. Sanderson, E.P. Jacobs, *Journal of Membrane Science* **2002**, *201*, 17–29.
8. J. Li, R.D. Sanderson, G.Y. Chai, *Sensors and Actuators B* **2006**, *114*, 182–191.
9. J. Mendret, C. Guigui, C. Cabassud, N. Doubrovine, P. Schmitz, P. Duru, J.Y. Ferrandis, D. Laux, *Desalination* **2006**, *199*, 373–375.

# Part III
# Process-oriented Monitoring Techniques

# 12
# Monitoring of Membrane Processes Using Fluorescence Techniques: Advances and Limitations

*Carla A. M. Portugal and João G. Crespo*

## 12.1
## Introduction: Why Use Natural Fluorescence as a Monitoring Technique?

This chapter discusses the use of fluorescence techniques for the monitoring of membrane processes, making use of the intrinsic fluorescence properties of the various components involved. This chapter deals with situations where none of these components is labeled with an external reporter, a fluorescence probe, as happens in many other applications described in this book (see Chapters 3, 4 and 8).

As previously discussed, when a target molecule is labeled with an external fluorescence reporter, the chemical character of the conjugate may differ from the original molecule and this change may induce a different behavior in this molecule when interacting with the membrane material. The opposite is also true, when the labeling procedure is applied to the membrane. Unlike radiolabeling, fluorescence labeling may induce significant changes in the chemical behavior of the original molecule, under specific conditions.

Therefore, the use of the intrinsic ("natural") fluorescence behavior of a given molecule may be an extremely powerful approach to investigate and monitor a specific process, because the molecule to be inspected is not labeled/altered by an external reporter. However, these techniques are restricted to the monitoring of compounds that exhibit natural fluorescence.

One of the most interesting features of natural fluorescence results from the fact that the fluorescence response of such a molecule depends very much on its microenvironment. Take, for example, the fluorescence behavior of a tryptophan residue in a protein molecule: its fluorescence response differs, depending on the position of such a tryptophan within the protein (e.g. more exposed at an outer hydrophilic environment or more buried inside a hydrophobic region of the protein). Also, for the same tryptophan example, the proximity or distance of potential quenchers influences its fluorescence response. This behavior can be explored in order to gather information about the molecular interaction between the target fluorescent compound and a membrane, as discussed in this chapter.

*Monitoring and Visualizing Membrane-Based Processes*
Edited by Carme Güell, Montserrat Ferrando, and Francisco López

ISBN: 978-3-527-32006-6

This chapter introduces and discusses different fluorescence techniques that do not require the use of external labeling: steady-state fluorescence (including 2D fluorescence), fluorescence anisotropy and time-resolved fluorescence; and it provides illustrative examples showing how these techniques may be used for the monitoring of membrane processes.

## 12.2 Natural Fluorescence Techniques

Fluorescence emission is a radiation de-excitation process of an excited molecule, corresponding to the transition from its lowest singlet excited electronic state to the ground state. The ability of an excited molecule to emit fluorescence light when decaying to its ground state is defined by its fluorescence quantum yield, $\Phi_F$. The fluorescence quantum yield corresponds to the fraction of absorbed energy that is lost in the form of fluorescence light.

Fluorescence emission of a molecule depends on its chemical structure and characteristics, on the environmental conditions (temperature, pH, solvent properties) and it is also sensitive to the interactions of the fluorophore with the environment in its vicinity. This behavior is characteristic of heterocyclic compounds containing nitrogen (e.g. indole, carbazole) or molecules with these substituting groups, as is the case of tryptophan. The fluorescence characteristics of tryptophan depend strongly on the polarity of its microenvironment (or solvent). This dependence is particularly important, as it has been used for identifying the structural alterations of proteins after submission to denaturating conditions (chemical denaturating agents [1], high temperatures or processing through porous media [1, 2]) where the tryptophan(s) belonging to the protein chain are used as structural reporters.

In addition, there are compounds (known as fluorescence quenchers) which are capable of reducing the fluorescence quantum yields of fluorophores, causing a decrease in the fluorescence emission. This effect is a consequence of energy, electron and proton transfer reactions occurring in the excited state, due to interactions between fluorophores and quenchers in close proximity. The efficiency of the quenching mechanisms depends on the concentration of the quencher, the stability of the intermediary complex formed in an excited state and the environmental conditions, such as temperature and solvent viscosity. Examples of effective fluorescence quenchers are hydroxyl groups (–OH), carboxyl groups (–COOH), oxygen (oxygen induces the oxidation of fluorescing species), halogen ions ($Cl^-$, $I^-$, $Br^-$), heavy metals, hydrogen bonds and disulfide bonds [3–5].

Fluorescence is a very sensitive technique capable of providing information on the structural status of the fluorophore or the molecules they are inserted in, making possible the monitoring of specific processes. In well defined and controled circumstances, it also allows for a quantitative determination of the fluorescence compounds, with detection limits within the parts per billion (ppb) range.

The detection limit of fluorescence techniques is strongly dependent on the efficiency of the fluorescence equipment. The type of information provided by these techniques depends on the fluorescence mode used – either steady-state (steady-state fluorescence) or time-resolved (time-resolved fluorescence) – and also whether the excitation is performed with natural or polarized light (steady-state or time-resolved fluorescence anisotropy) [3].

### 12.2.1 Steady-state Fluorescence

Steady-state fluorescence measurements provide information about physico-chemical events in equilibrium, in an excited state. Steady-state fluorescence measurements are performed by exposing a fluorophore (sample) to a continuous excitation light beam produced by a light source able to provide continuous radiation across a large spectrum range. The high-pressure Xe arc lamp fulfils these requisites, allowing a continuous intense light between ~250 nm and 1200 nm. The light provided by the light source is guided to an excitation monochromator or filter that only permits the passage of light with the selected excitation wavelength, $\lambda_{exc}$ (Figure 12.1). The excitation wavelength reaches the sample and light absorption occurs. Fluorescence light is emitted from the sample at different emission wavelengths ($\lambda_{em}$ are necessarily higher than $\lambda_{exc}$) and different intensities. Fluorescence emission then passes by an emission monochromator that splits the emitted light into the distinct $\lambda_{em}$ and the fluorescence signal is finally detected by a photomultiplier or a diode array detector [3].

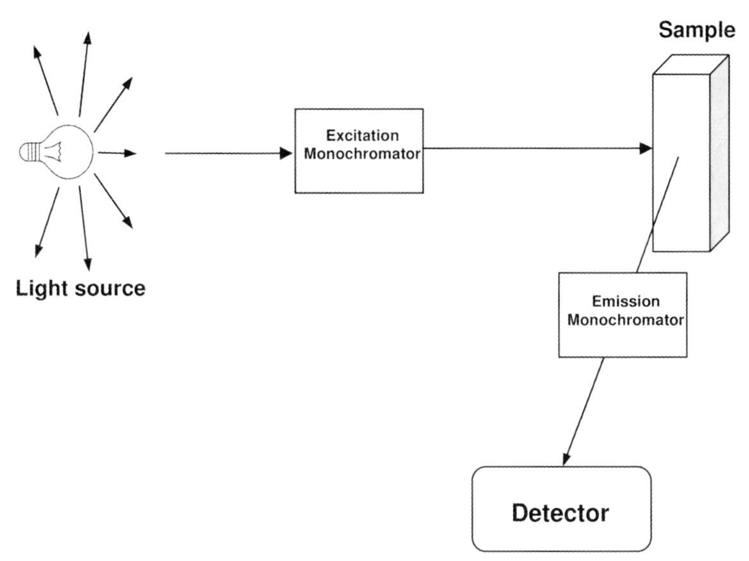


**Figure 12.1** Schematic diagram of the configuration of the excitation and emission components in a spectrofluorometer.

The fluorescence spectra obtained corresponds to the sum of the individual contributions of the emitting fluorophores present in the sample, capable of absorbing energy at the selected $\lambda_{exc}$. The information elicited through spectral analysis can be based on: (a) variation in fluorescence intensity, (b) shifts in maximum emission and (c) emission band broadening.

Variation in fluorescence intensity may be correlated with changes in the molecular structure of the fluorophore; it may indicate the proximity of a quencher or the formation of different chemical species resultant from possible physico-chemical events that may occur in the system. These chemical species appearing in the system may also be fluorophores, with their own fluorescence characteristics or fluorescence quenchers, contributing in a constructive mode (increasing fluorescence) or destructive mode (decreasing fluorescence), respectively, to the overall fluorescence spectra. Shifts in maximum emission may be correlated with changes in the polarity of the environment near the fluorophore. An example is tryptophan in proteins acting as a structural reporter of these molecules: structural changes of proteins may induce different exposures of a tryptophan to the solvent. The exposure of a buried tryptophan to an aqueous solvent decreases the fluorescence intensity and shifts the maximum emission to higher wavelengths (maximum emission red shift) [1, 2, 6]. Emission band broadening may be attributed to an increase of the structural heterogeneity in solution. Using again the example of tryptophan in proteins, if there are populations of proteins with different degrees of molecular unfolding in solution, one may expect to find tryptophans with different degrees of exposure to the solvent. Consequently, this will extend the fluorescence emission of tryptophans through a large emission wavelength range, leading to emission band broadening [1].

However, since steady-state fluorescence spectra result from the sum of the contribution of the individual species present in the system, the identification of individual species or chemical events (structural alteration, formation or disappearance of molecular species due to chemical reactions) occurring within complex systems may not be straightforward. This limitation may be overcome by using time-resolved fluorescence techniques [7].

### 12.2.2
### Time-resolved Fluorescence

Time-resolved fluorescence is a time-dependent technique allowing for the determination of events occurring during the excited state lifetime of the fluorophore, which may vary from a few picoseconds to nanoseconds.

The lifetime (or decay time) of the fluorophores can be obtained using two different time-resolved methods: pulse fluorometry (generally, using the single photon timing method) or phase modulation fluorometry. Pulse fluorometry is by far the most popular method; therefore the whole discussion developed in this chapter is based on the principles of this method.

In pulse fluorometry, the sample is excited by a pulsed light beam ($\delta$ pulse), provided by a mode-locked laser or by a flash lamp. The fluorescence response to

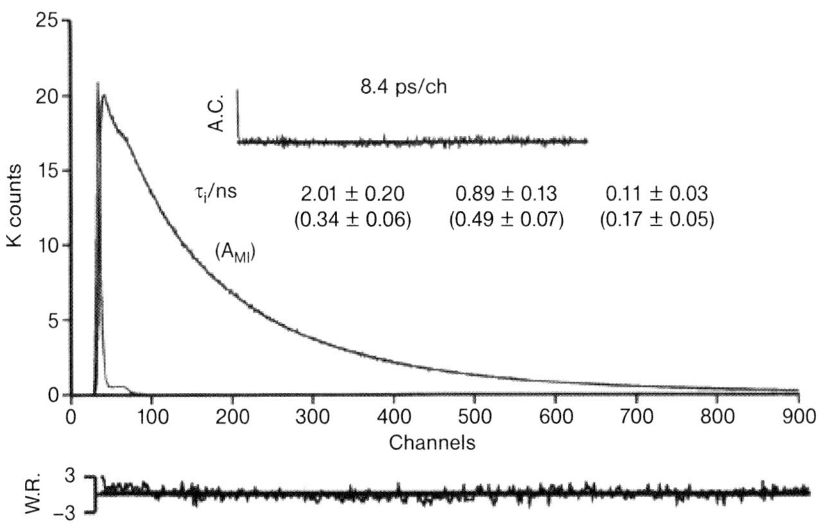


**Figure 12.2** Fluorescence decay of β-lactoglobulin in aqueous solution, at pH 3, fit with tri-exponential function. The fluorescence decay times ($\tau_i$) were acquired at $\lambda_{exc} = 290$ nm and $\lambda_{em} = 350$ nm. The quality of the fit was evaluated based on the values of $\chi^2$ obtained, and on the quality of the residual distribution (W.R.) [8].

the excitation pulse is most commonly expressed by a fluorescence decay curve (Figure 12.2). The fluorescence decay curves are usually acquired using vertical polarized light excitation and selecting the fluorescence at the magic angle (54.7°), to avoid polarization effects [3, 7].

The fluorescence decay curve accounts for the contribution of the individual decay times of all fluorescing residues (fluorophores in different environments) decaying to the ground state within the selected measuring time window. Therefore, through this technique it is possible to access to the individual contribution of each fluorescence species, *i*, in solution, to the overall fluorescence spectra by the signature of their decay times, $\tau_i$.

The deconvolution of the individual decay times can be obtained by fitting the fluorescence decay curve with a sum of exponentials (Equation 12.1). The number of fluorescing species in solution is equal to the number of exponential parcels needed to adequately fit the fluorescence decay. The quality of the fit is evaluated by the chi squared value ($\chi^2$) for goodness of fit and through analysis of the residuals distribution (W.R.):

$$I(\lambda, t) = \sum_i \alpha_i(\lambda) e^{-t/\tau_i} \tag{12.1}$$

where $I(\lambda,t)$ is the time evolution of fluorescence intensity (a.u.), $\alpha_i$ is the amplitude of the decay, $\tau_i$ is the fluorophore decay time (ns) and $t$ is the measuring time (ns).

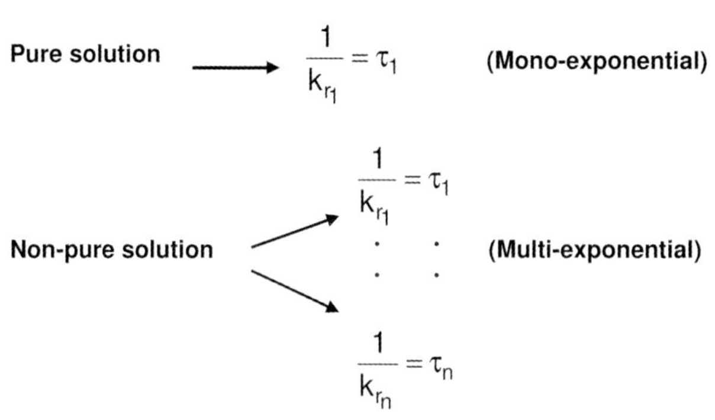


**Figure 12.3** Schematic diagram of the decay rate constants, $k_{ri}$, and respective decay times, $\tau_i$, obtained after fit of mono and multi-exponential fluorescence decays.

Fluorescence decays obtained for pure solutions, in the absence of fluorophore–solvent interactions, are perfectly described by mono-exponential functions (a single decay time is obtained). However, the fluorescence decay of multi-fluorophore solutions (usually the case of complex systems) requires the adjustment of multi-exponential functions (Figure 12.3). Note that a solution containing a single fluorophore in different environments should be considered as a mixture of fluorophores.

In some circumstances, complex systems such as biologic samples are characterized by fluorescence decays which are inadequately interpreted by using this mathematical discrete approach (Equation 12.1) is not valid in these cases). Instead, the fluorescence decays describing a multitude of events occurring simultaneously in an excited state (complex systems) obey a continuous distribution of decay times [9, 10]. According to this, fits of these fluorescence decays should be performed using Equation (12.2):

$$I(\lambda, t) = \int_{\tau} \alpha_t(\lambda) e^{-t/\tau} d\tau \qquad (12.2)$$

**Table 12.1** Values of the lifetime of NATA (N-acetyltryptophanamide) in dioxane/water solutions with variable percentages of water [6].

| Water (%) | Lifetime, $\tau$ (ns) |
|---|---|
| 0 | 5.2 |
| 20 | 5.1 |
| 40 | 4.8 |
| 60 | 4.1 |
| 80 | 3.6 |
| 100 | 3.1 |

**Table 12.2** Values of $\tau_i/\tau_0$ obtained for β-lactoglobulin upon addition of different concentration of $I^-$ (in the form of NaI) in 10 mM citrate/phosphate buffer solutions, at pH 5. $\tau_0$ is the decay time of β-lactoglobulin solutions in the absence of $I^-$ and $\tau_i$ are the fluorescence decay times of β-lactoglobulin solutions in the presence of different concentrations of $I^-$ (adapted from [8]).

| $I^-$ (M) | $\tau_1/\tau_0$ | $\tau_2/\tau_0$ | $\tau_3/\tau_0$ |
|---|---|---|---|
| 0 | 1 | 1 | 1 |
| 0.05 | 0.80 | 0.97 | 0.96 |
| 0.25 | 0.50 | 0.93 | 0.88 |
| 0.5 | 0.40 | 0.87 | 0.82 |

The value of the decay times ($\tau_i$) depends on the environmental conditions near fluorophores, such as: environmental polarity (Table 12.1) and the proximity of fluorescence quenchers (Table 12.2). For instance, a decrease in decay times can be ascribed either to the exposure of a fluorophore to a highly polar environment or to an increase in quenching efficiency. Consequently, analysis of the decay times allows for the detection of environmental changes in the vicinity of individual fluorophores, resulting from possible chemical events (structural changes, chemical reactions, chemical interactions between fluorophores or involving fluorophores and quenchers) occurring in the system.

### 12.2.3 Steady-state Fluorescence Anisotropy

When a sample is excited with polarized light (instead of randomly oriented natural light) only the molecules having a transition moment identical to the excitation light beam are excited. Light polarization is achieved by the placement of excitation and emission polarizers before the excitation light beam reaches the sample and after the sample compartment (i.e. before emitted light beam reaches the detector; Figure 12.4).

The rotational mobility of molecules in the excited state will induce the depolarization of light, causing excitation polarized light and fluorescence emission to have different angular displacements.

The fluorescence anisotropy is the fraction of polarized light emitted by a fluorophore when excited with polarized light, as calculated by Equations (12.3) and (12.4):

$$r = \frac{\text{Polarized light}}{\text{Total light}} = \frac{I_{||} - I_{\perp}}{I_{||} + 2I_{\perp}} = \frac{I_{VV} - I_{VH} \times G}{I_{VV} + I_{VH} \times G} \tag{12.3}$$

$$G = \frac{I_{HV}}{I_{HH}} \tag{12.4}$$

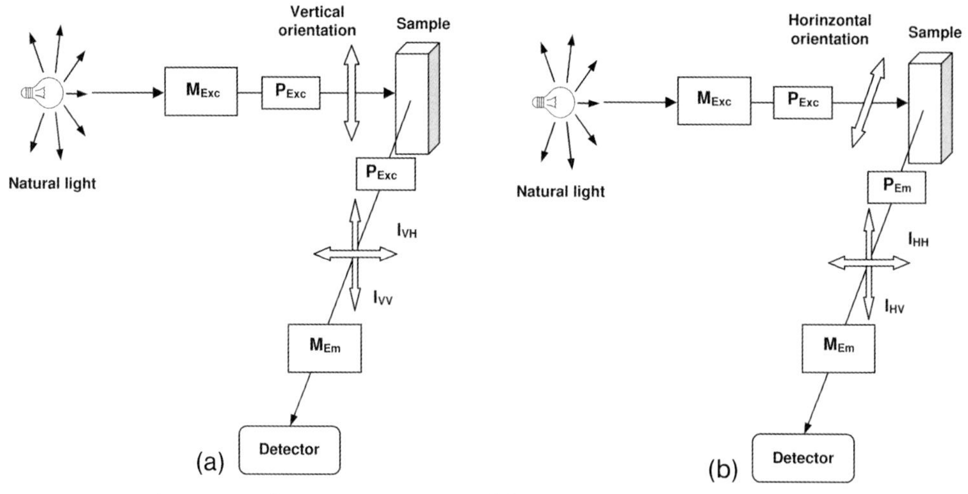


**Figure 12.4** Schematic representation of the configuration of excitation and emission components during acquisition of steady-state fluorescence anisotropy, comprising light source, detector, excitation and emission monochromators ($M_{Exc}$, $M_{Em}$) and polarizers ($P_{Exc}$, $P_{Em}$) and showing the four possible combinations of emission and excitation polarizers: (a) excitation light with vertical orientation, (b) excitation light with horizontal orientation.

where $I_{||}$ and $I_{\perp}$ account for the fractions of parallel ($I_{VV}$ and $I_{HH}$) and perpendicular ($I_{VH}$ and $I_{HV}$) polarized emitted light.

Fluorescence anisotropy can also be calculated by the quotient between the fluorescence intensities obtained at the four different combinations of the excitation and emission polarizer positions, as shown by Equation (12.3), $I_{VV}$ and $I_{HH}$ account for the fluorescence intensity obtained with both polarizers at the vertical and horizontal position, respectively. $I_{HV}$ corresponds to the fluorescence emission obtained with excitation polarizer horizontally and emission polarizer vertically oriented. In contrast, $I_{VH}$ corresponds to the fluorescence emission obtained with excitation polarizer vertically and emission polarizer horizontally oriented. The G factor (Equation 12.4) is a parameter that accounts for the sensitivity of the detection system for vertically and horizontally polarized lights [3].

Fluorescence anisotropy can be correlated with the rotational freedom of the fluorophores present in the sample. High anisotropy values correspond to a low depolarization of the fluorescence emitted light and are indicative of the presence of fluorophores with low rotational freedom. In contrast, low anisotropy values are a consequence of a high depolarization of the emitted light and are commonly ascribed to the presence of highly mobile fluorophores.

When a fluorophore is inserted within a matrix (fluorophores in solid supports or bound to molecular chains, such as polymers or proteins) the fluorescence anisotropy is composed by different rotational contributions: the self rotation of the fluorophore within the matrix and the overall rotation of the complex formed by the fluorophore and the matrix [11]. Consequently, the value of fluorescence anisotropy obtained in the steady-state corresponds to the weighted average fluorescence anisotropy of the system. The rotation of larger particles is generally associated to longer correlation times ($\phi_m$). Correlation times of spherical molecules correlate with their molar volume, $V$, temperature, $T$, and solvent viscosity, $\eta$, through the Stokes–Einstein equation (Equation 12.5):

$$\phi_m = \frac{\eta M_w V}{RT} \tag{12.5}$$

The correlation time parameter, $\phi_m$, is especially relevant in time-resolved anisotropy measurements, when the discrimination of the individual rotation ability of the different chemical species is possible.

### 12.2.4
### Time-resolved Fluorescence Anisotropy

Determination of fluorescence anisotropy decay involves the acquisition of fluorescence decays using distinct combinations of excitation and emission polarizers, in accordance with Equation (12.6):

$$r(t) = \frac{\text{Polarized light}}{\text{Total light}} = \frac{I_{||}(t) - I_{\perp}(t)}{I_{||}(t) + 2I_{\perp}(t)} = \frac{I_{||}(t) - I_{\perp}(t)}{I(t)} \tag{12.6}$$

The discrimination of the individual correlation times, $\phi_m$, and their assignment to the fluorophores, fluorophore aggregates or fluorophore–matrix assemblies can be obtained by solving Equation (12.1) and the following equations (Equations 12.7 and 12.8) [3, 7]:

$$I_{||}(t) = \frac{I(t)}{3}[1 + 2r(t)] = \frac{\sum_i \alpha_i e^{-t/\tau_i}}{3}\left[1 + 2\sum_i r_0 e^{(-t/\theta_{mi})}\right] \tag{12.7}$$

$$I_{\perp}(t) = \frac{I(t)}{3}[1 + 2r(t)] = \frac{\sum_i \alpha_i e^{-t/\tau_i}}{3}\left[1 - \sum_i r_0 e^{(-t/\theta_{mi})}\right] \tag{12.8}$$

Fluorescence anisotropy is also a very sensitive and powerful technique with an enormous potential in several application fields. This technique provides information concerning the mobility of the fluorescence probes, being extremely useful when studying processes that require an understanding of the organization of molecular and supramolecular assemblies, or the monitoring of molecular interactions (adsorption of molecules to surfaces, enzyme–substrate interactions). Fluorescence anisotropy may also find an application for probing structural/conformational changes of molecules (molecular aggregation, molecular unfolding, molecular bond cleavage) [1, 2, 8, 12] and gathering information about the stability and behavior of micelle and microemulsion systems [13].

### 12.2.5
### Steady-state Fluorescence versus Time-resolved Fluorescence Techniques

Steady-state fluorescence techniques are highly sensitive, non-invasive tools that can be used online and in situ, allowing information to be elicited in real time. Steady-state fluorescence offers convenience and speed, being useful in cases where a detailed knowledge of the molecular dynamic behavior is not the primary interest. The inability of steady-state fluorescence to discriminate the presence of distinct fluorophores in different environments is the major limitation of this technique. In this sense, time-resolved fluorescence constitutes a useful complementary tool, allowing for a better comprehension of the dynamic behavior of all the fluorophores present in the system (the participation of a fluorophore in a chemical event or the influence of this event over the structural and chemical properties of each fluorophore in solution). Time-resolved fluorescence is also a non-invasive and extremely sensitive method. However, since it is a time-dependent technique it may not be able to account for the fluorescence contribution of highly quenched fluorophores. Commonly, these fluorophores are characterized by very short decay times, which may lie below the experimental resolution. Identically, molecules or molecular assemblies presenting extremely low correlation times may be not detected by time-resolved techniques, since they may be out of the measuring time window. Steady-state fluorescence measurements can be affected by inner filter effects, which occur when the excitation light intensity is not homogeneously distributed through the whole cuvette and is essentially absorbed by the molecules presented at the cuvette walls. This effect may be caused by high fluorophore concentrations or the presence of other molecules that compete for light absorption. In contrast to steady-state fluorescence, time-resolved fluorescence measurements are independent of the absorbance of the solution, whereas light-scattering effects (frequently observed in turbid solutions) can be easily corrected during analysis of the results.

## 12.3
## Monitoring of Membrane Processes

### 12.3.1
### Steady-state Fluorescence for the Monitoring of Membrane Water Treatment Processes and Membrane Bioreactors

Several compounds exhibit intrinsic fluorescence behavior when exposed to an excitation source. Among the molecules that exhibit fluorescence behavior we may refer to a number of compounds which are intracellular fluorophores, such as tryptophan, NAD(P)H, riboflavin and pyridoxine, as well as a number of other compounds that frequently occur in water sources, included under the generic designation of humic acids (usually poorly characterized).

Taking into consideration that some of these intracellular fluorophores correlate well with the observed behavior of biological systems (growth, activity), a significant amount of work has been dedicated to developing techniques for the monitoring of biological reactors. Traditionally, fluorescence monitoring concentrated on NAD(P)H, as its fluorescence can be used as an indicator for cellular activity and metabolic state [14]. Similarly, tryptophan fluorescence was found to be a reliable indicator of cell concentration in bioreactor systems [14, 15]. However, one of the main problems in fluorometry is that fluorescence signals may be influenced by a whole range of factors, both external (environmental) and intrinsic to the fluorescence method. These factors include: (a) photodegradation of the excited molecule, which can be avoided or reduced by stirring the sample during measurement, (b) quenching of the excited state due to the presence of diverse interacting species within the microenvironment of the reporting fluorophore, (c) absorption of fluorescence either by external components or by the fluorophore itself (inner filter effect), (d) light scattering enhanced by the presence of cells or particles and (e) excitation of other compounds by the radiation emitted by a particular fluorophore (cascade effect) [16]. The combined effect of these phenomena may lead to considerable alterations in the fluorescence spectra of a given fluorophore, making quantitative use of this method extremely difficult.

Only when the concept of multi-fluorophore monitoring was introduced [14] and subsequently extended to two-dimensional (2D) fluorometry [15, 17] were some of the inherent drawbacks of fluorometry overcome. Instead of using the single excitation–emission approach, where the pair of excitation–emission wavelengths corresponding to the maximum emission response is used, the 2D fluorometry approach simultaneously scans a range of selected excitation and emission wavelengths which results in a three-dimensional (3D) fluorescence map (where the coordinates are the excitation wavelength, the emission wavelength and the resulting emission intensities). These 3D maps can then be transformed into 2D plots (surface-projected plots) where excitation–emission pairs with the same emission intensity are linked with iso-intensity contour lines (Figure 12.5). This approach easily takes account of fluorescence shifts and spectra alteration across the whole wavelength range monitored.

These fluorescence maps can therefore be regarded as overall fluorescence "fingerprints" of the system under study. They capture, in a convoluted mode, all the information that a defined system is giving, when inspected with UV/visible radiation. Each fingerprint represents the sum of responses by a series of fluorophores present in the sample and their complex interactions in their specific environment. Therefore, such fingerprints are rather sensitive to any changes, not only in terms of the fluorophore composition of the media, but also in terms of the environmental conditions the fluorophores are exposed to (pH, ionic strength, salt composition). This complexity may be regarded as a problem, because it makes difficult the deconvolution of such embedded information. But, we should recognize that these fingerprints are extremely rich due to the diversity of "responses" they capture. The challenge is how to recover and use the information in these fluorescence maps for process monitoring and for the development of expert control systems.

The second major challenge involves the development of 2D fluorescence probing devices that may allow for in situ, online and non-invasive monitoring of

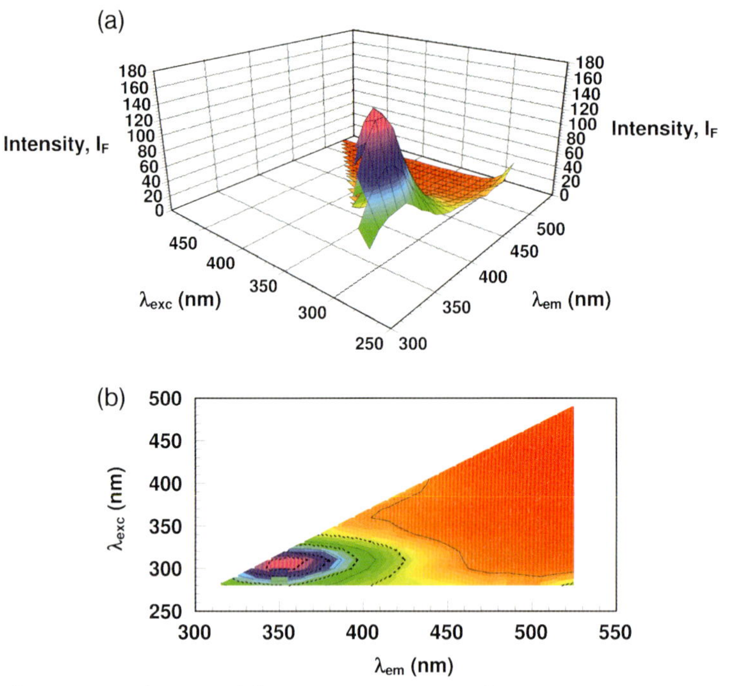


**Figure 12.5** Tri-dimensional fluorescence maps: (a) surface contour plot and (b) projected surface plot, highlighting the iso-intensity contour lines.

the system under study, whether it is a membrane water treatment process, a membrane bioreactor or any other process where fluorophores are present. Operation under these conditions further enhances the potential of the fluorometry technique, as changes in the system can be detected automatically, in real time and without disrupting the process.

Figure 12.6 shows how a simple optical fiber cable can be used for delivering excitation light and capturing emission light [18]. The fluorescence probe is attached to the outer wall of the membrane module or the bioreactor vessel, making it possible to monitor membrane-attached biofilms/biofouling or situations where suspended cells predominate. Fluorescence scans can be taken at different fixed positions along the module or the reactor, accounting for heterogeneities in the biofilm/biofouling structure and the respective physiological status, as well as heterogeneities inside the bioreactor vessel.

The loss of light transmission, inherent to the use of optical fibers, constitutes the major limitation to the application of this technology. This problem is generally minimized through the use of optical fiber bundles, which comprise a large number of excitation- and emission-dedicated optical fibers that merge at the top end of the bundle (the top makes contact with the system to be monitored) in a randomly organized assembly at the tip surface to avoid problems due to system heterogeneities.

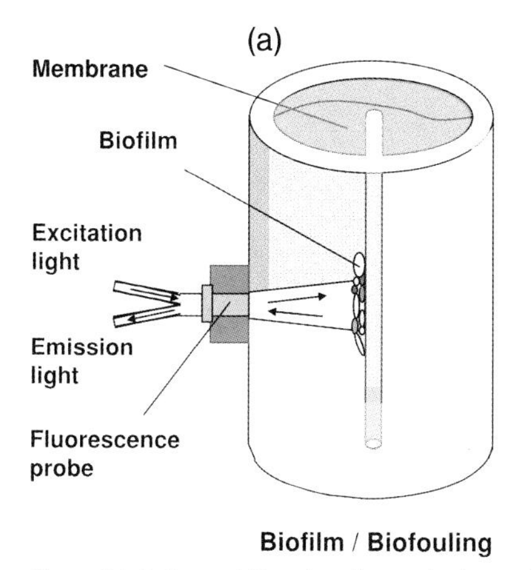


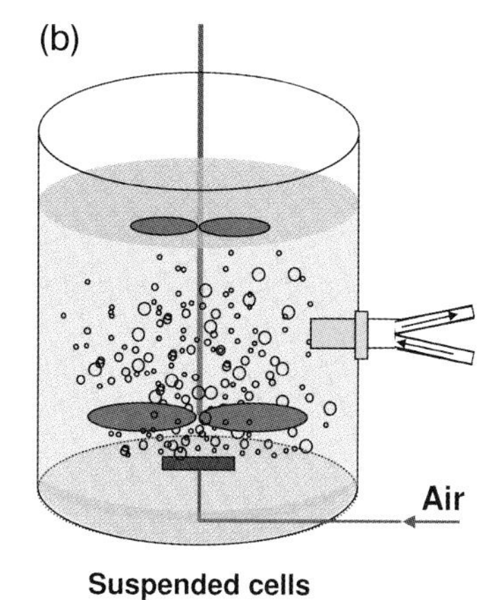


**Figure 12.6** Optical fiber bundle applied in online and in situ 2D fluorescence measurements for monitoring: (a) the biofilm growth at the membrane surface and (b) the performance of membrane bioreactor for water treatment processes.

Two-dimensional steady-state fluorometry can therefore be applied for the characterization of different water sources, such as spring water, surface and groundwater supplies and wastewater streams. As can be easily seen in Figure 12.7, different water streams originate rather different fluorescence maps, where it is possible to identify multiple fluorescence regions and, ultimately, associate some of these regions with the presence of defined constituents.

From Figure 12.7, the presence of humic acids can undoubtedly be identified in the surface water sample inspected. This approach may be associated with liquid chromatography techniques, which allow for sample fractionation, identification and quantification of specific fluorophores that correlate with defined regions of the fluorescence map [19].

Two-dimensional fluorescence can be also applied for the monitoring of membrane bioreactors, due to the ability of this technique to perceive fluorescence differences between feedwater streams, bioreactor media and treated water permeates (Figure 12.8).

The challenge, however, is still how to obtain quantitative data from the fluorescence maps acquired. Although 2D fluorescence maps are able to gather and register the complete fluorescence response in a large region of wavelengths, the information acquired is rather convoluted, due to superposition of the different phenomena mentioned above. Spectra deconvolution techniques have been applied by several authors, with spectra subtraction being the simplest and most generalized [18, 20, 21]. Figure 12.9, shows how spectra subtraction can elicit the response of a membrane bioreactor when exposed to a change in the composition of a target pollutant which is being used as carbon source by a mixed culture able to degrade it. In this example, the concentration of the pollutant 3-chloro-4-methylaniline (3C4MA) in the feed stream was increased from 250 mg/L to 500 mg/L.

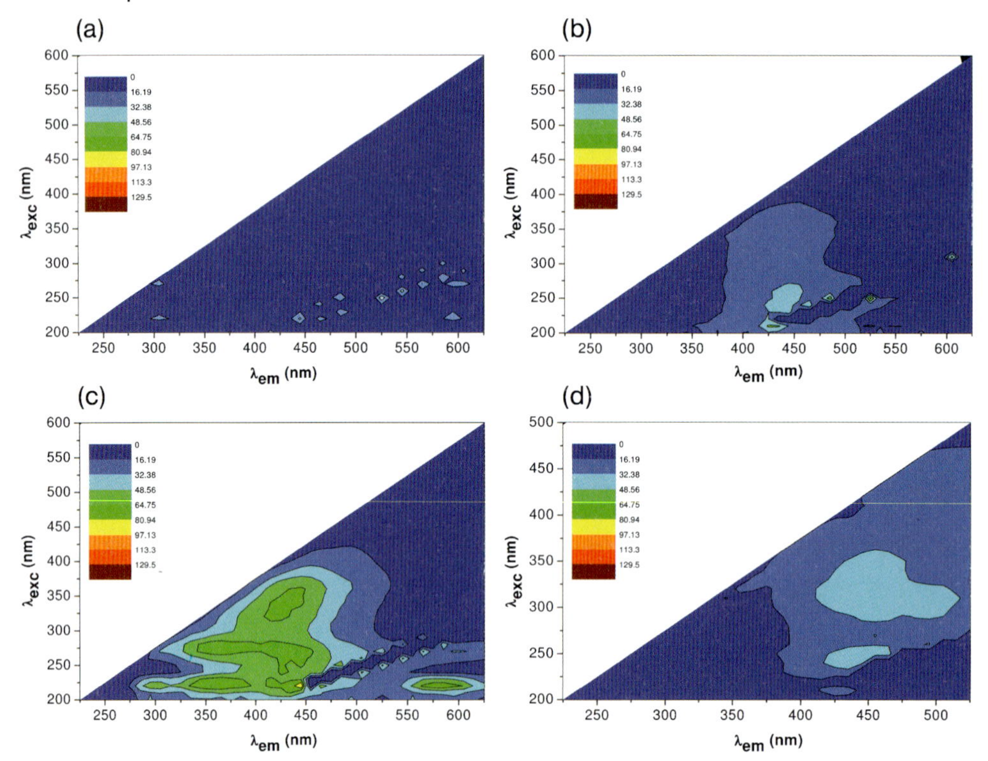


**Figure 12.7** Identification of the presence of humic substances in different water streams through comparison of 3D fluorescence maps acquired online for different streams: (a) spring, (b) surface and (c) wastewaters with the 3D fluorescence map obtained for (d) pure humic acids aqueous solutions.

By subtracting the fluorescence maps obtained under these two feeding concentrations, it was possible to clearly identify an increase in the fluorescence response associated with tryptophan (excitation ~290 nm; emission ~350 nm), which reflects an increase in biomass when using a higher carbon source concentration. Identically, the fluorescence increase at excitation ~360 nm, emission ~430 nm may be attributed to an overall increase in the NAD(P)H signal due to the higher metabolic activity of the cells when exposed to a higher concentration of the carbon source.

However, not all shifts in a fluorescence map can be explained so satisfactorily, which makes clear the limitations of the spectra subtraction method. A general drawback of spectra subtraction, applied in this context, is its qualitative, interpretative character at best. Hence, an improved spectra interpretation method would aim at deconvoluting fluorescence data such that they may be used for quantitative monitoring of system performance, thereby utilizing as much information as possible from the fluorescence maps.

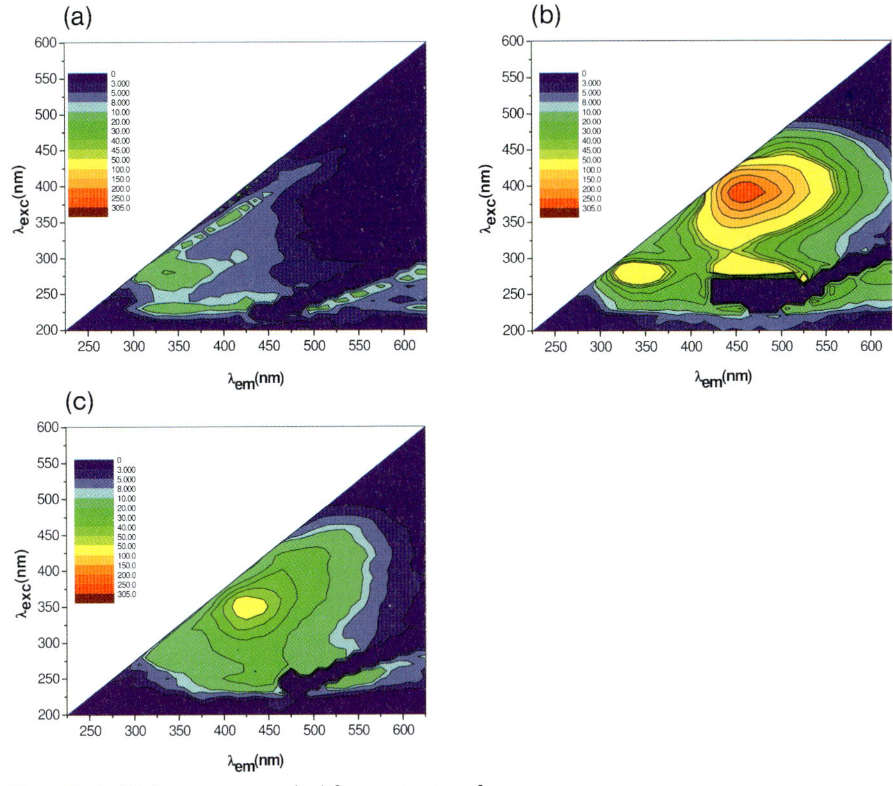


**Figure 12.8** 2D fluorescence applied for monitoring of membrane bioreactor in wastewater treatment. 3D fluorescence maps were acquired for: (a) feedwater stream, (b) bioreactor media and (c) treated water permeates.

It was shown that individual analysis of fluorophore peaks may fail to utilize a large part of the convoluted and contextual information contained in the fluorescence maps. Another difficulty is that the maps may be distorted due to changes in environmental conditions and other effects, as discussed above. To overcome these obstacles, it is necessary to apply a holistic approach to fluorescence spectra interpretation, which analyzes the entire fluorescence fingerprint in a contextual manner, without particular regard to the individual fluorophores involved in the process, which in fact does represent an "educated bias" approach that restricts full analysis of the data. This new approach utilizes much more information from the fluorescence map, as it includes not only all fluorescence peaks but also regions outside the immediate peak areas. The transition from qualitative analysis to quantitative monitoring is achieved through the association of all fluorescence maps acquired with relevant process performance parameters determined independently, such as time-evolving membrane permeability data (or instantaneous membrane flux) and the concentration of target compounds in the permeate (or rate of conversion of target compounds in the system).

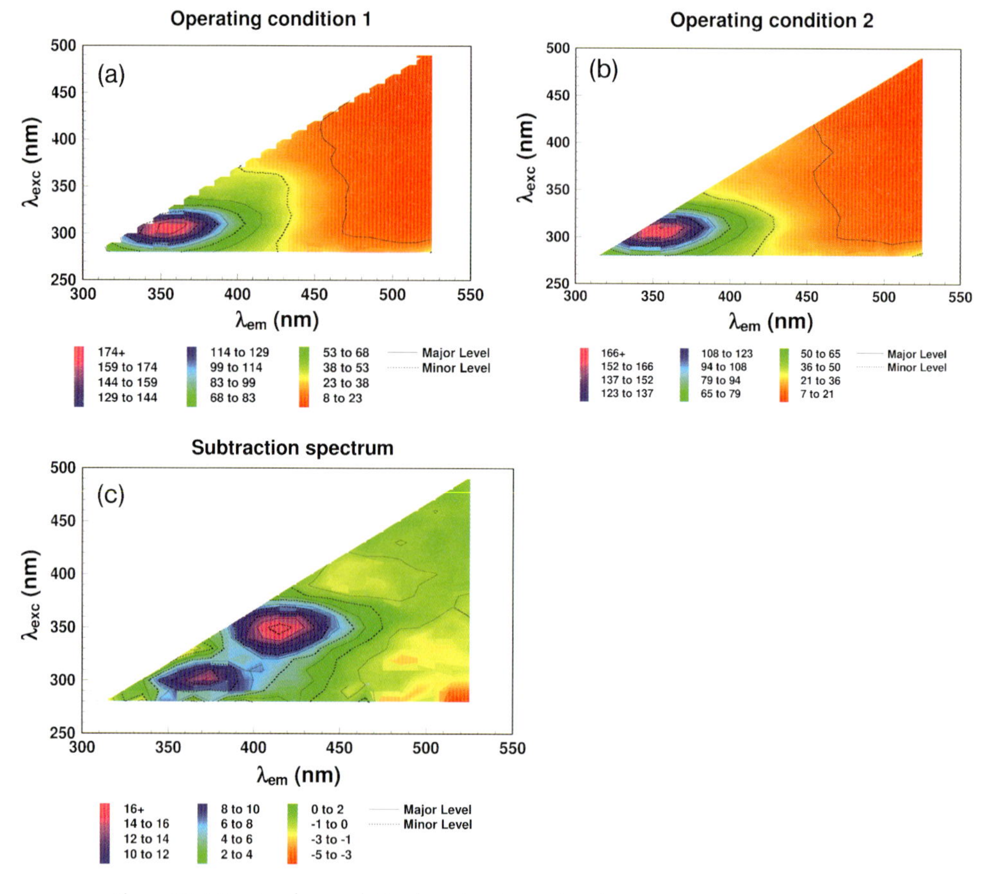


**Figure 12.9** Spectra deconvolution by using the subtraction technique to elicit the response of membrane bioreactor when exposed to different concentrations of pollutant (3-chloro-4-methylaniline). Fluorescence spectra acquired: (a) in the presence of 500 mg/L of pollutant, (b) in the presence of 250 mg/L of pollutant and (c) subtraction fluorescence spectrum.

The mathematical tools to perform such task may use the operating parameters of the system under study as input variables (pH, temperature, feed composition, feed flow rate, aeration rate, etc.) together with the fluorescence maps acquired, correlating them with relevant output process performance variables (such as membrane permeability, concentration of target compounds in the permeate, see above). Different mathematical tools have been proposed, such as stepwise multiple regression [15, 17] principal component analysis [22], backpropagation artificial neural networks [23] and feedforward artificial neural networks [18]. This latter approach is represented in Figure 12.10. As shown in this work [18] by using the

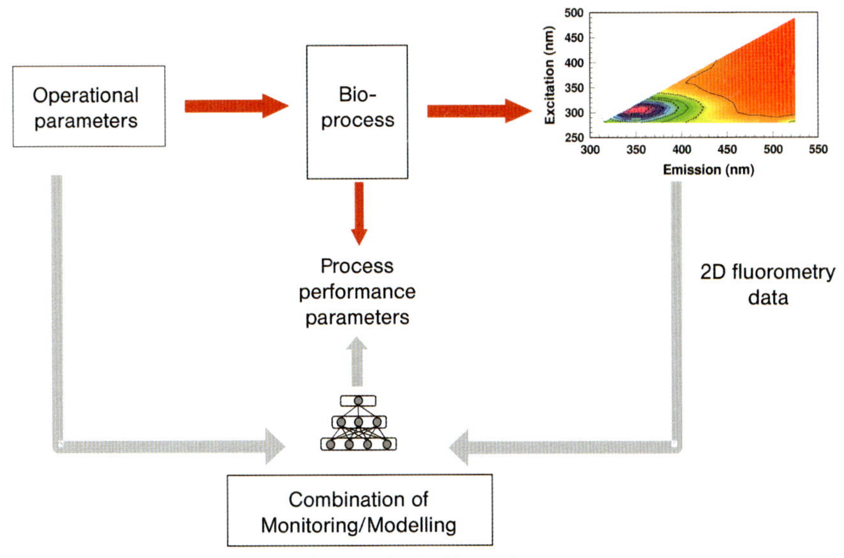


**Figure 12.10** Schematic diagram showing the feedforward artificial neural networks approach applied for eliciting information from 3D fluorescence maps.

validated artificial neural networks as a softsensor, this methodology has the potential of replacing offline membrane biofilm/biofouling monitoring by online 2D fluorescence monitoring, as process performance parameters determined offline can be inferred from the fluorescence maps. Similar approaches can be developed by using other mathematical tools such as projected latent structures (PLS).

## 12.3.2 Natural Fluorescence Techniques for Monitoring the Membrane Processing of Biological Molecules

As briefly mentioned in Section 12.1, one of the most interesting features of natural fluorescence results from the fact that the fluorescence response of a given molecule depends very much on its microenvironment. This feature can be used in order to gather information about the structure of complex molecules such as polypeptides and proteins, which may integrate several fluorescent amino acids residues, such as tryptophan, tyrosine and phenylalanine. Among these, tryptophan exhibits the highest quantum yield, which makes it a good candidate to be used as an intrinsic fluorescence reporter.

The basic concept is the use of the fluorescence response of tryptophan residues, embedded in the polypeptide/protein structure, which is sensitive to changes in its microenvironment. Therefore, if a given protein interacts with a membrane surface and, due to this interaction, changes its 3D structure, it can be anticipated that the relative position of the tryptophan residue(s) may be altered. Such changes may include:

1. A higher exposure of a buried tryptophan to the surrounding solvent (usually water) due to unfolding processes, which leads to contact with a more hydrophilic environment.
2. The opposite process, involving the movement of a tryptophan residue to a more buried, hydrophobic environment.
3. An increase in tryptophan residue mobility, which occurs when the protein assumes a more unfolded conformation in the region where the tryptophan is located.
4. A change in the relative position of a tryptophan residue towards internal protein quenchers, such as aspartate residues and disulfide bonds, which may occur as a result of processes of folding/unfolding.

All of these alterations, which may be induced when a protein molecule interacts with a membrane surface either during convective (filtration) processes or under diffusive conditions, can be reported by intrinsic fluorescence probes such as tryptophan residues and detected by using appropriate fluorescence techniques. In fact, the use of these techniques and the correct interpretation of their response is only possible when the number of tryptophan residues present in the protein is relatively low (one or two) because, otherwise, it becomes extremely difficult to assign a given fluorescence response to the corresponding tryptophan, limiting the interpretation effort.

During permeation, proteins are exposed to different processing conditions, such as shear stress (e.g. when permeating the membrane pores) and interaction with the membrane and the membrane pore surface, which may induce reversible or irreversible changes in the protein structure. In fact, the permeation of protein molecules through membranes with a molecular weight cut-off lower than their mass has been reported. This permeation may be attributable either to pore size distribution (e.g. the presence of pores with a higher diameter than the nominal cut-off), or to the shape and orientation of the protein, which may favor its passage through the membrane pores [24]. However, since distinct conformational states of proteins possess small energetic differences, the hypothesis of protein structural alteration (e.g. molecular elongation) during the permeation process (which facilitates their passage through the membrane pores) has to be considered.

The impact of parameters such as protein mass to membrane cut-off ratio ($\lambda$) or the effect of the membrane material on their structure/function during the permeation process is currently unknown and constitutes an essential requisite for the full development and implementation of membrane processes for the selective fractionation of proteins.

Permeation studies have been performed using membranes of different materials (regenerated cellulose, RC; polyethersulfone, PES), with a distinct cut-off, at different transmembrane pressures (TMP) [1, 2, 25]. β-Lactoglobulin and horseradish peroxidase were used as model proteins for studying the impact of

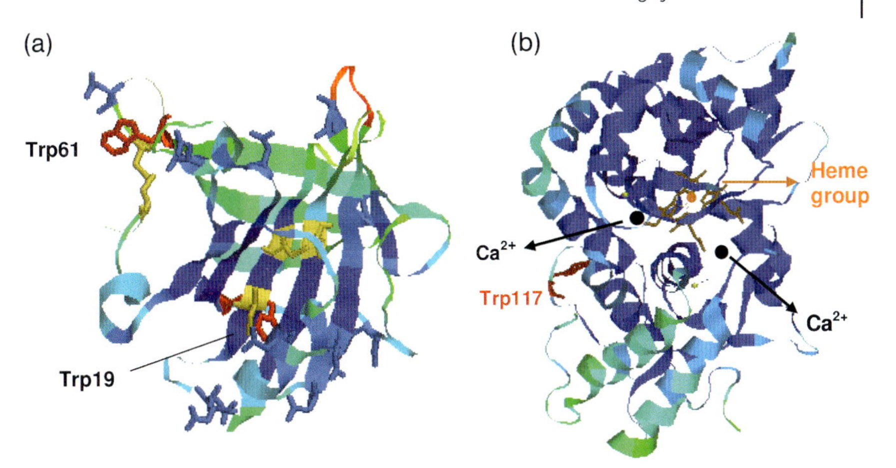


**Figure 12.11** (a) Schematic representation of β-lactoglobulin molecule highlighting the relative positions of tryptophans, Trp19 and Trp61 (red color) and the quenchers aspartate (at blue color) and disulfide bonds (orange color). (b) Schematic representation of the horseradish peroxidase (HRP-4C) molecule highlighting the relative positions of tryptophan, Trp117 (red color), heme group and the two structural calcium ions, $Ca^{2+}$ (black color).

membrane processing by ultrafiltration on possible changes in their structural characteristics and, eventually, their function/activity. The structural alterations induced were monitored by using complementary fluorescence techniques previously described (steady-state/time-resolved fluorescence, steady-state fluorescence anisotropy), using tryptophan as a structural reporter. Steady-state fluorescence and time-resolved fluorescence allow for the identification of changes of the environment in the vicinity of tryptophans (e.g. changes in the environment polarity), while changes in fluorescence anisotropy can be correlated with alterations in tryptophan mobility.

β-Lactoglobulin is a globular protein with unique structural characteristics, which have been very well described, making it an excellent model for processing studies. β-Lactoglobulin has two tryptophans (Trp19, Trp61) located in very distinct environments. Trp19 is located within the protein core (hydrophobic region), while Trp61 is placed at the exterior surface of β-lactoglobulin, in close contact with two strong quenchers: an aspartate residue and a disulfide bond (Figure 12.11a). β-Lactoglobulin solutions contain a mixture of monomers, dimers and higher-order oligomers whose proportion changes depending on the treatment applied, the pH and the ionic strength of the solution and the protein concentration.

In Figure 12.12, it is interesting to observe that no noticeable fluorescence changes are observed for the retentate solutions obtained after ultrafiltration with membranes having distinct cut-offs. These results seem to indicate that the rejected β-lactoglobulin molecules do not undergo significant structural changes.

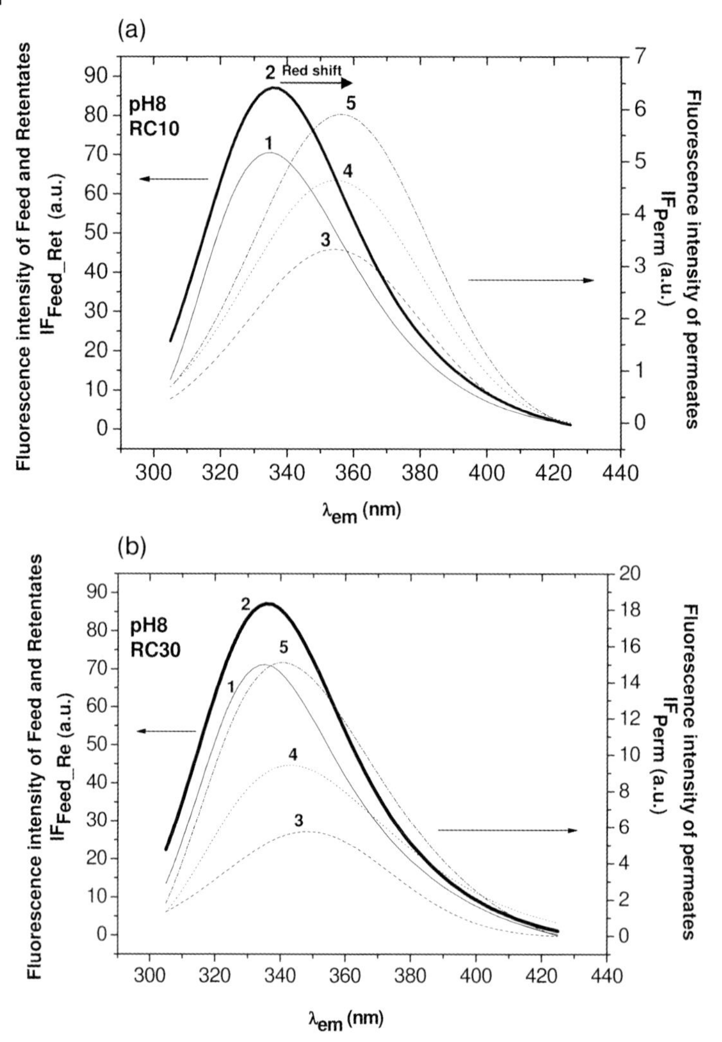


**Figure 12.12** Fluorescence emission spectra of β-lactoglobulin solutions, at pH 8, in a 10 mM phosphate buffer, acquired at $\lambda_{em} = 290$ nm, for: (1) feed, (2) retentate and permeate solutions collected after ultrafiltration with RC membranes of (a) 10 kDa (RC10) and (b) 30 kDa (RC30) at different TMP values: (3) 0.1 bar (10 kDa) or 0.025 bar (30 kDa), (4) 0.5 bar and (5) 1.0 bar.

However, the emission fluorescence spectra obtained for the β-lactoglobulin permeate solutions (acquired upon ultrafiltration using the 10 kDa membrane) reveals a substantial broadening of the emission band accompanied by a red shift in the fluorescence emission maximum. In contrast, the permeation of β-lactoglobulin solutions using membranes of the same material with 30 kDa cut-off leads to a less substantial broadening of the fluorescence emission bands, without noticeable red shifts in the emission maximum. These results indicate that the permeation of β-lactoglobulin using membranes with smaller membrane pores induces more significant alterations in the structure of β-lactoglobulin. It is also interesting to note that the magnitude of the emission band broadening and the red shift of the emission maximum, obtained for the permeates collected upon ultrafiltration with the 10 kDa membrane, are constant for all TMP values.

The fluorescence decays obtained for feed solutions of β-lactoglobulin are well fit by tri-exponential functions [8].

The time decay constants determined by time-resolved fluorescence can be attributed to the two tryptophan residues of β-lactoglobulin: Trp 19 more buried in the protein structure, in a more hydrophobic environment although less quenched, is mostly responsible for the two longest decay times, while Trp 61 located in a more exposed hydrophilic region but highly quenched, contributes the most to the shortest time decay observed [8]. The fluorescence decays of permeates obtained after ultrafiltration with 30 kDa membranes and retentates are still acceptably fit by the sum of three exponentials. In addition, the values of the decay times obtained for these solutions do not change appreciably (Table 12.3). This provides evidence that the structural characteristics of β-lactoglobulin are not significantly affected by ultrafiltration with 30 kDa membranes (at least in the vicinity of the fluorescence probes used: Trp19, Trp61). However, a different conclusion is drawn from the fluorescence decays obtained for permeates collected after ultrafiltration with 10 kDa membranes. Acceptable fits of the fluorescence decays of these permeate solutions are only achieved with the sum of four exponentials. The presence of a fourth longest component is attributed to the increase of structural heterogeneity of the protein solutions. That is, the structural alterations induced on β-lactoglobulin during permeation through 10 kDa membranes, generate the formation of a mixture where folded and unfolded β-lactoglobulin structures coexist [1].

These results are quite consistent with the behavior observed by steady-state fluorescence and strongly suggest that, when β-lactoglobulin is processed with a membrane with larger pores, the impact on the protein structure is minimal while, when a membrane with tighter pores is used, this impact may become relevant.

Measurements of steady-state fluorescence anisotropy were also performed for feed, retentate and permeate solutions of β-lactoglobulin obtained with both membranes (10 kDa, 30 kDa) at different TMP values. The analysis of the fluorescence anisotropy data (Figure 12.13) shows, in comparison with the fluorescence anisotropy obtained for the feed solutions, a decrease in fluorescence anisotropy observed at all applied TMP values.

This behavior is compatible with an increase in the mobility of tryptophans upon permeation using both membranes. The decrease in fluorescence anisotropy observed for the permeate solutions collected with the 10 kDa membrane clearly indicates that permeation through the 10 kDa membrane induces molecular unfolding. The decrease in fluorescence anisotropy was considerably lower upon permeation of β-lactoglobulin through a 30 kDa membrane, indicating, in

**Table 12.3** Fit of the fluorescence decays obtained for solutions of β-lactoglobulin in feed, retentate and permeates collected after ultrafiltration at different TMP values, using regenerated cellulose (RC) membranes with a nominal cut-off of 10 kDa and 30 kDa. $\tau_i$ and $A_{Ni}$ correspond to decay times and amplitudes of the different tryptophan residues, while $\chi^2$ accounts for the quality of the fits [1].

| | $\tau_i$ (ns) | $\tau_2$ (ns) | $\tau_3$ (ns) | $\tau_4$ (ns) | $A_{N1}$ | $A_{N2}$ | $A_{N3}$ | $A_{N4}$ | $\chi^2$ |
|---|---|---|---|---|---|---|---|---|---|
| Feed solution | 0.11±0.03 | 0.89±0.13 | 2.01±0.20 | – | 0.17±0.05 | 0.49±0.07 | 0.34±0.06 | – | 1.08 |
| TMP (bar) | | | | | | | | | |
| | Permeation using RC membrane with 10 kDa cut-off | | | | | | | | |
| 0.025 | 0.15 | 0.66 | 2.25 | 5.43 | 0.16 | 0.28 | 0.47 | 0.09 | 1.26 |
| 0.5 | 0.11 | 0.58 | 1.91 | 4.44 | 0.16 | 0.28 | 0.43 | 0.13 | 0.99 |
| 1 | 0.24 | 0.74 | 2.10 | 4.9 | 0.16 | 0.28 | 0.45 | 0.11 | 1.19 |
| | Permeation using RC membrane with 30 kDa cut-off | | | | | | | | |
| 0.1 | 0.09 | 0.81 | 2.17 | – | 0.25 | 0.52 | 0.23 | – | 0.94 |
| 0.5 | 0.10 | 0.72 | 1.93 | – | 0.15 | 0.52 | 0.33 | – | 1.02 |
| 1 | 0.20 | 0.91 | 2.17 | – | 0.17 | 0.60 | 0.23 | – | 1.08 |
| Retentate | 0.22 | 0.97 | 2.14 | – | 0.19 | 0.55 | 0.26 | – | 1.13 |

agreement with that found for steady-state and time-resolved fluorescence, the presence of less significant structural alterations.

In addition, as shown in Figure 12.13 during permeation with 10 kDa membranes, the variation of the fluorescence characteristics of β-lactoglobulin (i.e. the structural alterations undergone by β-lactoglobulin) is independent from the applied TMP. In contrast, the profile of fluorescence changes obtained for permeates collected using the 30 kDa membrane reveals a relationship between the fluorescence data and TMP. In this case, the changes in fluorescence (steady-state fluorescence, steady-state fluorescence anisotropy) became more substantial with the decrease in TMP (i.e. increase in the length of permeation time: the average time needed for a protein to cross the membrane pores). The decrease in the shear stress inside the membrane pores, as a consequence of lower pore constriction (30 kDa membrane), leads to a more visible contribution of the effects associated with membrane–protein interactions to the structural changes of β-lactoglobulin. The possibility of membrane–protein interactions increases at higher permeation time lengths, that is the contribution of this effect is more relevant at lower TMP values, thus explaining the higher structural alterations observed under these conditions.

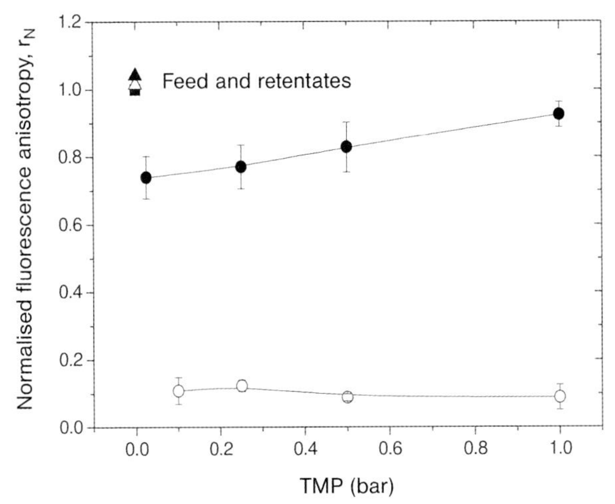


**Figure 12.13** Variation of the normalized steady-state fluorescence anisotropy, $r_N$, acquired at $\lambda_{exc} = 290$ nm and $\lambda_{em} = 350$ nm for β-lactoglobulin solutions, obtained after ultrafiltration with RC membranes of 10 kDa (○) and 30 kDa (●), at different TMP values, with the respective retentates (△,▲), in respect to feed (■).

The relevance of membrane–protein interactions was particularly evident during studies performed with horseradish peroxidase [25]. Peroxidases are heme enzymes that participate as catalysts in oxidative reactions where hydrogen peroxide is commonly the electron acceptor. The C isoenzyme of horseradish peroxidase is a monomeric glycoprotein with a molecular mass of ~44 kDa. This protein contains a prosthetic heme group buried in the central region of the protein (Figure 12.11b). The heme group is constituted by a protoporphyrin ring coordinated by an iron ion whose formal charge changes between +3 and +5, along the course of enzymatic reactions. Moreover, this enzyme contains four disulfide bridges and two binding $Ca^{2+}$ ions, proximal and distal to the heme, that do not directly participate in catalyzed reactions, but play an important role in sustaining the protein structure, thus assuring the enzymatic activity of horseradish peroxidase. The HRP-4C isoenzyme is a single tryptophan protein (Trp117), and therefore, it is an interesting model for investigating the impact of processes (such as ultrafiltration) on its 3D structure using fluorescence techniques.

As can be seen in Figure 12.14, membranes with the same nominal cut-off of 30 kDa, but made of different hydrophilic materials (RC, PES), have a quite distinct impact when used for processing HRP-4C solutions by ultrafiltration.

Processing with a 30 kDa RC membrane induces a strong red shift in the steady-state fluorescence spectra of the permeates collected. This red shift may be interpreted as a result of a higher exposure of Trp 117 to the aqueous environment

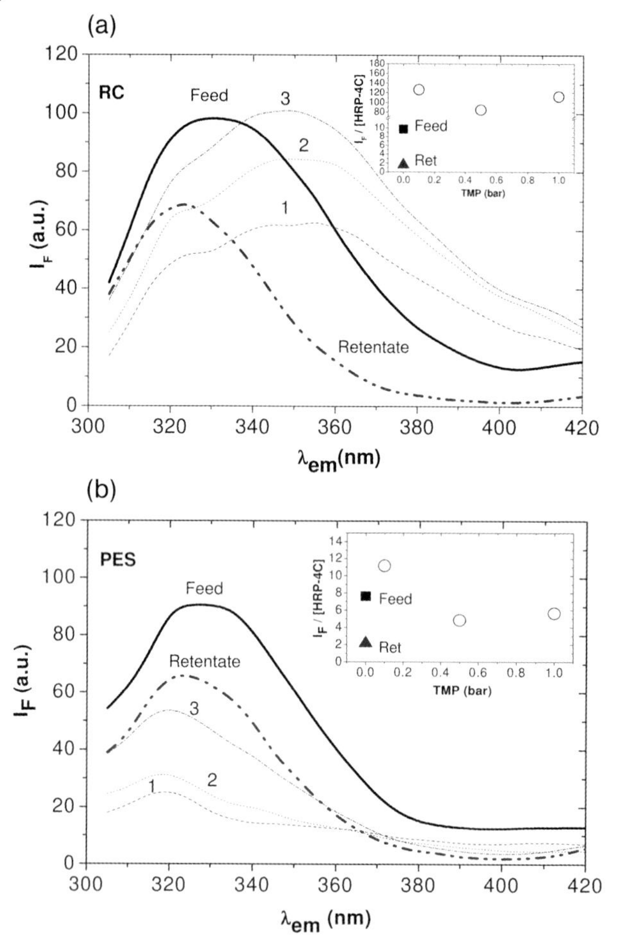


**Figure 12.14** Fluorescence emission spectra of HRP-4C solutions acquired at $\lambda_{exc} = 290$ nm, for (—) feed, (–·–·–) retentate and permeate solutions collected after ultrafiltration with (a) RC and (b) PES membranes, both with a cut-off of 30 kDa, at different TMP values: (1) 0.1 bar, (2) 0.5 bar and (3) 1.0 bar. Insets Variation of concentration-normalized fluorescence intensities ($I_F$/[HRP-4C]), obtained, at $\lambda_{em} = 350$ nm, in the feed (■), retentates (▲) and permeates (O) collected at different TMP values.

due to protein unfolding, which is confirmed by the decrease in fluorescence anisotropy observed, which is associated with a higher mobility of Trp 117 (Figure 12.15a). Also, in the inset of Figure 12.14, it is possible to see an increase in fluorescence intensity, which may result from a decrease of the quenching efficiency over Trp 117 due to unfolding of HRP-4C, after permeation.

When we observe the steady-state fluorescence spectra (Figure 12.14b) and the fluorescence anisotropy response (Figure 12.15a) of HRP-4C, after processing with the 30 kDa PES membrane, we may conclude that the permeating proteins remain

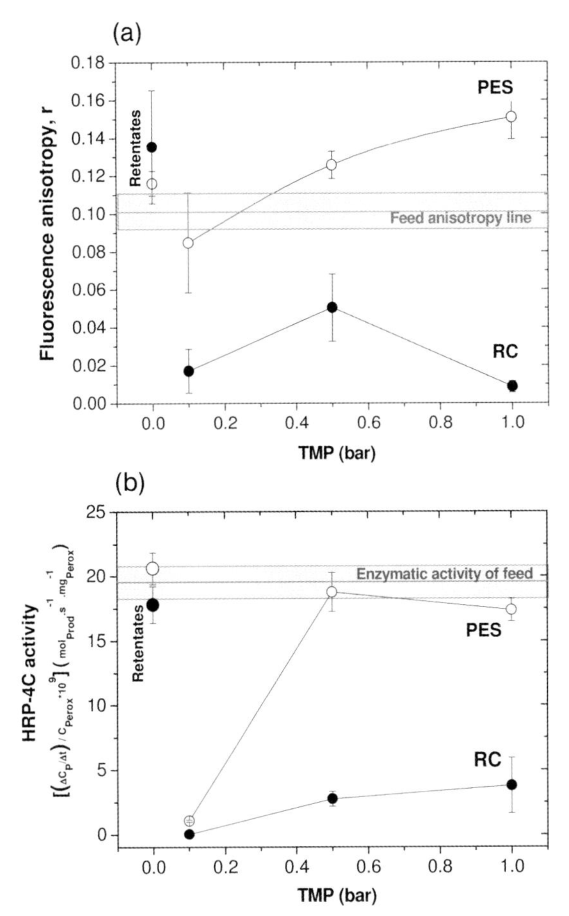


**Figure 12.15** (a) Variation in the steady-state fluorescence anisotropy, r, acquired at $\lambda_{exc} = 290\,\text{nm}$ and $\lambda_{em} = 350\,\text{nm}$ for horseradish peroxidase (HRP-4C) solutions obtained after ultrafiltration with PES (O) and RC (●) membranes, both with a cut-off of 30 kDa, at different TMP values, with the respective retentates. (b) Variation of the HRP-4C activity obtained after ultrafiltration using PES (O) and RC (●) membranes, at different TMP values. The fluorescence anisotropy, r, and the HRP-4C activity of the feed are represented by the gray line (the associated error is expressed by the gray box).

essentially unaffected, exhibiting no remarkable changes in their fluorescence spectra and anisotropy (an inversion to this fluorescence behavior is only observed for the first permeate solution, obtained at 0.1 bar). It is extremely interesting to notice that the activity of this enzyme follows very closely the results observed by fluorescence (Figure 12.15b). The horseradish peroxidase permeates processed with the 30 kDa PES membrane exhibit no change of enzymatic activity when compared with the corresponding feed solutions (permeate obtained at 0.1 bar is maintained as the exception), while the permeates obtained with the 30 kDa RC membrane display a significant reduction of activity, which was found to result from the loss of $Fe^{3+}$ affecting the function of the heme group of the enzyme [25]. Since the heme group is buried within the protein core, molecular unfolding is a necessary condition to allow interactions between the membrane surface and this ion. This fact, together with the different chemistries of the membrane surface (that regulates the affinity of $Fe^{3+}$ towards each membrane) explains the absence of structural/enzymatic activity alterations found for the retentates obtained with both membranes (RC, PES). The opposite behavior found for permeate at 0.1 bar was due to the higher depletion of $Fe^{3+}$ before saturation of the PES membrane, during the first stage of operation.

The fluorescence approach discussed provides information that may contribute to the development of new membrane materials more suitable for protein permeation, in terms of their surface chemistry and structural characteristics with the purpose of minimizing the negative impact of processing on protein function.

## 12.4
## Concluding Remarks

The case studies presented in this chapter are very much based on the personal experience and research interests of the authors and they should illustrate, not limit, the high potential of using natural fluorescence techniques for the monitoring of membrane processes. The examples selected reflect, however, the authors' perception about the areas of membrane research where natural fluorescence techniques may contribute to a better monitoring of membrane processes and, ultimately, to a better comprehension of the phenomena taking place when complex media interact with membranes.

The use of 2D fluorescence techniques for the monitoring of membrane processes, namely the ones involving water treatment and biological systems such cell culture/membrane bioreactors, may be adopted soon. The applications envisaged (some of them under study for which results were not shown) involve:

1. The 2D monitoring of membrane bioreactors at multiple locations, including the feed stream inlet, the membrane bioreactor itself at various positions, the permeate stream and membrane inspection during membrane autopsy procedures;
2. Online, early detection of water contamination with microbial cells in membrane drinking water processes;
3. Monitoring of cell culture, including animal cell cultures and tissue engineering, making use of the large variety of intracellular fluorophores present.

The combination of 2D fluorometry with adequate mathematical tools, which make possible the correlation of complex and convoluted fluorescence maps with process variables, is an essential step for a powerful use of the information which has been captured but is not easily accessible. The ultimate goal is the development of expert systems, softsensors, which make possible a more effective monitoring and process control, online, in real time and in a non-invasive mode.

The combined use of different natural fluorescence techniques, such as steady-state fluorometry, fluorescence anisotropy and time-decay fluorescence, has been revealed to be quite powerful. The use of these techniques in an integrated mode for the monitoring of membrane–protein interactions is only in its infancy. These techniques offer not only the possibility to study the interaction of proteins with membranes, under convective and diffusive conditions, but also they may be easily extended to studies involving proteins and other porous materials such as chromatography media. The areas of application of these techniques will range from polypeptide and protein fractionation to the monitoring of hemodialysis systems.

## References


1 C.A.M. Portugal, J.G. Crespo, J.C. Lima, *J. Membrane. Sci.* **2007**, *300*, 211–223.
2 C.A.M. Portugal, J.C. Lima, J.G. Crespo, *J. Membrane. Sci.* **2008**, *321*, 69–80.
3 B. Valeur, *Molecular Fluorescence, Principles and Aplications*, Wiley-VCH, Weinheim, **2002**.
4 M. Eftink, in: J.R. Lakowicz (ed.), *Topics in Fluorescence Spectroscopy, Vol. 2*, Plenum, New York, **1991**.
5 R.J. Lakowicz, *Principles of Fluorescence Spectroscopy*, Plenum, New York, **1983**.
6 T.Q. Faria, J.C. Lima, M. Bastos, A.L. Maçanita, H. Santos, *J. Biol. Chem.* **2004**, *279*, 48680–48691.
7 J. Beechem, L. Brand, *Annu. Rev. Biochem.* **1985**, *54*, 43–71.
8 C.A.M. Portugal, J.G. Crespo, J.C. Lima, *J. Photochem. Photobiol. B Biol.* **2006**, *82*, 117–126.
9 J.R. Alcala, E. Gratton, F.G. Prendergast, *Biophys. J.* **1987**, *51*, 597–604.
10 J.R. Alcala, E. Gratton, F.G. Prendergast, *Biophys. J.* **1987**, *51*, 925–936.
11 J.R. Lakowicz, B.P. Maliwal, H. Cherek, A. Balter, *Biochemistry* **1983**, *22*, 1741–1752.
12 J.M.G. Martinho, A.M. Santos, A. Fedorov, R.P. Baptista, M.A. Taipa, J.M.S. Cabral, *Photochem. Photobiol.* **2003**, *78*, 15–22.
13 N.E. Levinger, *Curr. Opin. Colloid Interface Sci.* **2000**, *5*, 118–124.
14 J.-K. Li, A.E. Humphrey, *Biotechnol. Bioeng.* **1991**, *37*, 1043–1049.
15 B. Tartakovsky, M. Sheintuch, J.-M. Hilmer, T. Scheper, *Biotechnol. Prog.* **1996**, *12*, 126–131.
16 G. Wolf, J.G. Crespo, M.A.M. Reis, *Environ. Sci. Biotechnol.* **2002**, *1*, 227–251.
17 B. Tartakovsky, L.A. Lishman, R.L. Legge, *Water Res.* **1996**, *30*, 2941–2948.
18 G. Wolf, J.S. Almeida, C. Pinheiro, V. Correia, C. Rodrigues, M.A.M. Reis, J.G. Crespo, *Biotechnol. Bioeng.* **2001**, *72*, 297–306.
19 N. Her, G. Amy, D. McKnight, J. Sohn, Y. Yoon, *Water Res.* **2003**, *37*, 4295–4303.
20 C. Lindemann, S. Marose, H.O. Nielsen, T. Scheper, *Sens. Actuactors B* **1998**, *51*, 273–277.
21 S. Marose, C. Lindemann, T. Scheper, *Biotechnol. Prog* **1998**, *14*, 63–74.
22 R. Ferrer, J. Guiteras, J.L. Beltrán, *Anal. Chim. Acta* **1999**, *384*, 261–269.
23 T.J. McAvoy, H.T. Su, M.S. Wang, M. He, J. Horvath, H. Semerjian, *Biotechnol. Bioeng.* **1992**, *40*, 53–62.
24 J.G. Crespo, M. Trotin, D. Hough, J.A. Howell, *J. Membrane Sci.* **1999**, *155*, 209–230.
25 C.A.M. Portugal, J.C. Lima, J.G. Crespo, *J. Membrane Sci.* **2006**, *284*, 180–192.

# 13 Membrane Emulsification Processes and Characterization Methods*

*Gabriela G. Badolato, Barbara Freudig, Ping Idda, Uwe Lambrich, Helmar Schubert, and Heike P. Schuchmann*

## 13.1 Introduction

Many liquid products are emulsions, systems of at least two or more immiscible or poorly miscible liquid phases, for example oil and water. Examples of these kinds of products are milk, butter, mayonnaise and cosmetic creams.

The preparation of an emulsion by mechanical processes comprises basically three steps: bringing the immiscible liquids together (premix), breaking up the premixed droplets by high energy input and stabilizing them, using additives like thickeners and emulsifiers [1, 2].

Many types of emulsification equipment are widely applied in industry, such as high pressure homogenizers and rotor–stator systems. In these machines the premix droplets are deformed and disrupted in the flow field of the emulsification device [1]. In addition to these techniques, alternative methods for the production of emulsions using microporous devices have been developed since the early 1990s.

In membrane emulsification processes, one phase (future disperse phase) is usually pressed through the pores of a membrane into the other phase (continuous phase) [3]. A different approach is the disruption of large droplets by pushing an emulsion premix through the pores of a membrane [4]. Furthermore, special shapes of the pore outlet of microchannel modules allow the production of small droplets due to a special detachment mechanism [5–7].

Membrane emulsification have several advantages compared with conventional emulsification processes, such as the low energy input required to produce emulsions [8] and the possibility to get emulsions with extremely narrow droplet size distributions [9]. However, there are also some problems related to membrane emulsification, like membrane fouling and low product output.

* A list of nomenclature is given at the end of this chapter.

*Monitoring and Visualizing Membrane-Based Processes*
Edited by Carme Güell, Montserrat Ferrando, and Francisco López

ISBN: 978-3-527-32006-6

Membrane emulsification processes can be directly visualized by microscope as well as by the use of high-speed cameras. In this case, information can be obtained about droplet disruption [10, 11] and fouling of the membrane. An indirect characterization method is the (inline) measurement of the emulsion characteristics. The emulsion is mainly characterized by its droplet size and droplet size distribution [2]. These influence important product characteristics like structure, mouthfeel, color and appearance, texture and viscosity [12, 13].

This chapter gives some information about membrane emulsification technology. Moreover, as the main topic of this work, specific membrane and microporous emulsification processes and methods for their characterization are presented.

## 13.2 Emulsification Technology

### 13.2.1 Emulsions

The basic types of emulsions are *oil in water* (o/w) emulsions, where oil droplets are dispersed in a continuous aqueous phase and *water in oil* (w/o) emulsions containing water droplets dispersed in oil (Figure 13.1). By dispersing an emulsion into a further continuous phase, a multiple or double emulsion can be obtained. A w/o emulsion dispersed into a second water phase leads to an *water in oil in water* (w/o/w) emulsion, whereas an o/w emulsion can be used to produce an *oil in water in oil* (o/w/o) emulsion by dispersing it into oil.

An important characteristic of an emulsion is its physical stability. This can be achieved by use of emulsifiers and stabilizers.

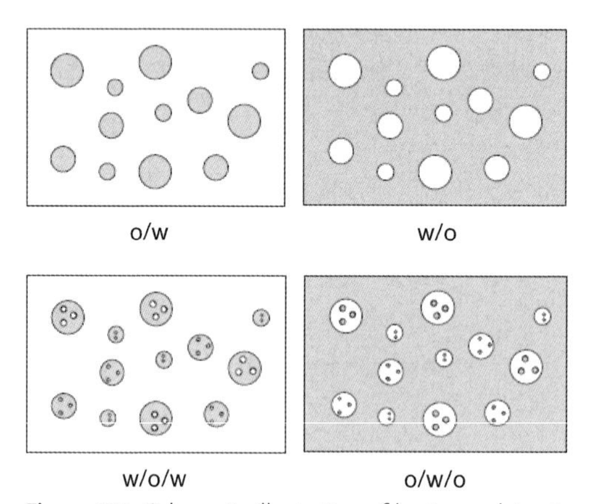


**Figure 13.1** Schematic illustration of basic emulsion types. The white color represents the aqueous phase, for example water and the gray color represents the oily phase, for example oil.

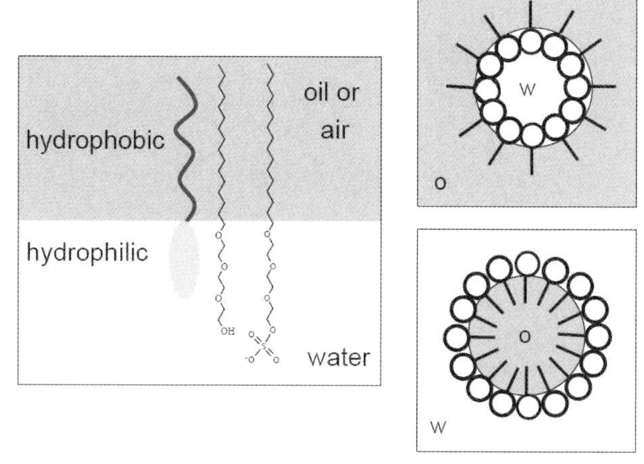


**Figure 13.2** Schematic illustration of an emulsifier molecule and emulsifiers adsorbed at the interface between the phases.

Emulsifiers are surface-active molecules having both a hydrophilic and a hydrophobic part in the molecule (Figure 13.2).

Due to this structure they tend to adsorb at the interfaces between a polar and a nonpolar fluid, for example water and oil. Emulsifiers reduce the surface tension and stabilize the surface by steric, electrostatic or hydrodynamic (Gibbs-Marangoni) effects [14]. Droplet coalescence (flowing of one or more droplets together) can thus be reduced or prevented. Some emulsifiers can be characterized by their hydrophilic/lipophilic balance (HLB) value that provides information on the ratio of hydrophilic to lipophilic character of the surfactant molecule. The HLB value helps to determine the phase in which the emulsifier is soluble. Usually, the emulsifier used is soluble in the continuous phase (Bancroft rule). Furthermore, the HLB value gives a first hint whether the emulsifier is suitable for the production of an o/w (HLB value 8–18) or a w/o emulsion (HLB value 4–6) [15].

Thickeners are applied in order to increase the viscosity of the continuous phase of the emulsion. Thus, the droplet movement is decreased, resulting in a reduced coalescence rate and sedimentation. At the same time the emulsion texture is changed.

Some materials, like proteins or emulsifying starches, act simultaneously as both emulsifier and thickener [2].

Besides coalescence and sedimentation, other mechanisms can cause instability of emulsions. They are summarized in Figure 13.3.

Sedimentation/creaming, ascent or decent of the droplets due to differences in mass density, and aggregation/flocculation of the droplets are reversible processes. In these cases, the original state of the emulsion can be re-obtained by mild shaking or stirring. Coalescence describes the process of two or more droplets combining and forming a new, larger droplet. It can occur directly after the breakup of a droplet, when the interface is not yet sufficiently occupied by emulsifier

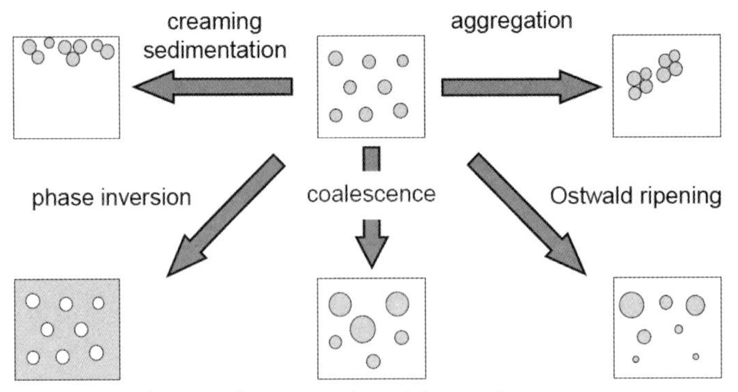


**Figure 13.3** Schematic illustration of instability mechanisms.

molecules, as well as during long-term shelf-life. High disperse phase fractions, unsuitable emulsifiers and variations in temperature can also lead to phase inversion (the continuous phase is transformed to the disperse phase and vice versa). The process of further shrinking of small droplet while larger droplets increase is called Ostwald ripening. It takes place when small fractions of the dispersed phase are soluble in the continuous phase. The driving mechanism is the high capillary pressure in the small droplets. Due to the lower capillary pressure in the larger droplets, the disperse phase molecules diffuse into those droplets and thus increase their volume.

## 13.3 Membrane Emulsification Processes

### 13.3.1 Membranes

In emulsification technology, a wide range of membranes has been used, such as: tubular membranes (Figure 13.4a) made from ceramics like aluminium oxide [8], special porous glasses like Shiratsu porous glass (SPG) membranes [9], polymers like polypropylene [8] and flat filter membranes (Figure 13.4b) made of PTFE [4]. Membranes can also be produced by micro engineering (Figure 13.4c) [6, 7, 11, 16]. The arrangement of the pores on the membrane, the pore shape and size and the porosity can be exactly set by this technology. Furthermore, the properties of the material allow very thin active layers, thus reducing the pressure drop but maintaining the mechanical stability [10].

Microchannel plates (Figure 13.4d) are microengineered devices made of silicon wafers. They were originally developed to investigate the rheology of blood cells [5]. Two cavities are used for the continuous and the disperse phase, respectively; and these are separated by walls. Small channels with a trapeze-shaped cross-section are etched in those walls. By covering the whole system with a glass plate, a channel is created that the liquid can be pushed through. The exit of this channel

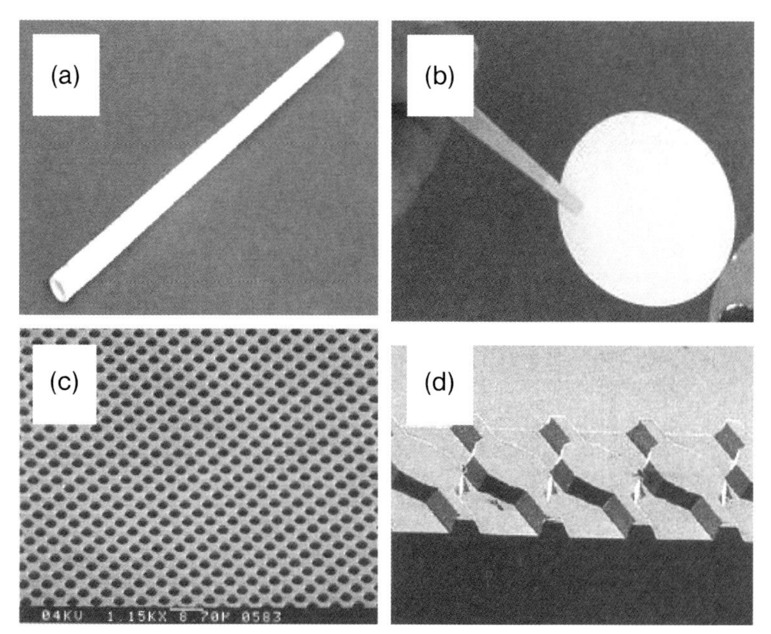


**Figure 13.4** Membranes used for emulsification technology. (a) Tubular membrane, (b) flat filter membrane, (c) microengineered membrane, (d) microchannel.

leads to the so-called terrace, an area with a strongly noncylindrical cross-section, the same height as the microchannel but several times its width [17]. In some investigations, the authors use so-called straight-through microchannels, which are, in a way, membranes with pores having an oblong cross-section [18].

### 13.3.2 Conventional or Direct Membrane Emulsification

By direct membrane emulsification (conventional membrane emulsification), the phase to be dispersed has to be pressed through a microporous membrane. Small droplets are formed and detached from the membrane by a flow of the continuous phase (Figure 13.5). For an appropriate droplet formation, the surface of the membrane has to be wetted by the continuous phase, for example a hydrophilic membrane has to be used to produce an o/w emulsion.

The pressure required for this process is determined by two effects. The first one is the pressure drop resulting from the flow of the disperse phase through the pores. It can be estimated by the Hagen-Poiseuille law (Equation 13.1) [8]. The pressure drop, $\Delta p$, depends on the viscosity of the disperse phase, $\eta$, the diameter and length of the pores, $d_p$ and $l_p$, and the flow velocity, $v$, in the pores:

$$\Delta p = \frac{32 \cdot \eta \cdot l_p \cdot v}{d_p^2} \qquad (13.1)$$

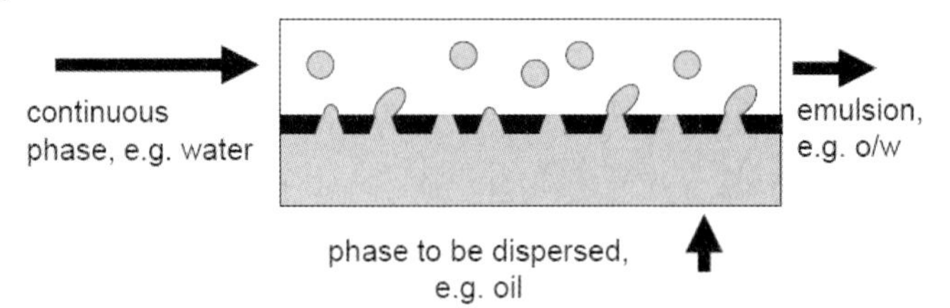


**Figure 13.5** Droplet formation by direct membrane emulsification process.

The second effect is the capillary pressure. In order to create a new interface at the cylindrical pore opening, the capillary pressure according to Equation (13.2) has to be overcome [8]:

$$p_c = \frac{4 \cdot \gamma \cdot \cos\delta}{d_p} \tag{13.2}$$

The capillary pressure, $p_c$, depends on the interfacial tension, $\gamma$, and the contact angle, $\delta$, and it is inversely proportional to the pore diameter, $d_p$.

By using a pressure above the capillary pressure, $p_c$, a droplet starts to grow and is detached by the shear force of the continuous phase after reaching a critical droplet diameter.

Among different forces interacting at the droplet in this state, $F_\gamma$, is dominant, holding the droplet at the pore opening and can be calculated with Equation (13.3) [8]. $F_\gamma$ is determined by the time-dependent interfacial tension, $\gamma(t)$, and the droplet diameter. Since the interfacial tension is influenced by the emulsifier dynamics, droplet detachment can be influenced by the emulsifier used [8]:

$$F_\gamma = \pi \cdot d_p \cdot \gamma(t) \tag{13.3}$$

The detaching forces acting on the droplet are mainly caused by the flow of the continuous phase. Dominant among these are the flow resistance force and the dynamic buoyant force.

Depending on the process conditions, the ratio of emulsion droplet size to membrane pore size is within a range of 2–10. Schröder for example reported a ratio between three and four [8].

Whereas only a few pores are active at a disperse phase pressure of about the capillary pressure, the volume flow through the pores as well as the fraction of active pores is increased at higher pressures. Since the probability for neighboring pores forming droplets at the same time rises, the risk of coalescence is higher at a high fraction of active pores. Consequently, low membrane porosity can decrease the risk of coalescence at the membrane surface. Absence of coalescence, due either to a low porosity or a disperse phase pressure of about the capillary pressure, enables one to produce emulsions with narrow droplet size distributions with direct membrane emulsification [8].

A detailed study of the influence of process parameters on the droplet sizes of the emulsion and the flux of the disperse phase in membrane emulsification have been subject of numerous investigations. An excellent overview is given by

Schröder [8]. Furthermore, Lambrich et al. [19] and Joscelyne et al. [20] give overviews of the literature available in this field of emulsification technology.

### 13.3.3 Membrane Emulsification in the Jetting Regime

Similar to the process described in Section 13.3.2, the disperse phase in membrane emulsification during the jetting regime is pushed through the pores of a membrane. However, in this case, the velocity of the disperse phase at the pore opening has to be higher. For this purpose, membranes of a low pressure drop, like the micro engineered membranes mentioned in Section 13.3.1, are very appropriate. This leads to liquid jets emerging from the pore (Figure 13.6).

The continuous phase flowing over the membrane surface deflect the jet and stretch it, depending on the ratio between the phase velocities. At a certain length, the jet becomes instable and breaks into single droplets, which are carried away by the cross-flowing continuous phase. The reason for the break-up is an effect called Rayleigh instability [21], which has been investigated since the nineteenth century because of his great importance in the fields of for example spraying or ink-jet printing.

In those applications, a liquid is dispersed into a quiescent gas phase, whereas membrane emulsification includes two liquid phases, resulting in a viscosity ratio of another order of magnitude. Furthermore, the two phases are flowing in a cross-flow pattern, which means that the flow of the surrounding fluid has to be kept in perspective, too. Investigations by Cramer et al. [22, 23] on drop formation in a co-flowing channel are close to the case of membrane emulsification in the jetting regime. In these investigations, the disperse phase emerges from a capillary in the center of a flow channel, whereas in membrane emulsification, the jet from a membrane pore has to cross the flow in the boundary layer and is deflected by the cross-flow. Since the flow velocity of the continuous phase is, according to Cramer, of great importance for the development of the jet, it is rather important to know the flow profile above the membrane and the distance between the membrane

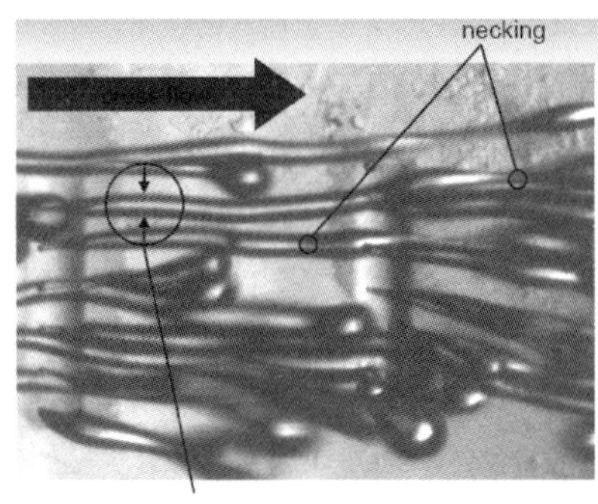


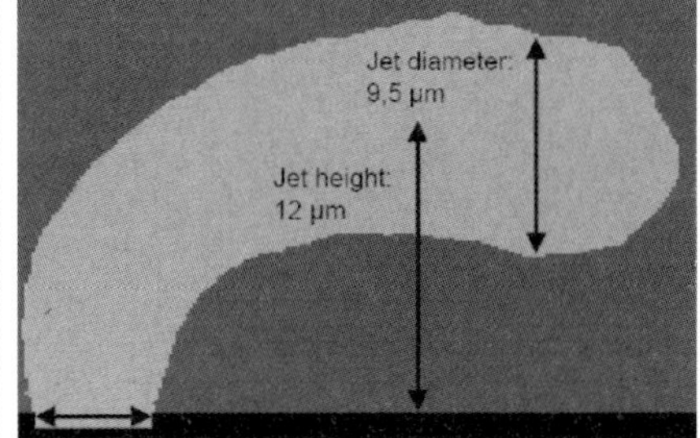


**Figure 13.6** Jetting at the membrane surface (video imaging and flow simulation; fluent) [10, 24–26].

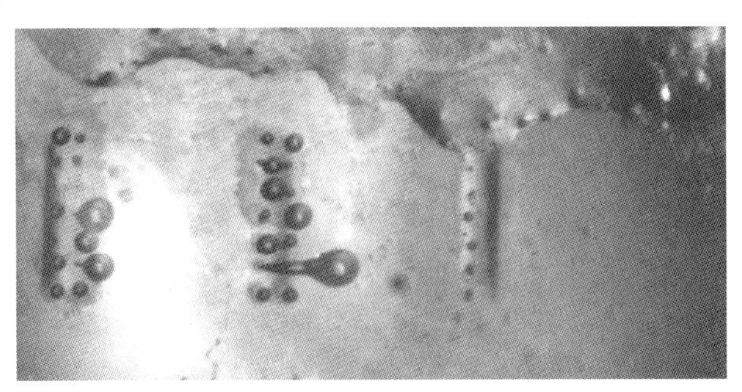

**Figure 13.7** Observation of membrane pores during an emulsification process.

surface and the deflected jet. This can be either measured by inline videotaping or calculated by flow simulations [10, 24].

Since this approach to membrane emulsification is rather new, only little information on the effect of process parameters on the development of jets is available. In this work it was monitored using a microscope (Olympus SZ 12, Olympus, 3000 Hz) with a high-speed video camera (speedCam Lite, Weinberger, Karlsruhe, Germany). In Figure 13.6 on the left hand we see a picture of rape seed oil jets (dispersed phase of the emulsion). The continuous phase is composed of demineralized water containing 1% Tween 80 (E443, polyoxyethylene (20) sorbitol monoolein; Carl Roth, Karlsruhe, Germany) with HLB value of 15.

The pore size of the microengineered membrane (Aquamarijn, The Netherlands) was 10 μm, the flow velocity was $v = 0.9\,\mathrm{m\,s^{-1}}$ and the transmembrane pressure $p_{tm} = 3$ bar.

It can be seen that nearly all pores produce liquid jets. Some of them become unstable within the range of the picture, whereas in other cases, necking takes place in areas out of the focus of the microscope. Compared wit the pore diameter of 10 μm, the jet diameter was four to five times larger (in the range of 40–50 μm).

In Figure 13.6 on the right hand we can see a flow simulation (Fluent). It was simulated for a system composed as water ($\eta = 1\,\mathrm{mPa\,s^{-1}}$) as continuous phase and oil ($\eta = 70\,\mathrm{mPa\,s^{-1}}$) as disperse phase. The interfacial tension was $\gamma = 27$ $\mathrm{mN\,m^{-1}}$, the contact angle was $\delta = 0°$. A membrane pore size of 5 μm was entered. The velocity of the continuous phase of $v = 0.6\,\mathrm{m\,s^{-1}}$ and transmembrane pressure $p_{tm} = 2$ bar were assumed.

Besides the observation of jets by use of a microscope with a high-speed video camera, it is also possible to determine the ratio of active pores, the flux, and the start of membrane pore blocking (fouling; Figure 13.7).

## 13.3.4
## Premix Membrane Emulsification

### 13.3.4.1 Process Principles

Premix membrane emulsification was developed by Suzuki [4]. In this process, an emulsion premix containing large droplets instead of the pure disperse phase is

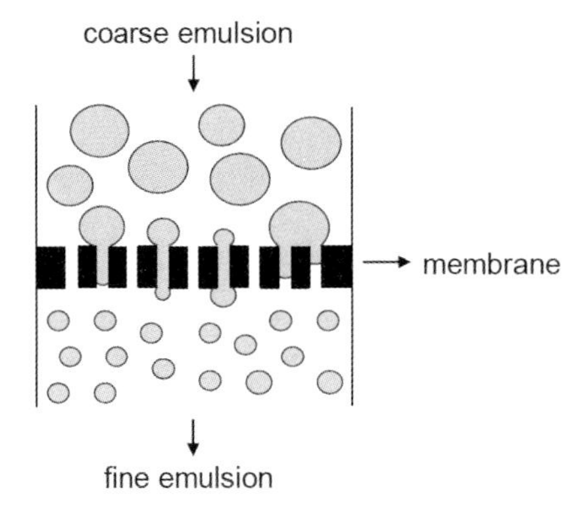


**Figure 13.8** Schematic illustration of premix membrane emulsification.

pressed through a membrane (Figure 13.8) at pressures below 15 bar. While passing the membrane pores, the droplets of the emulsion are disrupted and a fine emulsion is obtained.

The dispersed phase fraction of the resulting emulsion is determined by the concentration of the dispersed phase in the coarse emulsion, only. This means that high dispersed phase fractions can be obtained without recirculation of the emulsion through the membrane, especially if it is combined with a phase inversion during the passage through the membrane [27].

The key factor for the success of this process is the wettability of the membrane surface. A rule of thumb states that the continuous phase of the resulting fine emulsion has to wet the surface of the membrane as for direct membrane emulsification. Thus, for the production of an o/w emulsion, the membrane surface has to be hydrophilic, regardless whether the process is combined with a phase inversion or not [2].

#### 13.3.4.2 **Influence of Process Parameters**

For premix membrane emulsification, the factors influencing the emulsification results are the transmembrane pressure, repeated processing and the pore size of the membrane. Furthermore the composition of the emulsion, like disperse and continuous phase viscosity, emulsifier and emulsifier concentration have also an influence on the emulsification results [28–31].

One method to characterize the process is by measurement of the emulsion droplet size distribution. In this work, two methods were applied. One is a direct measurement using laser diffraction with polarization intensity differential scattering (PIDS) technology in a Coulter LS 230 (Beckman–Coulter, Krefeld, Germany). The second is an indirect one, based on the photometric measurements using a pulse laser photometer (Optimags Dr. Zimmermann, Karlsruhe, Germany). This was specially used for inline control of the emulsion quality during the process.

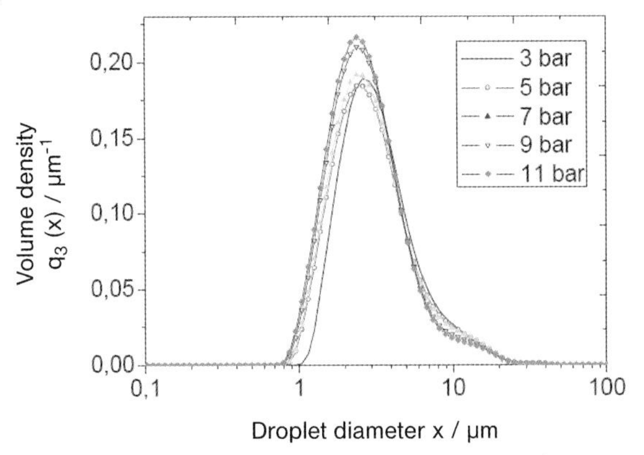


**Figure 13.9** Droplet size distributions in premix membrane emulsification at different transmembrane pressure differences.

**Influence of Transmembrane Pressure Difference** In principle, the higher the pressure difference, the smaller emulsion droplet sizes are obtained. However, the transmembrane pressure difference has only a minor influence on the emulsification results.

Figure 13.9 gives an example of droplet volume density distributions of emulsions obtained by pressing an emulsion premix through a membrane at transmembrane pressure differences varying from 3 bar to 11 bar. These pressure differences are 7.5- to 27.5-fold the minimum pressure difference required (capillary pressure). A hydrophilic polyamide membrane with a mean pore size of 0.8 µm was used. The emulsion premix consisted of 20% dispersed phase (vegetable oil). As continuous phase water containing emulsifier Tween 80 at a concentration of 2% was used. The Sauter diameter of the emulsion premix was $x_{3,2} = 25\,\mu m$.

For this system, irrespectively of the transmembrane pressured difference applied, the fine emulsion still has some big droplets after one pass through the membrane. Narrower droplet size distribution can be obtained by repeating the process.

**Influence of Repeated Processing** Repeated processing results in smaller droplets and narrower droplet size distributions of the fine emulsions. Figure 13.10 depicts an example of the volume density distributions of an emulsion with a dispersed phase of 30% after the first, second and third pass at 9 bar through a membrane with a mean pore size of 0.8 µm. In this case, at least two passes through the membrane are required in order to obtain a monomodal droplet size distribution.

**Influence of Membrane Pore Size** The smaller the mean pore size of the membrane, the smaller is the droplet size of the fine emulsion and the bigger is

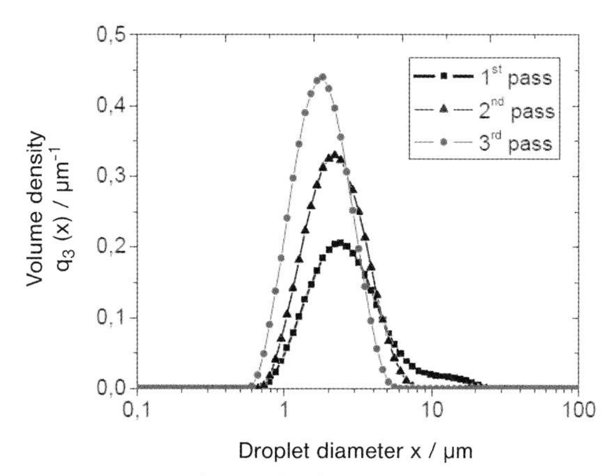


**Figure 13.10** Droplet size distributions in premix membrane emulsification after the first, second and third passes trough a membrane.

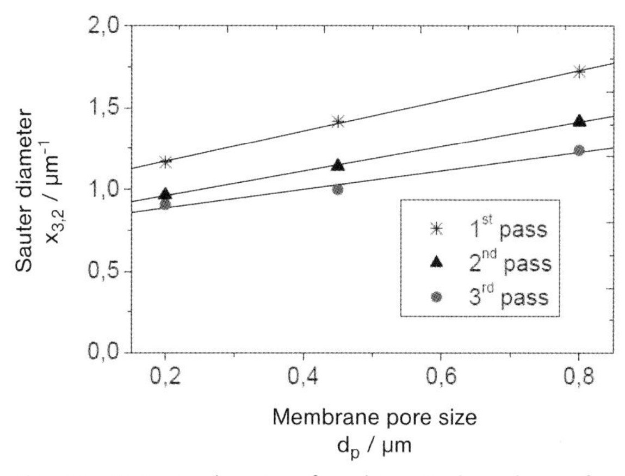


**Figure 13.11** Sauter diameter of emulsions in dependence of the mean pore size of the membrane after the first, second and third passes. Disperse phase fraction $\varphi = 5\%$; emulsifier: 2% Tween 80; transmembrane pressure difference: 12 bar.

the ratio between emulsion droplet diameter and membrane pore size diameter, $x/d_p$.

In the example given in Figure 13.11, polyamide membranes with different mean pore sizes were used (0.2 µm, 0.45 µm, 0.8 µm). A small disperse fraction of 5%, and a high emulsifier concentration (2% of emulsifier Tween 80) was chosen in order to avoid coalescence. Figure 13.11 gives Sauter diameters of the emulsions produced with membranes of different mean pore sizes after the first, second and third pass at 12 bar through the membrane. The Sauter diameter of the emulsion premix was 12 µm.

After three passes through the membranes with pore sizes of 0.2 μm, 0.45 μm and 0.8 μm, Sauter diameters of, respectively, 0.9 μm, 1.0 μm and 1.2 μm could be obtained. The ratios between droplet size and membrane pore size, $x/d_p$, were 4.5, 2.2 and 1.5, respectively. The smaller the pore size of the membrane, the bigger the capillary pressure (Equation 13.2) that has to be overcome to produce emulsions and consequently the bigger the ratio $x/d_p$ [32].

**Influence of Emulsifier Concentration** Emulsions of small droplet sizes and narrow droplet size distributions can be obtained at high emulsifier concentration. Figure 13.12 shows the volume density distributions of emulsions of a disperse phase fraction of $\varphi = 72\%$ and two different emulsifier (Tween 80) concentrations, 2.4% and 4.6%. The production parameters were: trans-membrane pressure difference of 12 bar, three passes and membrane mean pore size of 0.8 μm. In both cases the emulsifier concentration is above the critical micelle concentration (CMC).

Nevertheless, an increase in the emulsifier concentration results in smaller droplet diameters. At this high disperse phase fraction chosen, coalescence of new droplets leaving the membrane is found. The higher the emulsifier concentration, the shorter the stabilization time and, as a consequence, the less droplets coalesce after leaving the membrane.

**Influence of Disperse Phase Fraction** With increasing dispersed phase fraction, droplet collisions and thus droplet coalescence frequencies increase. Therefore, usually, the lower the disperse phase concentrations, the smaller are the mean emulsion droplet sizes found. An example is given in Figure 13.13. In this example, an emulsion premix with 40% of dispersed phase and a Sauter diameter of

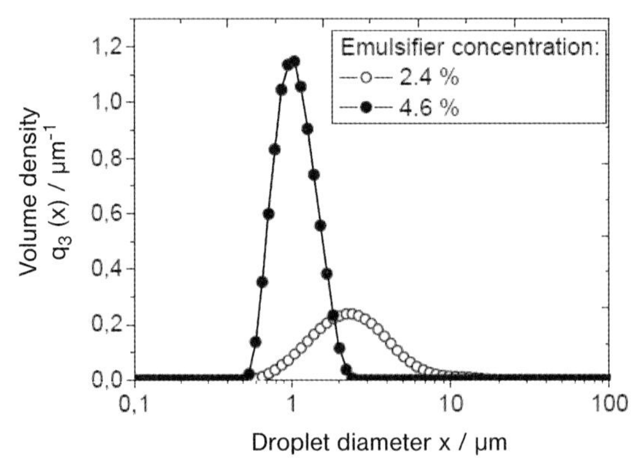


**Figure 13.12** Volume density distributions for emulsions with different emulsifier concentrations. Disperse phase fraction $\varphi = 72\%$; emulsifier: Tween 80; transmembrane pressure difference: 12 bar; three passes; membrane mean pore size: 0.8 μm.

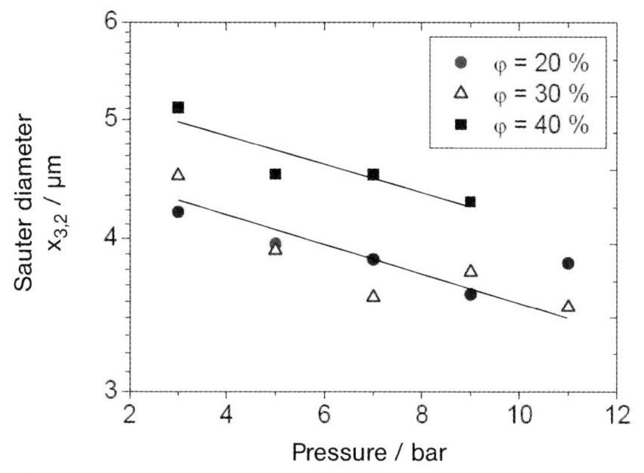


**Figure 13.13** Sauter diameter of emulsions with different disperse phase concentrations as a function of the pressure difference. Membrane mean pore size: 0.8 µm, one pass.

about 25 µm was produced. Part of this premix was diluted with water with 2% Tween 80 in order to obtain emulsions with exactly the same droplet size but different dispersed phase fractions (30%, 20%). Membranes with mean pore sizes of 0.8 µm were used for emulsification. The results given as a function of the transmembrane pressure difference depict only a slight influence of the transmembrane pressure difference, as already stated. However, the coalescence effect is more pronounced. Emulsions with slightly smaller Sauter diameter are obtained using a low dispersed phase fraction and high pressure.

#### 13.3.4.3 **Process Flux**

Figure 13.14 shows the premix membrane emulsification flux for different process and product parameters. The product flux ($\Delta p = 12$ bar, $d_p = 0.8$ µm, one pass) decreases from 28 $m^3 m^{-2} h^{-1}$ to 5 $m^3 m^{-2} h^{-1}$ with increasing dispersed phase concentration from 30% to 80%.

The flux in premix membrane emulsification is approximately ten times higher than the flux of conventional membrane emulsification, as measured for example by Schröder [8] considering membranes with similar mean pore diameters.

#### 13.3.4.4 **Inline Measurements**

The previous experiment was also inline monitored by photometric measurements. The pulse laser photometer used is well suited to characterize physical properties of emulsions like dispersed phase content or average droplet size [33]. The following results demonstrate that the pulse laser can not only be used in laboratory experiments but also for monitoring continuous emulsification processes.

Figure 13.15 shows the results in voltage as a function of time. The voltage corresponds to the turbidity of the liquid observed. The results agree with the results obtained by the measurement of the droplet size distribution using laser

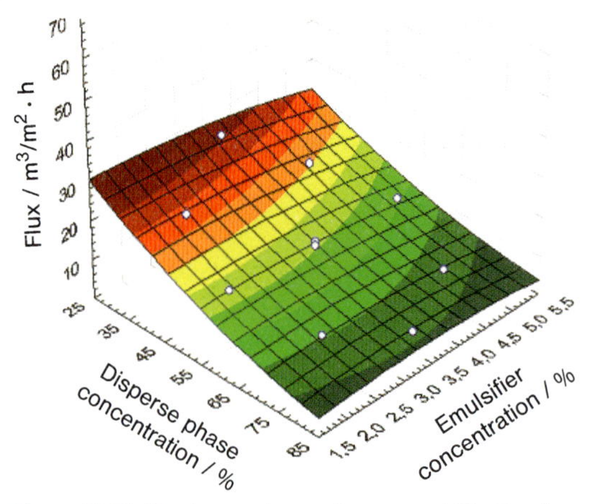


**Figure 13.14** Flux in premix membrane emulsification for different process and product parameters ($\Delta p = 12$ bar, $d_p = 0.8\,\mu$m, one pass).

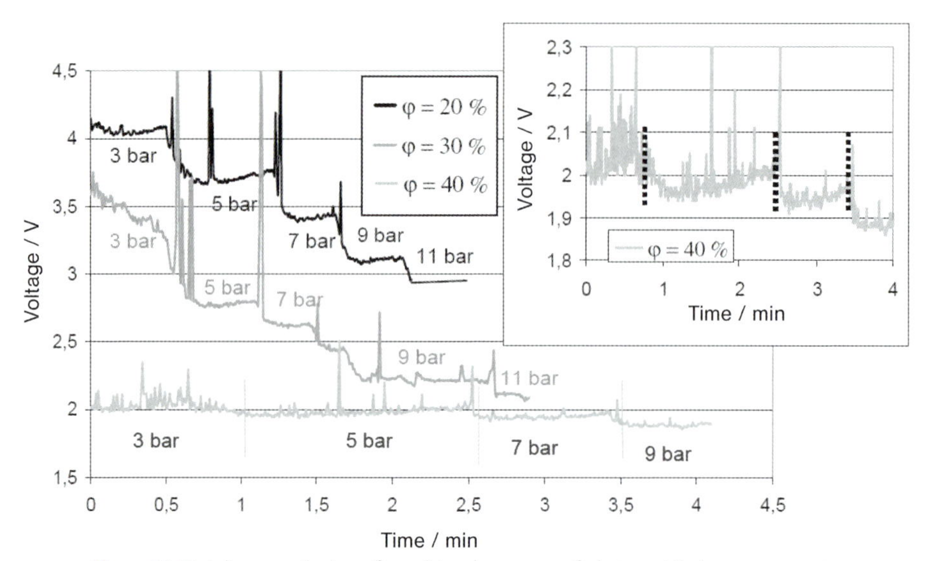


**Figure 13.15** Inline monitoring of emulsion by means of photometrical measurements.

diffraction. The higher the disperse phase fraction, the lower the voltage signal is, and the higher optical densities of the emulsions are measured. Furthermore increasing the pressure difference results in smaller droplets and consequently in higher turbidity. Thus, by means of pulse laser photometry both dispersed phase concentration and mean droplet sizes can be determined inline and used as base for a process control system. In case undesirable changes in droplet sizes are detected, the pressure difference may be readjusted during the processing [26].

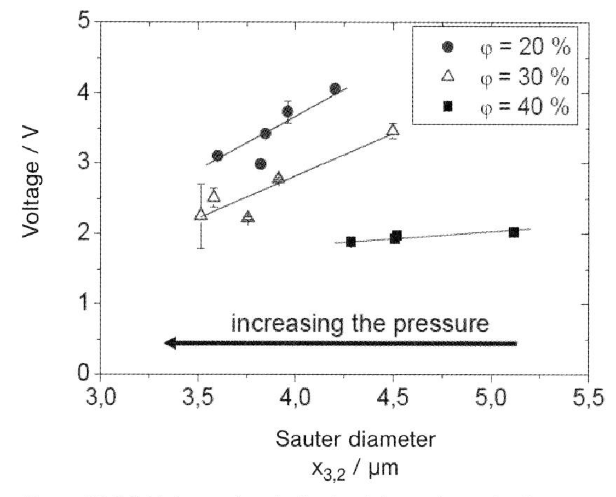


**Figure 13.16** Voltage signal obtained from the pulse laser photometer by Inline monitoring of emulsion against Sauter diameter measured by the Coulter LS 230.

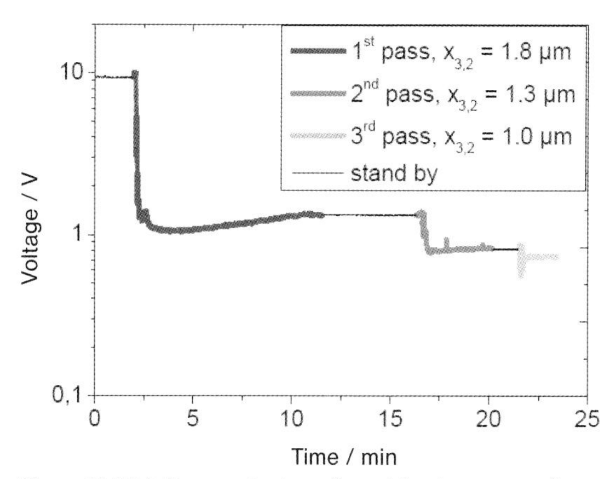


**Figure 13.17** Inline monitoring of emulsion by means of photometrical measurements. Example of monitoring of repeated process.

Figure 13.16 gives the voltage signal obtained from the pulse laser photometer against the Sauter diameter measured by the Coulter LS 230. The peaks between the different pressures are due to air bubbles and for this reason they are neglected. It depicts the good correlation between mean droplet size and optical density of an emulsion.

The effect of repeating process can also be monitored by inline measurement, as shown in Figure 13.17. An o/w emulsion with 10% dispersed phase and 2% Tween 80 and Sauter diameter $x_{3,2} = 13\,\mu m$ was pressed three times by 12 bar through a membrane with a mean pore size of 0.45 µm.

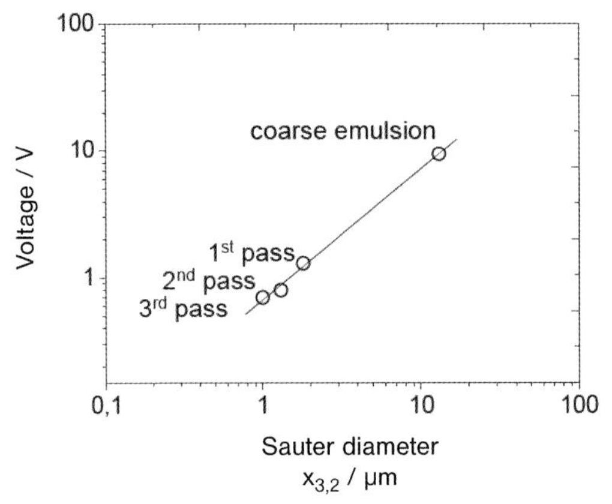


**Figure 13.18** Voltage signal obtained from the pulse laser photometer by Inline monitoring of coarse and fine emulsion against Sauter diameter measured by the Coulter LS 230.

As expected, droplet sizes decreased with increasing number of passes. As for the other measurements, the voltage signal during processing was in good agreement with the Sauter diameters determined by off-line laser diffraction (Figure 13.18).

### 13.3.5 Microchannel Emulsification

In microchannel emulsification the mechanism of droplet formation is totally different. Here, single droplets are formed in the absence of any shear forces.

A strongly noncylindrical geometry at the channel exit flattens the emerging droplet. The pressure difference that arises when the tip of the fluid leaps over the terrace into the well is caused by different Laplace pressures on the terrace and in the well [34]. On the terrace, the Laplace pressure that usually is determined by the two main radii, $R_1$ and $R_2$, and the interfacial tension, $\gamma$, is governed by the small radius, whereas the larger one can be neglected (see Equation 13.4). The wetting properties of the terrace and the glass plate and the distance between them determine the former radius. In the well, an increasing droplet diameter causes a decreasing Laplace pressure leading to a pressure gradient within the disperse phase on the terrace and in the well. Therefore, the fluid flows into the well causing a further grow of the droplet and the pressure gradient.

$$p_{\mathrm{Lap}} = \frac{1}{\gamma}\left(\frac{1}{R_1} + \frac{1}{R_2}\right) \tag{13.4}$$

As the volume flow from the terrace into the well is larger than the supply rate of the disperse phase, necking near the end of the terrace takes place below a critical velocity of the disperse phase in the microchannel. Due to the small radius in

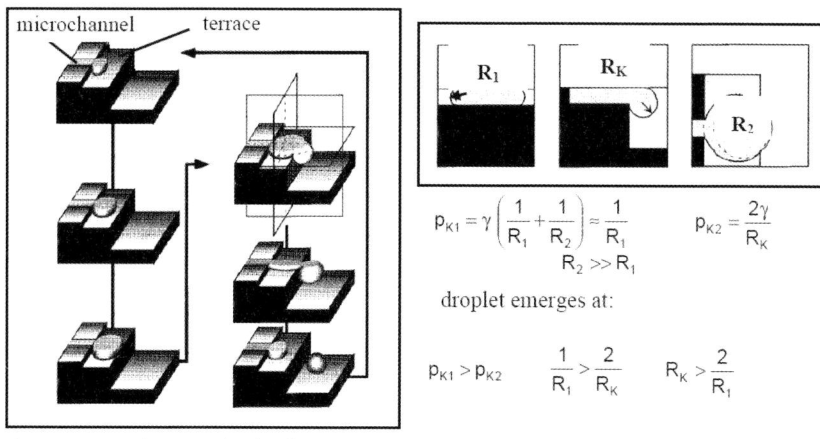


**Figure 13.19** Schematic droplet formation mechanism in microchannel emulsification [18].

the neck, the pressure in this area is higher than in the growing droplet and in the remaining liquid on the terrace. This effect, described by the equations in Figure 13.19, leads to the Laplace instability of the disperse phase causing the detachment of the droplet. Therefore, the geometry of the microchannel plate and its wetting properties determine the droplet size in this flow regime, only. Monodisperse emulsions can be produced, as long there is no change in pressure difference, dispersed phase velocity, or membrane pore geometry (as found in fouling processes). The schematic droplet formation mechanism is illustrated in Figure 13.19.

Above this critical velocity, a higher flow rate of disperse phase reduces the necking. Furthermore, the dynamic pressure on the terrace can not be neglected anymore and counterbalances the high pressure caused by the neck near the edge of the terrace. Thus, the mechanism of droplet detachment described above is prevented as the Laplace instabilities can not arise. Additionally, stabilizing effects caused by the emulsifiers influence necking. A detachment of droplets takes place due to other effects (e.g. cross-flow of the continuous phase) and the emulsion produced is no longer monodisperse [34].

For the production of double emulsions of the w/o/w or o/w/o type, mechanical stress during the second emulsification step can lead to a loss of dispersed phase of the inner or primary emulsion. Therefore, a process is aimed for that can be used to produce emulsions in the absence of shear forces. For that reason, microchannel emulsification seems to be a suitable technique. In order to prove the suitability of microchannel emulsification for the production of a multiple emulsion, we performed the second step of the production of a w/o/w emulsion using microchannel emulsification. The inner emulsion with a Sauter diameter of 0.25 μm was produced with a high pressure homogenizer, which is a conventional emulsification technique. Water was the disperse phase, purified rape seed oil containing 10% of PGPR (E476, Polyglycerinpolyrhicinoleat, Danisco Ingredients, Quickborn, Germany) as emulsifier represented the continuous phase. This emulsion was dispersed into demineralized water containing various

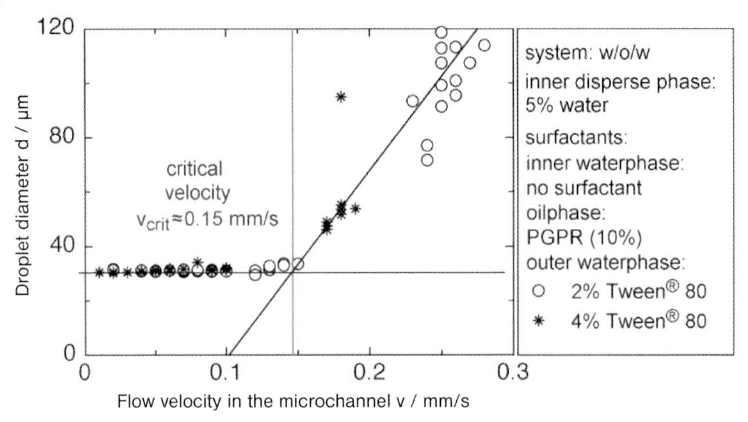


**Figure 13.20** Effect of surfactant concentration on the critical velocity.

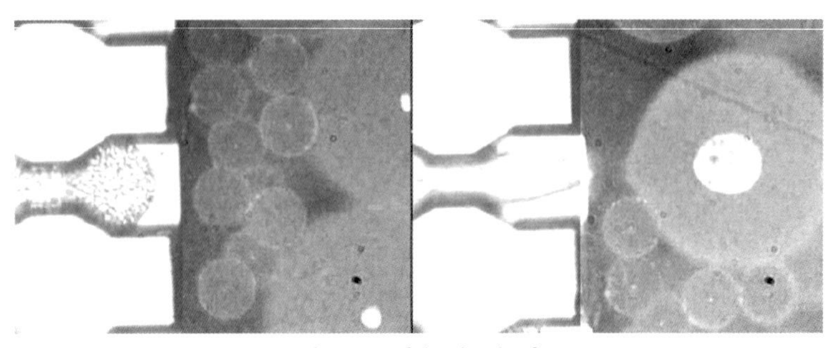

**Figure 13.21** Microscopic visualization of the droplet formation at dispersed phase velocities below (left) and above (right) the critical velocity.

concentrations of the (outer) emulsifier Tween 80. The effect of increasing the dispersed phase velocity above the critical value can be seen in Figure 13.20. As soon as the critical velocity is exceeded, an increase in the mean droplet diameter is detected. It is connected with broader droplet size distributions.

This effect can also be visualized using a microscope. Figure 13.21 shows, on the left, the droplet formation at velocities smaller than the critical velocity, and on the right, the droplet formation at velocities above the critical value. Above the critical velocity, differences in droplet formation on the terrace and droplets of bigger size in the continuous phase are monitored.

## 13.4 Summary and Conclusions

Membrane and microporous emulsification processes are associated with specific advantages and limitations that can be traced back to the nature of droplet formation taking place in the process. Since single droplets are formed at the exit of a pore or a microchannel, the disperse phase throughput in membrane emulsification and microchannel emulsification is rather low. Therefore, the implementation of membrane emulsification is mostly limited to high value products to date. Particularly in microchannel emulsification, the critical velocity represents a limit that does not allow an increase of the throughput per channel above a certain value. For the lack of investigations on scaling-up the system, the technique is reported to be used only in laboratories so far.

Membrane emulsification in the jetting regime offers a possibility of increasing the throughput of disperse phase per pore. In order to allow a high pore density at the membrane surface, the jets have to break-up close to the pore in order to prevent interactions between two or more jets. Further investigation in this field will help to explore the potential of this technique.

From the present point of view, premix membrane emulsification is the technique of choice for producing emulsions with high dispersed phase fractions at reasonable flux (see Figure 13.14) [32].

The effects of important process parameters are summarized in this work. The transmembrane pressure difference has little influence on the emulsification results. The higher the ratio between transmembrane pressure difference and capillary pressure, the smaller are the emulsion droplet sizes found, or the smaller is the ratio between emulsion droplet sizes and membrane pore sizes, $x/d_p$. Repeating the process (increasing the number of membrane passes) results in smaller droplets and narrower droplet size distributions in the resulting emulsions. For the systems studied at least two passes through the membrane are required in order to obtain a monomodal size distribution. The smaller the mean pore size of the membrane, the smaller the droplet size of the fine emulsion and the bigger the ratio between emulsion droplet diameter and membrane pore size diameter, $x/d_p$. With increasing emulsifier concentration, the stabilization time and, as a consequence, the coalescence rates of the droplets both decrease after leaving the membrane. Therefore, smaller mean droplet sizes are found in the product. No investigations on scale-up have been published yet.

Characterization of membrane processes is of high importance to understand the process and the mechanisms behind them, to determine product properties and to develop process-controlling systems. In this work, two kinds of characterization methods have been applied. The first is a direct one, visualizing the membrane using a high-speed camera. In this case the mechanisms of droplet formation and the effect and kinetics of pore-blocking (fouling) of the membrane can be directly observed. The second process characterization method, an indirect one, is based on the measurement of emulsion characteristics. Droplet size is measured using either off-line laser diffraction with polarization intensity differential scattering (PIDS) technology (diluted) or inline photometric measurements (pulse laser photometer, nondiluted). The photometric signal is proportional to the droplet size and dispersed phase concentration. It has been shown that pulse laser photometry is a method suitable for inline measurement as base for process control systems in emulsification. As this method is affordable in price, it is the method of choice in emulsion production.

## Acknowledgments

The authors wish to thank the Arbeitsgemeinschaft industrieller Forschungsvereinigungen (AiF; Projekt Inovationskompetenz II, PRO INNO II, KF0067201FK4), the Deutsche Forschungs-Gemeinschaft (DFG) in the framework of the Graduiertenkolleg 366: Grenzflächenphänomene in aquatischen Systemen und wässrigen Phasen, the Dechema/Max-Buchner-Forschungsstiftung Kennziffer 2484 and Verfahrenstechnik PRO 3 e.V. for the financial support of this work.

## Nomenclature

| | |
|---|---|
| $l_p$ | Pore length, m |
| $d_p$ | Pore diameter, m |
| $\eta$ | Viscosity of the disperse phase, Pa s$^{-1}$ |
| $v$ | Velocity, m s$^{-1}$ |
| $p_c$ | Capillary pressure, N m$^{-2}$ |
| $p_{tm}$ | Transmembrane pressure, bar |
| $\gamma$, $\gamma(t)$ | (Time)-dependent interfacial tension, N m$^{-1}$ |
| $\delta$ | Contact angle, – |
| $F_\gamma$ | Force caused by the interfacial tension, N |
| $p_{\mathrm{Lap}}$ | Laplace pressure, N m$^{-2}$ |
| $R_1$, $R_2$ | Radii of principal curvature, m |

## References

1 H.P. Schuchmann, T. Danner, *Chem. Eng. Tech.* **2004**, *4*, 364–375.
2 H. Schubert (ed.), *Emulgiertechnik, Grundlagen, Verfahren und Anwendung,* Behr's Verlag, Berlin, **2005**, pp 137–165, 369–431.
3 T. Nakashima, M. Shimizu, M. Kukizaki, *Key Eng. Mater.* **1991**, *61/62*, 513–516.
4 K. Suzuki, I. Shuto, Y. Hagura, *Food Sci. Technol. Int.* **1996**, *2*, 43–47.
5 T. Kawakatsu, Y. Kikuchi, M. Nakajima, *J. Am. Oil Chem. Soc.* **1997**, *74*, 317–321.
6 M.J. Geerken, *Emulsification with micro-engineered devices.* PhD thesis, University of Twente, **2006**.
7 I. Kobayashi, M. Nakajima, In: N. Kockmann (ed.), *Micro process engineering: fundamentals, devices, fabrication, and applications,* Wiley-VCH, Weinheim, **2006**, 149–172.
8 V. Schröder, *Herstellen von Öl-in-Wasser-Emulsionen mit mikroporösen Membranen,* PhD thesis, University of Karlsruhe, **1999**.
9 G.T. Vladisavljevic, U. Lambrich, M. Nakajima, H. Schubert, *Colloid Surface A* **2003**, *232*, 199–207.
10 H.P. Schuchmann, G.G. Badolato, U. Lambrich, H. Schubert, *Book of Abstracts 10th Aachen Membrane Colloquium,* **2005**, 171–184.
11 E. Zwan, K. Schröen, K. Dijke, R. Bomm, *Colloid Surface A* **2006**, *277*, 223–229.
12 H.P. Schuchmann, H. Schubert, *Chem. Eng. Technol.* **2003**, *26*, 67–76.
13 H.P. Schuchmann, In: U. Bröckel, W. Meier, G. Wagner (Eds.), *Product Design and Engineering: Best Practices, Vol. 1,* Wiley-VCH, Weinheim, **2007**, 63–93.

14 P. Walstra, in: P. Becher (ed.), *Encyclopedia of Emulsion Technology, Vol. 1*, Dekker, New York, **1998**, 52–127.

15 E. Dickinson, G. Stainsby, *Colloids in Food*, Applied Science, London, **1982**.

16 V. Schadler, E.J. Windhab, *Book of Abstracts 10th Aachen Membrane Colloquium*, **2005**, 585–591.

17 M. Nakajima, *Riken Review* **2001**, *36*, 21–23.

18 I. Kobayashi, S. Mukataka, M. Nakajima, *J. Colloid Interf. Sci.* **2004**, *279*, 277–280.

19 U. Lambrich, G.T. Vladisavljevic, *Chem. Eng. Tech.* **2004**, *76*, 376–383.

20 S.M. Joscelyne, G. Trägårdh, *J. Membrane Sci.* **2000**, *169*, 107–117.

21 L. Rayleigh, *Proc. R. Soc. Lond.* **1879**, *29*, 71–97.

22 C. Cramer, B. Berüter, P. Fischer, E.J. Windhab, *Chem. Eng. Technol.* **2002**, *25*, 499–506.

23 C. Cramer, P. Fischer, E.J. Windhab, *Chem. Eng. Sci.* **2004**, *59*, 3045–3058.

24 U. Lambrich, PhD dissertation, Universität Karlsruhe, **2008**, in preparation.

25 U. Lambrich, P. Idda, H. Schubert, H.P. Schuchmann, *Tropfenbildung durch Strahlzerfall – ein neuer Ansatz für das Membranemulgieren*, GVC/DECHEMA, Frankfurt, **2006**.

26 Abschlussbericht PRO INNO II: *Entwicklung einer Membrananlage zur Herstellung von nahezu monodispersen Emulsionen mit direkter Inline-Kontrolle der Emulsionseigenschaften*. Projekt (Inovationskompetenz II (PRO INNO II) KF0067201FK4), **2007**.

27 K. Suzuki, K. Hayakawa, Y. Hagura, *Food Sci. Technol. Res.* **1999**, *5*, 234–238.

28 G.G. Badolato, G. Krug, H.P. Schuchmann, H. Schubert, *Study of Process Parameters for Emulsification Using Plate Membranes*. 6. Slaca: Simpósio Latino Americano de Ciência de Alimentos, Brazil, **2005**.

29 *Report of Pro-3 Verfahrenstechnik scholarship holder*, Mitarbeiter, Gastwissenschaftler in der LVT, Giulliana Krug, **2007**, www.lvt.uni-karlsruhe.de.

30 G.G. Badolato, G. Krug, H.P. Schuchmann, H. Schubert, *Abstract of "Max-Buchner-Forschungsstiftung"*, 2484, **2007**, www.dechema.de/Abgeschlossene_Projekte.html.

31 H.S. Ribeiro, L.G. Rico, G.G. Badolato, H. Schubert, *J. Food Sci.* **2005**, *70*, 117–123.

32 G.G. Badolato, G. Krug, H.P. Schuchmann, H. Schubert, *Premix Membrane Emulsification: Higher Flux and Higher Dispersed Phase Concentration in Membrane Process*, Congrès Mondial de l'Emulsion, Lyon, **2006**.

33 S. Oliczewski, R. Daniels, *Pulse Laser – an Innovative Tool for the Stability Assessment of Emulsions and Creams*. Proceedings of the International Meeting on Pharmaceutics, Biopharmaceutics and Pharmaceutical Technology, Nuremberg, **2004**.

34 U. Lambrich, S. van der Graaf, K. Dekkers, H. Schubert, R.M. Boom, *Production of Double Emulsions Using Microchannel Emulsification*, Proceedings of ICEF 9, France, **2004**.

# 14
# Towards Fouling Monitoring and Visualization in Membrane Bioreactors

*Yulita Marselina, Pierre Le-Clech, Richard M. Stuetz, and Vicki Chen*

## 14.1
## Introduction

The past 20 years have seen a significant change in the industry of wastewater treatment with the implementation of the membrane bioreactor (MBR) technology [1]. With its numerous advantages over conventional activated sludge processes (CASP), MBR technology has become an option of choice for the treatment of wastewater. These advantages include higher effluent quality, good disinfection capability, higher volumetric loading and smaller footprint. Although the operating and maintenance costs of MBR are higher than those of CASP, the growing concerns of water resource availability and the higher standard of environmental legislations have driven the recent development of MBRs worldwide. In 2005, the MBR market was estimated at around US$ 216 million and forecasted to rise to US$ 363 million by 2010 [2].

As with any membrane process, membrane fouling occurs in MBRs and remains one of the major considerations faced by MBR operators. Fouling is generally defined as the unwanted deposition and accumulation of soluble and particulate materials onto or into the membrane surface, resulting in a decrease in the hydraulic performance of the process. In an MBR, fouling mainly originates from biological materials and leads to the formation of biofilms on the membrane surface. Incidentally, a more significant degree of membrane fouling can lead to membrane cleaning or even membrane replacement [3, 4].

Fouling behavior and mechanism can be assessed through the traditional measurement of the permeate flux and the transmembrane pressure (TMP), the monitoring of rejection and the use of empirical models based on these parameters. Moreover, observation techniques (both invasive and non-invasive) can be used for direct or indirect fouling visualization, therefore proposing an improved understanding of the fouling behavior on the membrane surface. The specific nature of the fouling obtained in MBR processes presents significant challenges to the use of these techniques and the purpose of this chapter is therefore to evaluate

*Monitoring and Visualizing Membrane-Based Processes*
Edited by Carme Güell, Montserrat Ferrando, and Francisco López

ISBN: 978-3-527-32006-6

the potential of various imaging techniques which can be used to visualize the fouling layer in MBR systems.

## 14.2
## Factors Affecting Fouling in MBR

In MBR, membrane fouling can be affected by the nature of the feed, the membrane properties and the hydrodynamic or operating conditions [4]. Better understanding of those interactions is necessary to implement more efficient fouling control and removal strategies. A brief summary of recent research conducted in the MBR field may also allow us to further comprehend the chemical and physical natures of the fouling layer we aim to observe with the different techniques introduced later in this chapter.

### 14.2.1
### Nature of the Feed

Due to the presence of living microorganisms, extracellular polymeric substances (EPS), soluble microbial products (SMP) and other particulate, colloidal and soluble materials in the MBR mixed liquor, the nature of the membrane fouling is significantly complex and variable. The EPS and SMP composition is controlled by different microbial processes, such as active cell secretion, shedding of cell surface matcrial, ccll lysis and adsorption from environment [5]. In wastewater membrane filtration systems such as MBR, these biopolymeric substances have been reported as some of the major foulant contributing to fouling [6]. Although their exact role in the MBR fouling formation is still unclear, the participation of these substances in the formation of biofilms has been widely reported [7].

The feed particle sizes, which range from colloidal materials up to large flocs, also contribute to fouling in MBR. The particle size characteristics in the bioreactor are influenced by operating parameters such as the dissolved oxygen concentration and the addition of a flocculation agent. Dissolved oxygen concentration has been reported to affect the microbial characteristics, such as flocculation and settling properties [8–10]. Low concentration could lead to severe membrane fouling due to poor flocculation, causing the increase concentration of soluble components and specific cake resistance [9]. The addition of a flocculation agent generates higher binding forces between the SMP and the bacteria, producing a larger mean particle size of activated sludge in MBR [11, 12].

More fundamental research based on the effect of the feed on MBR fouling has been conducted by using model solutions to mimic the major foulants found in the mixed liquor. Examples of model solutions that have been previously used include:

- Alginate as microbial polysaccharides [13].
- Yeast (which contains cell debris and soluble materials) [14].
- Bovine serum albumin (BSA) as protein [13].

- Inert particles like bentonite (mean diameter of 2.7 μm) [15] or latex (mean diameters of 3.0, 6.4, 12.0 μm) [14].
- Mixture solutions like alginate–BSA [13] and yeast–BSA [16].

### 14.2.2
### Membrane Properties

Fouling formations can also be affected by properties of the membrane itself, such as pore sizes and their distributions, surface roughness and membrane configurations. As expected, the effect of pore sizes and their distributions on the fouling formation are strongly related to the particle size distribution of the feed solution [4]. Smaller membrane pores tend to reject a wider range of materials but the resulting cake layers are expected to have higher resistances and to form more reversible fouling compared to those of larger pores. In contrast, filtration with larger membrane pores allows some of the foulants to deposit inside the membrane pores and to cause internal fouling, which normally forms irreversible fouling [4]. Membranes with higher surface roughness (and therefore presenting more shelters from the shear force of liquid flow) tend to exacerbate the fouling behavior [17]. Membrane module configuration also affects the fouling formation in MBR, as filtration permeability in submerged MBR is reported to be twice as high as those obtained for sidestream configuration at given operating conditions [18].

### 14.2.3
### Operating Conditions

As mentioned previously, activated sludge characteristics can vary significantly with MBR operating conditions. Aeration rate, filtration mode, dissolved oxygen level or sludge retention time (SRT) are some of the operating conditions that can have a strong influence on the nature of fouling obtained.

Aeration provides complex turbulence near the membrane surface, which increases particle back-transport to the bulk [19]. In the case of hollow fiber membranes, the particle back-transport mechanism can also be enhanced by lateral fiber movements resulting from the rising of bubbles. Fouling formation can also be limited by applying filtration pressure relaxation [20] or periodic back-washing [21]. Under these operating conditions, the fouling propensity is limited by minimizing the continuous build-up of reversible fouling deposition.

As it controls the nature of the biomass population in the reactor, the SRT remains one of the most important parameters in MBR operation. Operating an MBR at higher SRT leads inevitably to an increase in the mixed liquor suspended solid concentration but this, in itself, may not necessary imply greater fouling. A recent literature review on fouling in MBR reports the contradictory trends found in scientific publications [4]. While some studies reported higher fouling rates observed at lower SRT, supposedly due to the greater level of SMP, other experiments could not correlate SRT with membrane permeability.

## 14.3
## Fouling by Biological Material

When it comes to fouling in biological active processes like MBR, the terminology used may appear slightly confusing. Updated definitions for biofouling and biofilm are therefore proposed in this section. Biofouling is described as the undesirable deposition of materials of biological origin on a surface [22, 23], which participates in the reduction of hydraulic performances in MBR systems. Biofouling can be further described as the initial attachments of SMP onto a surface through adhesive forces during either passive adsorption or filtration condition (Figure 14.1a). During the filtration of mixed liquor in MBR, bacteria and other colloidal particles may then attach by cohesive mechanisms to the membrane surface already covered by SMP (Figure 14.1b). As the mixed liquor filters through the fouled membrane, it provides nutrients and dissolved oxygen to the deposited bacteria. As a result, the immobilized bacteria assimilate to the surrounding environment by producing EPS and by forming a complex structure: the biofilm layer [24, 25].

Biofilm formation therefore relies on early bacterial attachments and subsequent growth of bacteria on the membrane surface (Figure 14.1c) [7, 26]. It has also been reported that biofilm develops with more difficulty if the operating conditions exceed a threshold of interference, such as a given amount of available nutrients for biomass growth and shear force at the membrane surface [27]. In the initial stage of formation, the biofilm may not uniformly cover the surface, but the colonization tends eventually to spread across the overall surface. Dislocation of sloughing biofilm may initiate a new colony in another section of the membrane area [28]. The established biofilm is highly heterogeneous, features high moisture content and consists of both non-water-permeable materials (cells, scale, debris) and water-permeable substances (biopolymers) [22, 29].

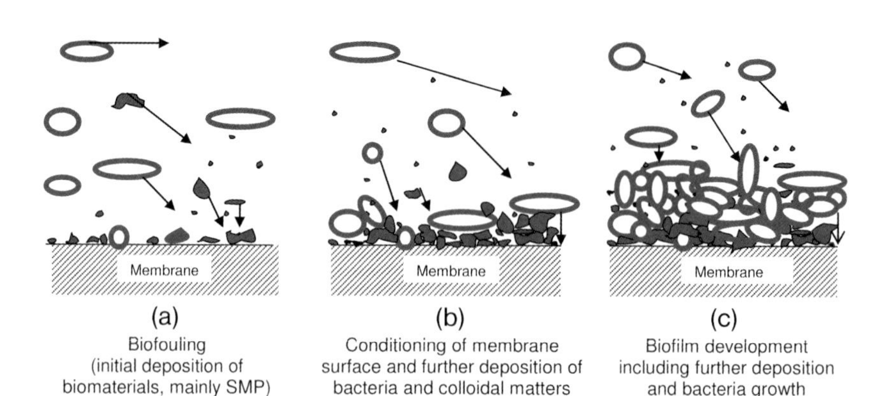


**Figure 14.1** Three stages during biofouling and biofilm formation (a–c) on a membrane surface. Black shapes indicate SMP, white ovoid shapes indicate bacteria surrounded with EPS.

The presence of dead and living bacteria in the biofilm has been showed to influence its physical characteristics. While non-starving cells provide higher adhesion characteristics and therefore an evenly distributed fouling layer, starving biomass tends to form a more irregular fouling layer on the membrane surface [7]. Significant differences are also observed when fouling mechanisms obtained during the filtration of biologically active particles are compared with those from filtration of inert-based compounds [30]. For example, in the early stage of biofouling layer formation, the deposition of bacteria occurs relatively quickly compared with the attachment of inerts onto the surface. This is due to the adhesive characteristics of the EPS present within the biological material. Once a layer of biofilm forms on the membrane surface, the membrane performance decreases due to further reversible or irreversible bacteria attachment.

## 14.4 Fouling Characterization Without Visualization Techniques

### 14.4.1 Permeate Flux and Transmembrane Pressure

Fouling propensity is commonly assessed by monitoring permeate flux and transmembrane pressure (TMP). Since membrane processes are generally operated either under constant TMP or constant permeate flux, a decrease in permeate flow rate or an increase of TMP is observed, respectively, once the fouling forms on the membrane. From these two parameters, calculation of hydraulic resistance ($m^{-1}$), membrane permeability ($L\,m^{-2}\,h^{-1}\,bar^{-1}$) or specific cake resistance ($m\,kg^{-1}$) is also possible and allows further assessment of fouling conditions. The hydraulic resistance of the filtration system can be quantified by correlating TMP and permeate flux during a clean water test. This correlation is described as the Darcy's law:

$$J = \mathrm{TMP}/R_{\mathrm{total}}\cdot\eta \tag{14.1}$$

With permeate flux ($J$, $m\,s^{-1}$) as a function of TMP (Pa), total resistance ($R_{\mathrm{total}}$, $m^{-1}$) and permeate viscosity ($\eta$, Pa s) [31]. $R_{\mathrm{total}}$ consists of the intrinsic membrane and fouling, both reversible and irreversible resistance. Furthermore, membrane permeability ($K$) is a function of flux on TMP and it also can be monitored to assess fouling.

$$K = J/\mathrm{TMP} \tag{14.2}$$

Specific cake resistance can be calculated as a function of flux, TMP and deposited mass [15, 32]. It is generally used to characterize the fouling cake structures, either in dead-end [32] or crossflow filtration [33]. Table 14.1 provides a list of specific cake resistance values for activated sludge, which was studied with different MBR operating conditions.

**Table 14.1** Specific cake resistance values obtained during biomass filtration in MBR.

| Varied operating conditions | Specific cake resistance ($\times 10^{-12}$ m kg$^{-1}$) | Ref. |
|---|---|---|
| Mixed liquor suspended solid concentration (4000–10 000 mg l$^{-1}$) | 3.5–13.0 | [87] |
| SRT (20–100 days) | 6–12 | [88] |
| Dissolved oxygen concentration (<0.1 mg l$^{-1}$, >3.0 mg l$^{-1}$) | 10–1000 | [10] |

### 14.4.2 Empirical Fouling Models

The mechanisms of membrane fouling can be predicted with empirical mathematical models, which are generally a function of TMP and flux [34, 35]. The fouling models, such as cake filtration, pore, standard, intermediate and complete blockage, are derived from Darcy's law [34, 36, 37]. Those fouling models can be applied for either constant flux [38, 39] or constant TMP [36, 40] operation.

Fouling models of pore blockage in the early stage of filtration followed by cake filtration were used to predict TMP profile during MBR operation, which was carried on under sub-critical constant flux mode [39]. The application of more rigorous fouling models was also studied to predict TMP during the treatment of municipal wastewater in MBR [38]. The implementation of adhesive forces and distribution of particles and membrane pore diameters were taken into account in the calculations of a more reliable prediction of the process behavior. The maximum error between the prediction and the experiments was around 7%, but the model slightly overestimated TMP at low flux and underestimated system performance at high flux.

### 14.4.3 Rejection Performance

During filtration, the rejected solutes or particulates can diffuse back to the bulk or accumulate on the membrane surface and inside the pore. This propensity is generally determined by measuring the level of pollutants in samples of the feed and permeates. However, it is especially relevant to study the specific rejection performances for the components that more prone to foul the membrane. A higher rejection value of these compounds would indicate their potential deposition on the membrane surface. Protein, carbohydrate and total organic content rejection can be monitored throughout the filtration process in MBR. During the filtration of activated sludge in MBR, protein and carbohydrate rejections were quantified as 20–65% and 75–95%, respectively [41]. However, more detailed fouling characterization can be obtained by extracting the fouling layer from the membrane. Solute rejections can then be measured in terms of SMP and EPS fractions [31]. It has been reported that SMP has a higher rejection value than EPS solution, indicating that SMP may cause more fouling than EPS.

## 14.5 Invasive Methods for Fouling Observation

Visualization with invasive methods in membrane application requires removing the membrane from its module. Losses, shrinkages and disturbances of the sample materials are likely to occur during the sample preparations and pose particular challenges for MBR applications. Each technique is discussed here for its potential to visualize and characterize the fouling layer in MBR applications.

### 14.5.1 Electron Microscopy

#### 14.5.1.1 Scanning Electron Microscopy

Electron microscopy is one of the main, traditional techniques for membrane characterization. Scanning electron microscopy (SEM) uses the high energy of an electron beam that scans across the sample. The interactions between electron beam and the surface produce an emission of back-scattered electrons and secondary electrons. The signals are detected, processed and then sent to a cathode ray tube scanning synchronously. SEM operates under vacuum condition ($<10^{-4}$ Pa) and the sample is required to be dried. Due to the presence of a high energy electron beam, the sample is also required to be coated with a thin layer of gold or carbon in order to disperse the surface charges [42]. But in practice, the high energy beams can still cause burning of polymeric samples even with a metal coating. SEM allows observation of the topographical surface and has a resolution of 1 nm under optimal sample conditions.

SEM has been utilized to observe the fouling layer morphology of a tubular membrane used to filter wastewater effluents. In this study, it was found out that the predominant fouling layer was composed of bacteria and EPS [43]. SEM was also used to visualize the cleaned fouled membrane in order to see the effectiveness of cleaning methods, such as sonication, chemical cleaning, backwashing and a combination of cleaning after membrane filtration with sludge in MBR [40].

The biofilm, or hydrogel, often formed in the MBR has to be dried prior to SEM observation. Hydrated fouling layer can be highly distorted by cracking and shrinkage during the drying process and high-vacuum operation, providing great limitations in the interpretation of images. Yet, SEM still can provide a broad survey of gross structures on the membrane surface. Other applications of SEM to characterization of membrane processes can be found in Chapter 3.

#### 14.5.1.2 Field Emission Scanning Electron Microscopy

Field emission scanning electron microscopy (FESEM) is adapted from SEM and combines the system of a field emission electron source smallest scanning probe to produce an ultra-high resolution of 0.7 nm. FESEM uses lower accelerating voltages than any other electron microscope. Thus it reduces beam-induced damage to the specimen and has been successfully used to characterize membrane structures down to the ultrafiltration (UF) range [44]. As with SEM, the specimen for FESEM analysis also requires a light coating with platinum or other metal to prevent charge build-up on the surface of the membrane.

The FESEM technique has been used mainly to characterize virgin membranes from UF to microfiltration (MF) pores sizes [44, 45]. Image analysis can provide information on the average pore distributions on the surface, which is not always readily available from the membrane manufacturers. A FESEM autopsy of fouled membrane used in wastewater revealed the presence of trapped bacteria in the fouling layer [46]. FESEM characterization of a fouled membrane, used to filter wastewater, can also give some indication of pore plugging or cake build-up by organic matter or colloidal particles [47, 48]. FESEM technology has only been used occasionally for MBR study, since it still has similar disadvantages as SEM techniques with regards to artefacts introduced by the drying process and high vacuum.

#### 14.5.1.3 **Environmental Scanning Electron Microscopy**

More recently, the introduction of environmental scanning microscopy provided an important alternative which avoided drying artifacts, particularly in the foulant layer. Environmental scanning electron microscopy (ESEM) can operate in either high-vacuum mode (dried sample) or low-vacuum mode (hydrated conditions). Fully saturated water vapour conditions as high as 7 kPa (50 Torr) and the elimination of any sample coating allow sample analysis without dehydration [42, 49].

Fouled membrane used in MF of wastewater was visualized with the ESEM technique to assess the effectiveness of lime-softening pretreatment prior to filtration. ESEM revealed that the fouling layer with pretreatment process was more porous, thus enhancing the membrane filterability [50]. In another study, the suitability of SEM and ESEM were compared to visualize colloidal materials on the membrane surface [42]. In this case, ESEM visualization was preferred due to its ability to perform observation in hydrated states. However, this technology has also been used in a limited way in MBR study, due to its lower image resolution than SEM.

#### 14.5.1.4 **Transmission Electron Microscopy**

In transmission electron microscopy (TEM), electrons are transmitted through the sample, allowing detailed visualization of the internal cross-sectional structure of individual microorganisms. TEM runs at high voltage (60–100 kV) and the penetrating power of electron beam is poor, thus TEM works better for thin samples [51]. Staining is required on the sample to provide image contrast during analysis.

In membrane application studies, imaging by TEM showed the presence of an irreversible biofilm on the fouled membrane used to filter surface water in reverse osmosis [22]. Then, quantitative analysis on the spatial arrangements and cellular ultrastructures of biofilm layer growth on a supporting surface was also clearly

**Table 14.2** Comparison of the different electron microscope techniques.

| | SEM | FESEM | ESEM | TEM |
|---|---|---|---|---|
| Sample preparations | Dried and coated with gold or carbon | Lightly coated with platinum | Not required | Thinly sliced and stained |
| Sample conditions | Dried | Dried | Dried or wet | Dried |
| Resolutions (nm) | 1 | 0.7 | 10–20 | 4 |
| Acceleration voltages (kV) | 5–45 | <5 | 10 | 100–300 |

visualized [49]. In another experiment, polystyrene particles were filtered with polysulfone membrane in a crossflow filtration [52]. Two different types of layers (one dense, the other more porous) were observed clearly when the membrane was cut tangentially. However, the TEM technique has not been used widely to characterize the fouled membrane in MBR due again to the need for substantial sample preparation and the potential of artefacts drying.

The differences of the electron microscope techniques (SEM, FESEM, ESEM, TEM) in terms of sample preparations, sample conditions, resolutions and acceleration voltage are listed in Table 14.2.

### 14.5.2 Atomic Force Microscopy

Atomic force microscopy (AFM) is a non-destructive technique that is able to produce 3D surface morphology images with a resolution of 2–10 nm. This technique is able to provide visualization at the atomic–molecular scale and is also employed to contour the sample surface. The probe is commonly made from silicon or silicone nitride, but it can also be coated with other materials, such as colloids (cellulose, latex) or microorganism cells (yeast, spores) [53]. AFM analysis can be done under hydrated conditions or carried out in fluids, avoiding sample damage during the drying process and dissipation of the electrostatic forces [54]. Three different types of scanning modes, such as contact, non-contact and tapping mode, can be specified, depending on the distance between probe and surface [55]. A more detailed description of the technique, as well as other applications to membrane characterization can be found in Chapters 5 and 6.

AFM analysis can provide an insight of sample morphology, giving important information on single colloid/cell interaction on a substrate [53]. The interaction forces between surface and probe can be measured and plotted as a function of its distance, giving an indication of the strength of adhesive forces [56]. The adhesion property of biofilm growth on a stainless steel substrate was studied with AFM under hydrated conditions [57]. Visualization under this condition enabled the observation of the distribution of EPS present in the biofilm layer.

Hydrated biofilm can also be observed with AFM, providing significant information of biofilm rheology and adhesion under repeated raster-scanning probe at a certain elevated load [58]. These repeated raster-scanning movements cause some displacement of material, from which its volume can be calculated as a function of raster-scan area and average depth in its abrasion. Then, the friction force can be quantified according to its AFM images as a number of image pixels within incremental friction force intervals. Thus, the biofilm cohesive energy per unit volume can be determined as a function of frictional energy force and volume of material detachments. The different thickness layers of fouled membrane, used in MBR at different ages of the biofilm, were studied for biofilm cohesive energy per unit volume [59]. It was found that biofilm cohesive energy correlated to polysaccharide concentration, which increased with the depth of biofilm layer under aerated condition.

The morphology visualization, adhesion and cohesive characteristics during biofilm analysis can provide a useful insight for fouling study in MBR. Only limited amounts of AFM analysis have been used on MBR surfaces; however this technique has the potential not only to provide imaging but also local force/rheological measurement of biofilm adhesion and structure.

### 14.5.3
### Confocal Laser Scanning Microscopy

While confocal laser scanning microscopy (CLSM) is an invasive optical technique, it is a mainstay in the biological and medical sciences in terms of imaging 3D structures. CLSM is a combination of an epifluorescence microscope and a laser light source, such as argon ion, helium–neon, krypton–argon, helium–cadmium or UV. These laser sources have different ranges of excitement wavelength, which are used to emit the specific fluorescent signal of specimens. The unique features of CLSM compared with other visualization techniques are the ability to identify fluorescent substances in the specimen and to construct a 3D structural image, obtained from optical sectioning [60]. Each different fluorescent substance can be clearly located and quantified. For instance, dead and living bacteria [61], cells [62] and foulant macromolecular materials, such as polysaccharides and proteins [9], can be stained within the membrane fouling layer (Table 14.3). The fluorescent markers are generally mixed with biofilm specimen, incubated at room temperature in the dark and the marker excess is then washed out with buffer solutions.

As mentioned in Section 14.3, biofilm is heterogeneous and contains dead and living microorganisms. The images obtained by CLSM showed that dead bacteria accumulated in the sublayer in thickened biofilm on a flat sheet hydrophilic polypropylene membrane, used in MBR [61]. In other MBR-based studies, CLSM was used to assess the fouled membrane at different dissolved oxygen concentrations [8–10]. The fouled membrane was stained and visualized for spatial distributions of polysaccharides, protein and nucleic cells (see Table 14.3 for equivalent fluorescent marker). Biofilm porosity, thickness, uniformity and heterogeneity were also quantified, allowing the determination of the membrane fouling characteristics for dissolved oxygen concentrations. The fouling rate at a high concentration of dissolved oxygen was found to be about 7.5 times slower than at low concentrations. However, the high dissolved oxygen level led to a thicker and more porous biofilm deposit layer [8–10]. Thus, membrane permeability is not necessarily a function of biofilm thickness but can also be characterized by the porosity of the fouled membrane. A low dissolved oxygen concentration has a higher fouling rate due to poor flocculation resulting in the presence of higher number of smaller particles.

**Table 14.3** Fluorescent markers, used in MBR fouling layer visualizations.

| Observations | Fluorescent marker | Ref. |
|---|---|---|
| Polysaccharides | Concavaline A | [8, 9, 11] |
| Protein | Hoechst 2495 | [9, 11] |
| Nucleic acid (cells) | Sybr green I | [8–12] |
| Living bacteria | Cytochrome 9 | [61] |
| Dead bacteria | Propidium iodide | [61] |

The strength of the CLSM technique lies in its potential to visualize cake structure and provide foulant characterization with minimal disruptions to the foulant cake. This technique is becoming rapidly adopted in the membrane community as a means to characterize biofilms formed during the filtration of biological material. Drawbacks still remain in the removal of the membrane sample from the feed solution and the potential for loosely bound layers to detach during the process. These loosely bound layers may still provide a significant hydraulic resistance that would not be captured in the CSLM analysis. More information on CSLM uses in membrane and membrane process characterization can be found in Chapter 4.

## 14.6 Non-invasive Observation Methods

A non-invasive observation method is able to monitor the changes during the filtration process and is performed without removing the membrane from its module. This technique can minimize sample destruction, allowing more accurate visualization analysis compared with the invasive observation method previously discussed.

## 14.6.1 Projector Technique

Projector technique (PT), one early application of optical techniques to membrane fouling visualization, utilizes a projector and screen to monitor fouling deposition. This technique was applied to measure biofilm thickness on the silicone rubber tubular membrane during wastewater filtration [63]. The thickness was then measured and photographed from direct visual observation on the projector screen.

From this direct visualization, the biofilm layer was shown not to be uniform, but featuring filaments and extensions [63]. The resolution of the PT technique is 1 mm; therefore it is unable to give information at the level of individual bacterial cell interactions. However, the PT technique can provide useful aggregate-level information about biofilm growing onto a tubular silicone membrane.

## 14.6.2 Microscope Observation

### 14.6.2.1 Direct Observation Through Membrane

To provide higher resolution, optical microscopy was utilized in a number of configurations. The direct observation through membrane (DOTM) setup consists of a microscope, a camera and a crossflow membrane module. Particle deposition on the membrane can be observed directly with DOTM thanks to the Anopore flat sheet membrane, which becomes transparent in wet conditions [14]. Particles larger than 1 μm deposited onto the membrane surface can be clearly identified with this technique. Submicron bacteria can also be visualized by using fluorescent microscopy and bacterial staining [30].

When yeast and latex particles were used as a model solution for DOTM experiments, it was found that the foulants were more likely to accumulate around existing depositions due to variations in the local pore size distribution [14]. The foulant deposition on the membrane could also be observed with direct visual observation (DVO), equipment similar to the DOTM setup, with the difference that the microscope focused on the feed side [64]. Both of these setups allowed the quantification of fouling surface coverage as a function of filtration time. Patchy foulant deposition was viewed after 5 min of filtration with 82% of surface covered with yeast [64] (see Chapter 2).

The foulant movements for either particles or bacteria can be summarized into three consecutive behavioral movements [14, 30]:

1. Foulants moved along the membrane without stopping (rolling or sliding).
2. Foulants then stopped on the membrane surface but most of them soon were moved away (momentary deposition).
3. Finally, permanent deposition started to occur.

Due to their simpler physiochemical properties, the deposition rates of model particles, which were observed with a direct observation technique were much lower in magnitude than those observed with yeast or bacteria cells [65]. Similarly, the bacteria fouling removal was much slower than non-biological colloids and took the shape of rolling flocs and detachment of large aggregate groups [30]. These techniques have been mainly used to show fouling coverage on the membrane surface while quantifications of cake thickness still remain difficult to determine.

#### 14.6.2.2 Direct Observation on Hollow Fiber Membrane

The DOTM experimental setup could be easily adapted to allow the direct observation (DO) of the fouling occurring on the surface of a hollow fiber membrane. The DO apparatus comprises a modified crossflow module, a microscope and a video camera (see in Figure 14.2). The membrane module can be horizontally mounted onto the optical bright-field microscope (Axiolab, Zeiss) with 10 times of magnification objective lens. During the visualisation, the hollow fibre membrane was observed as dark/black colour, and the camera could be easily focussed on the edge of the hollow fibre membrane.

The effect of bubbles passages on the fouling formation was studied with a DO setup [66]. The system did not capture the images of bubble passages onto the membrane surface due to camera speed limitation, but the consequences of the bubbles on the deposited cake were recorded. As expected, the introduction of bubbles into the system led to a limitation of the cake layer formation by enhancing particle back-transport and reducing the particle adhesion to the membrane surface.

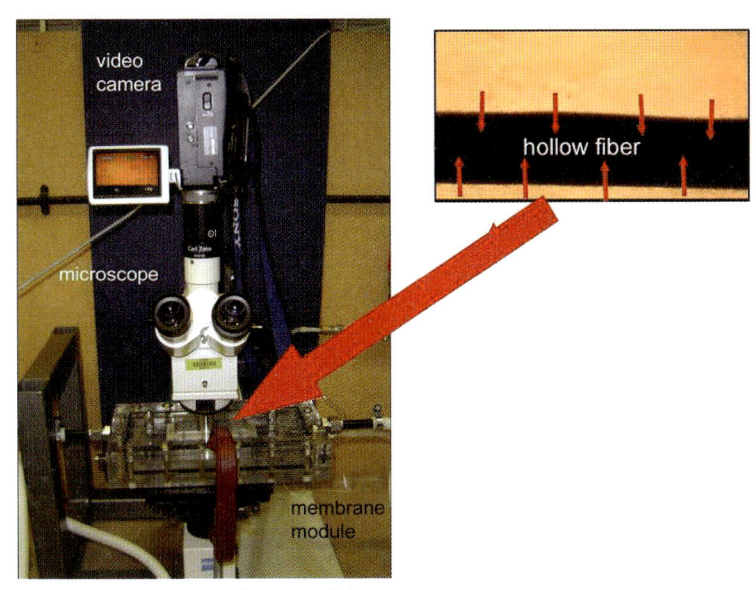


**Figure 14.2** The DO membrane module.

Fouling phenomena during the filtration of alginate and bentonite mixture were observed with the DO technique [67, 68]. However, detailed characterization of particulate material by using a bentonite suspension was also required to understand its fouling mechanisms. In order to assess the reliability of the DO technique, the experiments of 200 mg $l^{-1}$ of bentonite concentration, at 15 mm $s^{-1}$ have been repeated five times [69]. The TMP was measured simultaneously giving an indication of cake resistance during the cake formation process. The movement of bentonite particles near the membrane surface (up to 800 μm) were clearly visible under the microscope. The still images in Figure 14.3 were captured at different filtration times of 0, 15, 60 and 135 min. This DO setup provides in-situ observation tangentially to the membrane surface; the cake height and particle velocity can therefore be characterised. It was observed that the fouling deposition mechanism was characterised by the formation of both stagnant and fluidised cake layers. The particles gradually deposited on the membrane surface, the fouling thickness measured at 15, 60 and 135 min were 30, 70 and 180 μm, respectively. The fluidised cake height ($H_{fc}$) measured at the beginning of filtration was 5–10 μm and stabilised at 30–50 μm after 90 min of filtration.

At the end of filtration, the fouled membrane was cleaned by applying backwashing while the same CFV of 15 mm $s^{-1}$ was maintained. Figure 14.4 shows the temporal changes of fouling layer morphology during the cleaning period at time 1, 4, 6, 7, 8 and 9 min. During cleaning, it was observed that the fouling layer can be characterised into two different structures, expanded and fluidised layers. The height of the expanded cake ($H_{ec}$) gradually increased with cleaning time; its values at cleaning times of 4, 5 and 6 min were 230, 300 and 320 μm, respectively. As the $H_{ec}$ increased, the fouling structure was observed to become weaker and more porous compared to the fouling layer in the beginning of cleaning. The upper layer

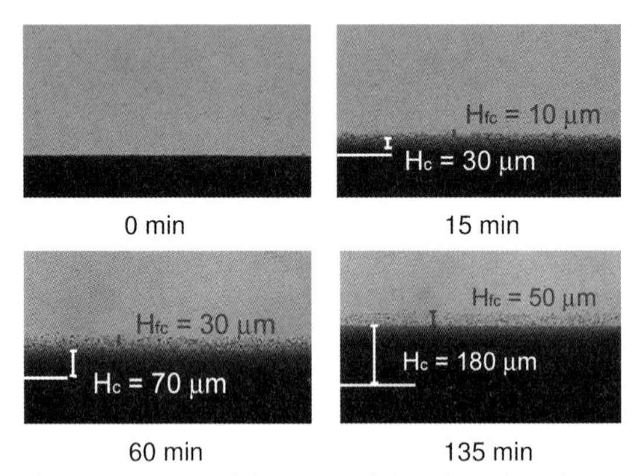


**Figure 14.3** Temporal changes in fouling deposition for 200 mg $l^{-1}$ bentonite filtration (CFV of 15 ± 0.2 mm $s^{-1}$, flux of 75 l $m^{-2}$ $h^{-1}$). The hollow fibre membrane is visible as the dark area below the white line. $H_c$ and $H_{fc}$ indicate stagnant cake height and fluidised cake height, respectively.

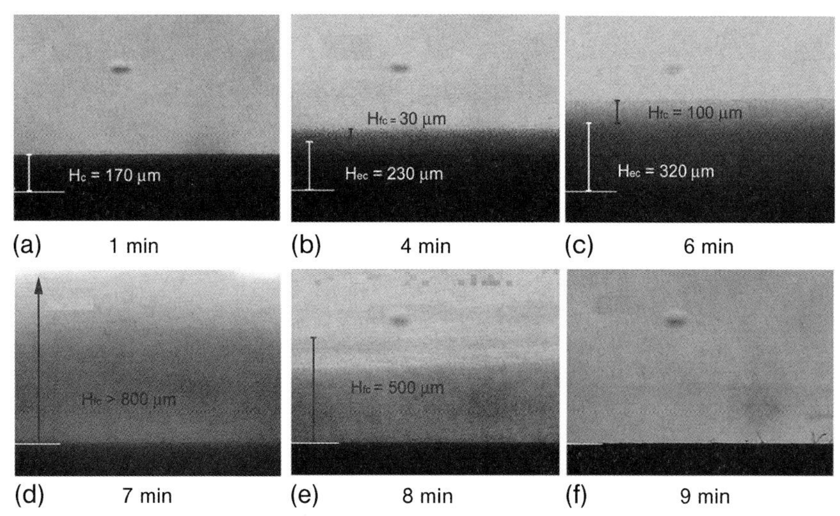


**Figure 14.4** Fouling removal (backwash and CFV 15 mm s$^{-1}$) mechanisms at cleaning times of 1 min (a), 4 min (b), 6 min (c), 7 min (d), 8 min (e) and 9 min (f). $H_c$, $H_{ec}$ and $H_{fc}$ indicate the height of stagnant fouling, expanded fouling and fluidised cake, respectively.

of the expanded cake was gradually eroded, creating a fluidised cake. The $H_{fc}$ during the cleaning also increased with cleaning time; the $H_{fc}$ values at cleaning time of 2, 4, 5 and 6 min were 10, 30, 40 and 100 µm, respectively. Finally, the porous fouling layer expanded to a critical point at which it became totally fluidised at 7 min, with its height greater than 800 µm. As a result, most of the fouling was easily removed after 8 min of cleaning through the crossflow and backwashing.

The standard deviation related to parameters of the results, such as TMP, fouling height and the cleaning time, obtained by the DO technique was approximately 20%. This technique allows monitoring multilayer deposit quantitatively in function of filtration time, as well as the assessment of local velocities near the cake surface by tracking particle velocities. The DO technique has significant potentials to be explored for MBR study, since filtration of activated sludge, effect of filtration mode and effect of bubbling would be assessed. However, as with all optical microscopies, solution turbidity is a limiting factor in its application.

#### 14.6.2.3 Membrane Fouling Simulator

A membrane fouling simulator (MFS) provides a similar visualization method to DOTM and DVO. This technique features microscopic visualization of the surface of a spiral wound membrane and a crossflow membrane module constructed from stainless steel plate. TMP during filtration is also measured, while fouling deposition of biomass on spacer is visualized simultaneously. MFS is easy to handle, simple, robust and small, with a flow capacity of 15–25 L h$^{-1}$ [70, 71].

## 14.6.3 Laser Applications

### 14.6.3.1 Laser Beam Excitation

In this technique, a laser beam of He–Ne focuses tangentially onto the surface of tubular membrane. A fraction of the laser light is absorbed by the deposited fouling layer and the signal reduction is monitored to translate the deposited thickness [72].

This technique was used to measure the fouling thickness deposited on a tubular membrane. Bentonite as a particulate model solution was filtered during fouling characterization experiments. The maximum CFV and solid concentration used for this analysis were $0.3\,m\,s^{-1}$ and $375\,mg\,L^{-1}$, respectively [72]. The limitations of the technique are the inability to visualize the cake structure and the requirement of light adsorption into the lumen, reducing the visualization potential for MBR study. Extended information can be found in Chapters 11 and 15.

### 14.6.3.2 Laser Excitation Near Infrared Region

Laser excitation near infrared, which is known as 3D femtosecond imaging (3DFI) can be used as non-invasive fouling visualization technique. The laser is installed in series with a microscope and camera to observe the fouling layer directly on the flat sheet membrane in a crossflow membrane module [73]. The 3DFI can be used in conjunction with fluorescent label, similar to CLSM; however it can only provide a laser range in the infrared region (see Chapter 8).

The fouling deposition of yeast was visualized by labeling it according to its cell wall, dead cells and living cells [73]. The limitation of this device was the inability to observe a fouling layer of more than 45 μm. Therefore, the application of a high concentration and other parameters, such as crossflow velocity (CFV) and length of filtration, was limited. However, use of the fluorescent technique has the potential to analyze the key foulant species within the fouling layer during filtration in MBR by fluorescent tagging without removal from the filtration cell.

## 14.6.4 Ultrasonic Time Domain Reflectometry

The ultrasonic time domain reflectometry (UTDR) technique uses sound waves to measure the fouling height on the membrane surface during crossflow filtration and cleaning process. This technique transmits sound waves onto a membrane module consisting of many layers, including the upper plate of the membrane module, fouling layer, membrane and bottom layer of the crossflow system. The sound waves are introduced without altering or damaging the clean or fouled membrane and are recorded in terms of amplitude versus time. The arrival time of the reflected signal decreases when fouling is formed on the membrane surface. The minimum thickness that can be detected with UTDR is 100–750 nm. Thus, the progressive deposition of materials onto the membrane surface can be clearly recorded by this technique.

The UTDR technique has been applied to monitor fouling growth and to estimate fouling density on the membrane surface by filtering paper mill effluent [74] and calcium sulfate [75] by flat sheet membrane and by tubular membrane module [76]. The fouling growths and removals during the filtration of paper mill effluent were observed according to its response signals for 20 h. It was observed that fouling thickness increased rapidly in the beginning of filtration, followed by a slow increase. The signal at 20 h of filtration had a sharper wave than the one obtained at the beginning of filtration. Subsequent SEM image showed that a sharper wave indicates a denser or more compressible fouling layer [74]. In another experiment, the fouled membrane, used to filter calcium sulfate for 7 h, was autopsied with SEM and its cross-sectional image confirmed the fouling thickness reading by UTDR [75].

UTDR has been successfully used to monitor the fouling growth and density on membrane surfaces, either on a flat surface or on a curved surface (tubular membrane). However, the UTDR result does not provide direct optical visualization and it requires to be confirmed by another visualization technique to identify the characteristics of peak amplitude signal and relate them to a specific foulant deposition. This technique can be applied to monitor fouling in MBR because it is sensitive enough to detect the changes in fouling layer without labeled species and optically transparent molecules; but it requires significant effort in signal analysis to correlate foulant thicknesses and densities (see Chapters 11 and 15).

### 14.6.5
### Electrochemical Shear Probe

An electrochemical probe can be used to measure the shear forces at the surface of a membrane in the presence of bubbles. The system consists of a cathode that is mounted flush to the outside of a Teflon tube as a test fiber and a reference anode. The cathode is called the shear probe and is generally made of platinum wire (0.5 mm diameter) which protrudes onto a test fiber. The electrochemical reaction between cathode and anode is provided under the influence of a constant electric potential of 250 mV. The obtained signal in voltage drop is then conditioned through an amplifier and low-pass filter. The relationship between mass transfer at the probe and the limiting diffusion current is used to convert the voltage values into shear forces [77, 78].

This electrochemical probe has been used to study the surface shear rate when dual-phase flow (air sparging) was introduced to the system [77]. More recently, high-speed video camera with a capture rate of 1000 Hz was synchronized with data acquisition from shear probe signals [78]. The results from this technique provided a good insight on how bubble shape varied with operating conditions, as well as surface shear. The use of an electrochemical probe can be used to optimize the fouling in MBR via gas sparging.

### 14.6.6 Photo-interrupt Sensor

The photo-interrupt sensor can be used to determine the fouling thickness distribution on a membrane surface. This non-invasive setup consists of a high-intensity infrared light emitting diode with a wavelength of 950 nm as the emitter and a high-gain transistor as the collector [79, 80] (see Chapter 15). This sensor is placed in the upper plate of a crossflow membrane module. When the distance between the sensor and the object is decreased, the reflective current of the light source increases. Subsequently, fouling thickness can be quantified according to its voltage response. The effective range of this technique's ability to measure fouling thickness is in the range 10–5000 μm. A 6% error was previously recorded for fouling thickness quantification for feed concentrations below 1% (w/w) [78]. In general, this analytical technique is excellent for its low cost and effective analysis of the fouling layer thickness distribution, but its application is limited to dilute feed concentrations.

### 14.6.7 Nuclear Magnetic Resonance

The Nuclear magnetic resonance (NMR) image is obtained from the map distribution of spin electrons in a magnetic field and detection of their resonances. The technique can produce detailed information about topography and dynamic movements in either a solution or solid samples. The sensitivity of the technique depends on the amount of magnetic field introduced into the sample and the sample state itself, such as its temperature. NMR technique is limited to the sample size, which has to fit within the magnetic coils (ranges up to 30 cm in diameter) [81].

NMR has been used to observe the concentration polarization phenomena of silica particle deposition at different CFVs in the lumen side of the tubular membrane (in/out filtration) [82]. The concentration polarization at low CFV showed non-uniform polarization layers, while a lower thickness of concentration polarization and more rapid steady-state condition were observed at higher CFV. In another experiment, NMR has also been used to map the flow rate distributions during the filtration of an oil/water emulsion in a membrane bundle containing five fibers [83]. In one study, NMR was used to indicate the uneven distribution of permeate flow through the packing of fibers; and this condition might influence membrane performance and efficiency. NMR analysis gives dynamic flow characteristics, such as concentration polarization and velocity distribution during membrane filtration. Due to the limitation of data acquisition time, its application to the MBR field is probably severely limited.

### 14.6.8
### Particle Image Velocimetry

Particle image velocimetry (PIV) is used to trace particle motion in a fluid flow or for direct observation of flow phenomena. The particles are illuminated with a plane sheet of light and images are taken with a known time separation [84]. The distance of particle in a controlled time difference can be calculated to provide a 2D velocity map.

A PIV application was used to observe particle movements/distributions in a crossflow module during filtration of yeast cells (without spacer) [85]. A particle velocity map was also defined to describe flow distribution along the rectangular crossflow module (without membrane) with spacers [86]. In another experiment, the flow velocity distribution within the fiber bundle (nine fibers in a $3 \times 3$ array) was observed to identify the presence of "dead zones" (indicated by particle movements) during bubble injections [84]. It was found that if small bubbles were introduced from the center of the bundle, a dead zone was formed and caused fiber blockage.

Fouling deposition in crossflow membrane filtration can be indicated by the deceleration of particles on the membrane surface. PIV was used to identify early fouling phenomena (particle decelerations) and the "dead zone" during membrane filtration [84]. PIV has the same limitation as other optical methods, like DOTM, as it requires optically transparent solutions. However, how this technique can be extended to real module with a large number of fibers is a challenge; and it may be limited to laboratory validation of bubbling and computational fluid dynamic models of MBR systems.

## 14.7
## Conclusions

Fouling monitoring in MBR can be done through the simple assessment of the flux and TMP. However, fouling visualization can provide more detailed characterization on its morphological structures. Fouling characterization techniques can be divided into two categories according to their destructive nature to the sample. Invasive (EM, AFM, CLSM) and non-invasive techniques (PT, microscope observation, laser beam applications, UTDR, electrochemical shear probe, photo-interrupt sensor, NMR, PIV) were studied for their potentials and when possible their effectiveness to monitor fouling in MBR applications. Each technique provides specific insights to characterize fouling but significant limitations exist due to the labile nature of most MBR foulants.

In MBR application, biofouling and biofilm originate from the complex nature of the feed. Biomass mixed liquor contains solids, inert, colloids, macromolecules and microorganisms in the soluble fractions. As in any biological process, the microorganisms involved in MBR tend to colonize any fixed support material or membrane surface. Moreover, once filtered, the biomass and its components are

Table 14.4 Applications and limitations of non-invasive techniques.

| Non-invasive technique | Applications | Limitations | Ref. |
|---|---|---|---|
| PT | Measurement of fouling thickness during the filtration of synthetic wastewater on tubular membrane | Unable to detect the individual bacterial cell interactions | [63] |
| DOTM | Measurements of cake coverage and visual observation of particle (yeast and latex) movements on the membrane surface | Unable to measure the thickness of cake height | [14] |
| DO | Measurements of cake height, visual observation (tangentially to the membrane), measurement of particle velocity | Difficult to focus on the edge of hollow fiber | |
| MFS | Direct visualization of fouling deposition on between spacers | Unable to measure the thickness of cake height | [70] |
| UTDR | Measurements of cake height, qualitative cake density (filtration of paper mill effluent) | Unable to give visual observation | [74] |
| Electrochemical shear probe | Measurements membrane surface shear rate in the presence of air sparging. Online measurement with high speed camera allows the observation of bubble shapes | No visualization of fouling layer | [77] |
| Photointerrupt sensor | Measurements of cake height | Unable to give visual observation | [79] |
| NMR | Measurements of concentration polarization layer, velocity distribution (filtration of silica particles) | Limitation on data acquisition time | [82] |
| PIV | Measurements of velocity distribution during yeast filtration | Requires optically transparent solutions | [86] |

known to create a fouling layer with complex structures. The fouling layer on the membrane surface is relatively unstable and it could be degraded during the autopsy process.

An invasive method is not preferred for cake visualization in MBR due to its destructive nature to the fouling layer during sample preparation and analysis. Dynamic movements during fouling deposition or removal at different operating parameters, as obtained with a non-invasive method, can explain the fouling mechanisms explicitly. Optical or laser techniques offer perhaps the fastest response time, although limited by the turbidity of the solution. Table 14.4 provides a brief summary of non-invasive techniques, with their relative advantages and limitations. Potential methods for fouling monitoring/characterization in MBR include:

- Analysis of different macromolecular substances in fouling layer (CLSM).
- Fouling deposition and removal (PT, photo-interrupt sensor and DO).
- Concentration polarization (NMR).
- Bubbling (electrochemical shear probe and PIV).

There is synergy between efforts to monitor shear distribution using electrochemical probes and foulant monitoring, as poor hydrodynamics eventually lead to foulant build-up. In situ monitoring of fouling in industrial MBR systems still remains elusive; however, a combination of small optical and shear sensing probes may provide long-term solutions.

## Acknowledgments

The authors gratefully thank the Australian Research Council for their financial support of this study, Dr. Hong-yu Li for her contributions to this paper and also Siemens Water Technology for supplying the PVDF hollow fiber membrane.

## References

1 W. Khongnakorn, C. Wisniewski, L. Pottier, L. Vachoud, *Sep. Purif. Technol.* **2007**, *55*, 125–131.

2 S. Judd, *The MBR Book: Principles and Applications of Membrane Bioreactors in Water and Wastewater Treatment*, Elsevier, Oxford, **2006**.

3 P. Le-Clech, A. Fane, G. Leslie, *Filtr. Separat.* **2005**, *June*, 20–23.

4 P. Le-Clech, V. Chen, T.A.G. Fane, *J. Membrane Sci.* **2006**, *284*, 17–53.

5 C.S. Laspidou, B.E. Rittmann, *Water Res.* **2002**, *36*, 2711–2720.

6 N. Jang, X. Ren, G. Kim, C. Ahn, J. Cho, I.S. Kim, *Desalination* **2007**, *202*, 90–98.

7 H.-C. Flemming, G. Schaule, *Desalination* **1988**, *70*, 95–119.

8 M.-A. Yun, K.-M. Yeon, J.-S. Park, C.-H. Lee, J. Chun, D.J. Lim, *Water Res.* **2006**, *40*, 45–52.

9 H.Y. Kim, K.-M. Yeon, C.-H. Lee, S. Lee, T. Swaminathan, *Sep. Sci. Technol.* **2006**, *41*, 1213–1230.

10 Y.-L. Jin, W.-N. Lee, C.-H. Lee, I.-S. Chang, X. Huang,

T. Swaminathan, *Water Res.* **2006**, *40*, 2829–2836.
11 B.-K. Hwang, W.-N. Lee, Pyung-Kyu Parka, C.-H. Lee, I.-S. Chang, *J. Membrane Sci.* **2007**, *288*, 149–156.
12 W.-N. Lee, I.-S. Chang, B.-K. Hwang, P.-K. Park, C.-H. Lee, X. Huang, *Process Biochem.* **2007**, *42*, 655–661.
13 Y. Ye, P. Le-Clech, V. Chen, A.G. Fane, B. Jefferson, *Desalination* **2005**, *175*, 7–20.
14 H. Li, A.G. Fane, H.G.L. Coster, S. Vigneswaran, *J. Membrane Sci.* **1998**, *149*, 83–97.
15 C. Gourgues, P. Aimar, V. Sanchez, *J. Membrane Sci.* **1992**, *74*, 51–69.
16 Y. Ye, V. Chen, *J. Membrane Sci.* **2006**, *265*, 20–28.
17 S. Percival, J.T. Walker, *Methods Enzymol.* **2001**, *337*, 187–200.
18 S.J. Judd, P. Le-Clech, T. Taha, Z.F. Cui, *Membrane Technol.* **2001**, *135*, 4–9.
19 Z.F. Cui, S. Chang, A.G. Fane, *J. Membrane Sci.* **2003**, *221*, 1–35.
20 S.P. Hong, T.H. Bae, T.M. Tak, S. Hong, A. Randall, *Desalination* **2002**, *143*, 219–228.
21 P.J. Smith, S. Vigneswaran, H.H. Ngo, R. Ben-Aim, H. Nguyen, *J. Membrane Sci.* **2005**, *255*, 99–106.
22 R. McDonogh, G. Schaule, H.-C. Flemming, *J. Membrane Sci.* **1994**, *87*, 199–217.
23 C. Sommariva, A. Comite, G. Capannelli, A. Bottino, *Desalination* **2007**, *204*, 175–180.
24 B. Tansel, J. Sager, J. Garland, S. Xud, L. Levine, P. Bisbee, *J. Membrane Sci.* **2006**, *285*, 225–231.
25 S. Tsuneda, H. Aikawa, H. Hayashi, A. Yuasa, A. Hirata, *FEMS Microbiol. Lett.* **2003**, *223*, 287–292.
26 I.-S. Chang, P. Le-Clech, B. Jefferson, S. Judd, *J. Environ. Eng.* **2002**, *128*, 1018–1029.
27 H.-C. Flemming, *Elements of an Integrated Anti-fouling Strategy with Emphasis on Monitoring*, in: Proceedings of the Membrane Biofouling Meeting – Causes and Control, Singapore, **2003**.
28 H.-C. Flemming, *Introductory Microbiology and Bioadhesion*, in: Proceedings of the Membrane Biofouling Meeting – Causes and Control, Singapore, **2003**.
29 J.S. Baker, L.Y. Dudley, *Desalination* **1998**, *118*, 81–90.
30 H. Li, A.G. Fane, H.G.L. Coster, S. Vigneswaran, *J. Membrane Sci.* **2003**, *217*, 29–41.
31 X.-M. Wang, X.-Y. Li, X. Huang, *Sep. Purif. Technol.* **2007**, *52*, 439–445.
32 B. Lodge, S.J. Judd, A.J. Smith, *J. Membrane Sci.* **2004**, *231*, 91–98.
33 A.A. McCarthy, P.K. Walsh, G. Foley, *J. Membrane Sci.* **2002**, *201*, 31–45.
34 B. Blankert, B.H.L. Betlem, B. Roffel, *J. Membrane Sci.* **2006**, *285*, 90–95.
35 G.R. Bolton, A.W. Boesch, M.J. Lazzara, *J. Membrane Sci.* **2006**, *279*, 625–634.
36 R. Jiraratananon, D. Uttapap, P. Sampranpiboon, *J. Membrane Sci.* **1998**, *140*, 57–66.
37 C. Duclos-Orsello, W. Li, C.-C. Ho, *J. Membrane Sci.* **2006**, *280*, 856–866.
38 A. Broeckmann, J. Busch, T. Wintgens, W. Marquardt, *Desalination* **2006**, *189*, 97–109.
39 S. Ognier, C. Wisniewski, A. Grasmick, *J. Membrane Sci.* **2004**, *229*, 171–177.
40 A.L. Lim, R. Bai, *J. Membrane Sci.* **2003**, *216*, 279–290.
41 A. Drews, J. Mante, V. Iversen, M. Vocks, B. Lesjean, M. Kraume, *Water Res.* **2008**, in press.
42 F.J. Doucet, L. Maguire, J.R. Lead, *Anal. Chim. Acta.* **2004**, *522*, 59–71.
43 H. Ivnitsky, I. Katz, D. Minz, E. Shimoni, Y. Chen, J. Tarchitzky, R. Semiat, C.G. Dosoretz, *Desalination* **2005**, *185*, 255–268.
44 K.-J. Kim, M.R. Dickson, V. Chen, A.G. Fane, *Micron Microsc. Acta* **1992**, *23*, 259–271.
45 Y. Ye, V. Chen, A.G. Fane, *Desalination* **2006**, *191*, 318–327.
46 S.B. Sadr-Ghayeni, P.J. Beatson, A.J. Fane, R.P. Schneider, *J. Membrane Sci.* **1999**, *153*, 71–82.
47 L.D. Nghiem, A.I. Schafer, *Desalination* **2006**, *188*, 113–121.

48 R. Fabris, E.K. Lee, C.W.K. Chow, V. Chen, M. Drikas, *J. Membrane Sci.* **2007**, *289*, 231.

49 S.B. Surman, J.T. Walker, D.T. Goddard, L.H.G. Morton, C.W. Keevil, W. Weaver, A. Skinner, K. Hanson, D. Caldwell, J. Kurtz, *J. Microbiol. Methods* **1996**, *25*, 57–70.

50 J. Zhang, Y. Sun, Q. Chang, X. Liu, G. Meng, *Desalination* **2006**, *194*, 182–191.

51 D.A. Brown, T.J. Beveridge, C.W. Keevil, B.L. Sherriff, *FEMS Microbiol. Ecol.* **1998**, *26*, 297–310.

52 V.V. Tarabara, I. Koyuncu, M.R. Wiesner, *J. Membrane Sci.* **2004**, *241*, 65–78.

53 W.R. Bowen, T.A. Doneva, H.B. Yin, *Desalination* **2002**, *146*, 97–102.

54 H.H.P. Fang, K.-Y. Chan, L.-C. Xu, *J. Microbiol. Methods* **2000**, *40*, 89–97.

55 R. Chan, V. Chen, *J. Membrane Sci.* **2004**, *242*, 169–188.

56 A. Razatos, *Methods Enzymol.* **2001**, *337*, 276–285.

57 I.B. Beech, J.R. Smith, A.A. Steele, I. Penegar, S.A. Campbell, *Colloids Surfaces B: Biointerf.* **2002**, *23*, 231–247.

58 F. Ahimou, M.J. Semmens, P.J. Novak, G. Haugstad, *Appl. Environ. Microbiol.* **2007**, *73*, 2897–2904.

59 F. Ahimou, M.J. Semmens, G. Haugstad, P.J. Novak, *Appl. Environ. Microbiol.* **2007**, *73*, 2905–2910.

60 J.R. Lawrence, T.R. Neu, *Methods Enzymol.* **1999**, *310*, 131–143.

61 J. Zhang, H.C. Chua, J. Zhou, A.G. Fane, *J. Membrane Sci.* **2006**, *284*, 54–66.

62 Z. Yang, X.F. Peng, M.-Y. Chen, D.-J. Lee, J.Y. Lai, *J. Membrane Sci.* **2007**, *287*, 280–286.

63 L.M. Freitas-dos-Santos, A.G. Livingston, *Biotechnol. Bioeng.* **1995**, *47*, 82–89.

64 W.D. Mores, R.H. Davis, *J. Membrane Sci.* **2001**, *189*, 217–230.

65 S.-T. Kang, A. Subramani, E.M.V. Hoek, M.A. Deshusses, M.R. Matsumoto, *J. Membrane Sci.* **2004**, *224*, 151–165.

66 S. Chang, A.G. Fane, *J. Chem. Technol. Biotechnol.* **2000**, *75*, 533–540.

67 P. Le-Clech, Y. Marselina, R. Stuetz, V. Chen, *Desalination* **2006**, *199*, 477–479.

68 P. Le-Clech, Y. Marselina, Y. Ye, R.M. Stuetz, V. Chen, *J. Membrane Sci.* **2007**, *290*, 36–45.

69 Y. Marselina, P. Le-Clech, R.M. Stuetz and V. Chen, **2008**, in preparation.

70 J.S. Vrouwenvelder, J.A.M.V. Paassen, L.P. Wessels, A.F.V. Dam, S.M. Bakker, *J. Membrane Sci.* **2006**, *281*, 316–324.

71 J.S. Vrouwenvelder, S.M. Bakker, M. Cauchard, R.L. Grand, M. Apacandie, M. Idrissi, S. Lagrave, J.A.M.V. Paassen, J.C. Kruithof, M.C.M.V. Loosdrecht, *Water Sci. Technol.* **2007**, *55*, 197–205.

72 M. Hamachi, M. Mietton-Peuchot, *Chem. Eng. Sci.* **1999**, *54*, 4023–4030.

73 D. Hughes, U.K. Tirlapur, R. Field, Z. Cui, *J. Membrane Sci.* **2006**, *280*, 124–133.

74 J. Li, D.K. Hallbauer, R.D. Sanderson, *J. Membrane Sci.* **2003**, *215*, 33–52.

75 J. Li, L.J. Koen, D.K. Hallbauer, L. Lorenzen, R.D. Sandersonc, *Desalination* **2005**, *186*, 227–241.

76 J. Li, R.D. Sanderson, G.Y. Chai, *Sensor Actuat. B Chem.* **2006**, *114*, 182–191.

77 P.R. Berube, G. Afonso, F. Taghipour, C.C.V. Chan, *J. Membrane Sci.* **2006**, *279*, 495–505.

78 C.C.V. Chan, P.R. Berube, E.R. Hall, *J. Membrane Sci.* **2007**, *297*, 104–120.

79 K.-L. Tung, S. Wang, W.-M. Lu, C.-H. Pan, *J. Membrane Sci.* **2001**, *190*, 57–67.

80 W.-M. Lu, K.-L. Tung, C.-H. Pan, K.-J. Hwang, *J. Membrane Sci.* **2002**, *198*, 225–243.

81 V. Chen, H. Li, A.G. Fane, *J. Membrane Sci.* **2004**, *241*, 23–44.

82 D. Airey, S. Yao, J. Wu, V. Chen, A.G. Fane, J.M. Pope, *J. Membrane Sci.* **1998**, *145*, 145–158.

83 S. Yao, M. Costello, A.G. Fane, J.M. Pope, *J. Membrane Sci.* **1995**, *99*, 207–216.

84 A.P.S. Yeo, A.W.K. Law, A.G. Fane, *J. Membrane Sci.* **2006**, *280*, 969–982.
85 J.S. Knutsen, R.H. Davis, *J. Membrane Sci.* **2006**, *271*, 101–113.
86 M.M. Gimmelshtein, R. Semiat, *J. Membrane Sci.* **2005**, *264*, 137–150.
87 J. Cho, K.-G. Song, K.-H. Ahn, *Desalination* **2005**, *183*, 425–429.
88 Z. Ahmed, J. Cho, B.-R. Lim, K.-G. Song, K.-H. Ahn, *J. Membrane Sci.* **2007**, *287*, 211–218.

# 15
# Monitoring Technique for Water Treatment Membrane Processes*

*Kuo-Lun Tung*

## 15.1
## Introduction

Membrane fouling and subsequent flux decline are undesirable but inevitable problems in membrane filtration processes for water treatment. To deal with these formidable obstacles, numerous methods have been adopted by experts in the discipline of water treatment membrane process: for example the development of low-fouling membranes, the design of high-efficiency modules, the selection of optimal operational strategies and several improvements in peripheral control, monitoring and cleaning techniques. However, the laggardly development and improvement of these methods has limited the competitiveness of the processes and their wide acceptance in industry during the past three decades. An early warning technique for fouling problems in a water treatment membrane process is crucial to the improvement of the membrane process operation as well as to the development of a fouling prevention strategy. Usually, the problems of severe flux decline or process failure are noticed if the quality of the product or produced water fails given standards. This is the most expensive method of fouling monitoring, but still widely adopted in practice. Towards the end of the last millennium, researchers began to be aware of the importance of in situ understanding as well as online monitoring of the fouling growth. Several in situ monitoring techniques have been developed in the laboratory to reveal the growth mechanism of fouling layer in membrane filtration processes. Nevertheless, process-oriented, reliable and predictable online monitoring techniques for onsite fouling analysis and control are still few. Brief introductions of the currently available fouling monitoring techniques for laboratorial research and onsite application are described in the next section. Finally, an integrated technique for online fouling monitoring with process-oriented capabilities of: (1) in situ measurement of the fouling layer thickness, (2) dynamic analysis of the fouling layer structure and

* A list of nomenclature is given at the end of this chapter.

*Monitoring and Visualizing Membrane-Based Processes*
Edited by Carme Güell, Montserrat Ferrando, and Francisco López

ISBN: 978-3-527-32006-6

(3) monitoring of membrane fouling potential in a membrane filtration process is introduced in the last section.

## 15.2
## Development of Fouling Monitoring Techniques

The traditional fouling problem in a practical water treatment membrane process is monitored on the basis of parameters such as flux decrease, pressure drop and conductivity or turbidity of feed stream and is diagnosed afterwards on destruction of the membrane. Fouling analysis is always examined based upon a trial and error approach. Moreover, the detection of fouling potential is in the water phase instead of on the membrane surface and it gives no information about the location, extent and composition of foulants [1, 2]. The lack of this information on the membrane surface could be a drastic barrier to the improvement of membrane process operation as well as to the development of fouling prevention strategy. Evidently, there is a keen need for better tools or measurements to monitor the membrane fouling process online, in situ, in real time and non-destructively. A comprehensive state of the art review of the development and assessment of these techniques has been reported in the literature [3–5].

### 15.2.1
### Requirements for a Successful Fouling Monitoring Technique

The analysis of membrane filtration processes involves consideration of phenomena in three regions: (1) the bulk fluid stream in the membrane module, (2) at the membrane–fluid interface and (3) in the membrane [3]. The majority of in situ monitoring techniques have been applied to analyze phenomena at the membrane–fluid interface, such as concentration polarization, cake formation and pore clogging. Therefore, Chen et al. [4] reviewed the development of in situ monitoring techniques by classifying them into two categories for concentration polarization analysis and fouling phenomena analysis, respectively. Chen et al. [3] reviewed them by classifying those techniques into two categories: optical and non-optical techniques. Unfortunately, most of these techniques have been developed only for fundamental researches at laboratory-scale or have found application in special sectors of technical systems, even though they might have potential for wider use [5]. From the information described in the review literature, the requirements for a successful fouling monitoring technique are generalized and listed in Table 15.1.

### 15.2.2
### Classification of Fouling Monitoring Techniques

In the list of requirements in Table 15.1, two of these requirements (in situ fouling assessment, foulant composition analysis) are the main difficulties for process-oriented applications. The online, in situ and in real-time analysis of compositions

**Table 15.1** Requirements for a successful fouling monitoring technique.

| Main items | Requirements |
|---|---|
| Performance measurement | Pressure drop, flux, conductivity or turbidity |
| Fouling assessment | Online, in situ, non-invasive, foulant composition analysis |
| Information acquirement | Real time, representative, accurate, reproducible, automatic |
| Device requirement | Reliable, user friendly, robust, low cost |

**Table 15.2** Classification of monitoring device levels [5].

| Level | Description |
|---|---|
| Level 1 monitoring devices | Devices which can detect the kinetics of deposition and changes in thickness of a fouling layer but cannot provide the information of foulant compositions. |
| Level 2 monitoring devices | Devices which can distinguish between inorganic and organic compositions of a given foulant. |
| Level 3 monitoring devices | Devices which can provide detailed information about the chemical composition of the foulant or directly address microorganisms. |

of the foulants in the full-scale membrane module is also still not available. In order to identify the validity of the developed monitoring techniques for process-oriented purposes, Flemming [5] distinguished these techniques by classifying them into three categories according to the level of fouling information they can provide (as indicated in Table 15.2).

Devices in level 1 can detect the kinetics of deposition and changes in thickness of a fouling layer but cannot provide the information of foulant compositions; devices in level 2 can distinguish between inorganic and organic compositions of a given foulant; while devices in level 3 can provide detailed information about the chemical composition of the foulant or directly address microorganisms. Devices to be classified into level 2 and level 3 are useful for fundamental research in the laboratory due to the limitations of device performance at current status, such as the difficulty with in situ examination and the need for removing samples from the system for analysis. Moreover, the analyses sometimes need to prepare the sample in advance: and the analyses are usually time-consuming. Devices in level 1 are more maturely developed for online and in situ process applications. However, the feasibility of in situ measurement is further dependent on the types of membrane module and/or process to be monitored and the measurement device adopted. Comparisons of available monitoring techniques for fouling detection at the membrane–fluid interface region in water treatment membrane processes are listed in Table 15.3. Some parts of the comparisons have been reported by Chen et al. [3]. A full comparison of the available level 1 monitoring techniques for all

**Table 15.3** Non-invasive monitoring techniques applied to membrane processes [2].

| Technique | | Real time[a] | Approximate resolution | Applicable module[b] | Applicable process[c] | Model-based[d] | Complex[e] | Cost[e] | References |
|---|---|---|---|---|---|---|---|---|---|
| Optical | Direct observation through membrane (DOTM) | R | >0.5 μm | F | MF; UF | N | L | L | [6–8] |
| | Laser triangulometry | R | 3 ~ 5 μm | F, H | MF; UF | N | L | M | [9, 10] |
| | Optical laser sensor | R | 5 μm | F | MF; UF | N | L | M | [11, 12] |
| | Interferometry | D | 20 μm | F | MF; UF, NF, RO | N | M | M | [13–15] |
| | Photosensor | R | 10 μm | F | MF; UF | N | L | L | [16–20] |
| Non-optical | Ultrasonic time-domain reflectometry (UTDR) | S | 5–10 μm | F, T, H, S | MF; UF; NF; RO | Y | M | L | [21–25] |
| | Impedance spectroscopy | S | <1 μm | F | ED | Y | H | L | [26–29] |
| | NMR imaging | D | 10 μm | F, T, H, S | MF; UF | N | H | H | [30–32] |
| | Small-angle neutron scattering (SANS) | D | 0.1 nm | F, T, H, S | MF; UF; NF; RO | Y | H | H | [33–36] |

a) R = rapid real time; S = relatively slow; D = depends on equipment or beam source.
b) Specified for application in commercial module: F = flat plate; S = spiral wound; T = tubular; H = hollow fiber.
c) MF = microfiltration; UF = ultrafiltration; NF = nanofiltration; RO = reverse osmosis; ED = electrodialysis.
d) Y = yes; N = no.
e) L = low; M = moderate; H = high.

regions can be found in the review literature [3, 4]. Some devices in level 2 and 3 can be found in the literature [5]. Nine techniques in level 1 category listed in Table 15.3 for comparison include the direct observation through membrane (DOTM) method, laser triangulometry, optical laser sensor (see Chapter 11), interferometry, photosensor, ultrasonic time-domain reflectometry (UTDR; see Chapter 11), impedance spectroscopy (see Chapter 9), NMR imaging and small-angle neutron scattering (SANS). They are compared based on the characteristics needed to be considered in their application, such as: real time, resolution, applicable membrane modules and membrane processes, model-based and cost. For monitoring applications in commercial membrane module, all the optical techniques are limited to a flat plate module due to the difficulty of installing the optical sensor and are also limited to application on MF and UF processes due to the limitation of resolution. The non-optical techniques have a wide applicability in different membrane modules; however, the relatively slow response time to signal change is their main limitation for application. NMR and SANS are powerful tools for fundamental research in the laboratory; however, their extremely high cost and high complexity limit their application in full-scale membrane filtration processes. Evidently, there is a keen need to improve or develop better devices or techniques to monitor the membrane fouling process online, in situ, in real time and non-destructively in a full-scale system.

### 15.2.3 Installation of a Process-oriented Fouling Monitoring System

From the review of currently available monitoring techniques, it is found that almost all the mentioned methods have been used in a specially designed membrane cell or module for membrane fouling study. Relatively few methods can be applied to commercial modules. Although the ultimate goal of membrane fouling monitoring is to mount the measuring device or to install the technique directly into a commercial membrane module, this still has practical difficulty with the currently available monitoring techniques. Therefore, a bypass flow system has been widely used in practice. Vrouwenvelder et al. [37] developed a practical tool: the membrane filtration simulator (MFS) for fouling prediction and control with a specially designed bypass flow module. Besides the requirements for an ideal monitoring technique mentioned in Section 15.2.1, the representativeness of the specially designed test module for a practical system is another important issue. The MFS tool was designed for monitoring membrane fouling in a spiral wound membrane module for nanofiltration (NF) and reverse osmosis (RO) processes. With a representative specially designed test module, the MFS can be applied for early warning, characterization of the fouling potential of the feed stream, evaluation and selection of an optimal pretreatment method and evaluation of fouling control strategies. The MFS tool provides a useful model of a bypass flow monitoring installation; however, further integration with real-time numerical analysis or precision monitoring devices to develop a more informative technique, for example to provide fouling layer structure or chemical composition of foulants in real time, is still needed. The next section presents in situ fouling layer thickness measurement devices integrated with an online dynamic analysis to provide real-time information on the fouling layer structure for membrane filtration processes.

## 15.3
## Dynamic Analysis of Online Fouling Monitoring

The growth of a fouling layer due to the deposition of undesirable materials on the membrane is a persistent problem in water treatment membrane processes. Particulate fouling, that is the deposition of suspended solids, colloids and microbial cells onto or into the membranes, is an especially delicate issue in the membrane filtration operation. Its complete removal by intensive pretreatment of the feed water is not always feasible. A technique for early warning and fouling monitoring is the desire of all engineers to achieve the long-term and stable operating performance of a membrane process. To fulfill this requirement, an integrated technique is proposed for the online fouling monitoring of a water treatment membrane filtration process. This online monitoring technique provides dynamic and real-time information about a fouling phenomenon and includes the process-oriented capabilities of (1) in situ measurement of fouling layer thickness, (2) dynamic analysis of fouling layer structure, and (3) monitoring of membrane fouling potential in membrane filtration processes for water treatment applications. The dynamic analysis was originally proposed by Lu and Hwang [38] and was modified to analyze the formation and compression of the particulate fouling layer during microfiltration. To perform this online dynamic monitoring technique, data is needed for: (1) pre-estimated surface porosity of a fouling layer and (2) real-time variation in the fouling layer thickness. The surface porosity of a fouling layer can be obtained by means of a low-head filtration system [39]. The main difference between the analysis originally proposed by Lu and Hwang and this study is that data for the permeate flux versus time was replaced with data for the real-time variation in the fouling layer thickness. How to measure the real-time variation of a fouling layer thickness is suggested and compared in the next section.

### 15.3.1
### Online Measurement of Fouling Layer Thickness

To perform this dynamic monitoring technique, either the time course of permeate volume data or the time course of the fouling layer growth data can be used for analysis. For fouling monitoring purpose, to use the data of real time variation of fouling layer thickness for dynamic analysis is more feasible than to use the permeate volume variation data [40], since it can provide a local fouling behavior at each membrane surface area with a reasonable number of measurement device installations. Two reported methods for measuring fouling layer thickness variation in a water treatment membrane process, one method in the category of optical technique and the other in the category of non-optical technique, are introduced to demonstrate their potential in the application of process-oriented online dynamic analysis. Other techniques with similar capabilities are also feasible for this analysis, for example laser triangulometry [9, 10], optical laser sensor [11, 12], interferometry [13–15] and so forth.

#### 15.3.1.1 Optical Method

The first suggested method to measure the thickness of a growing fouling layer for online dynamic analysis uses a photointerrupt sensor [18]. The base sensor is a high-sensitivity reflective type subminiature photointerrupter sensor (model GP2L22; Sharp, Japan). The internal connection diagram of the reflective type photointerrupter sensor that contains a high-intensity infrared light-emitting diode (LED) with a wavelength of 950 nm as the emitter and a high-gain silicon photo-Darlington transistor as the collector is depicted in Figure 15.1. When an object is close to the sensor, light from the emitter is reflected from the surface of the object into the collector. Decreasing the distance between object and sensor results in an increase in reflective current. This is the basic principle of measurement.

A schematic diagram of thickness measurement of a growing cake on the membrane surface during cross-flow filtration of a particulate suspension is

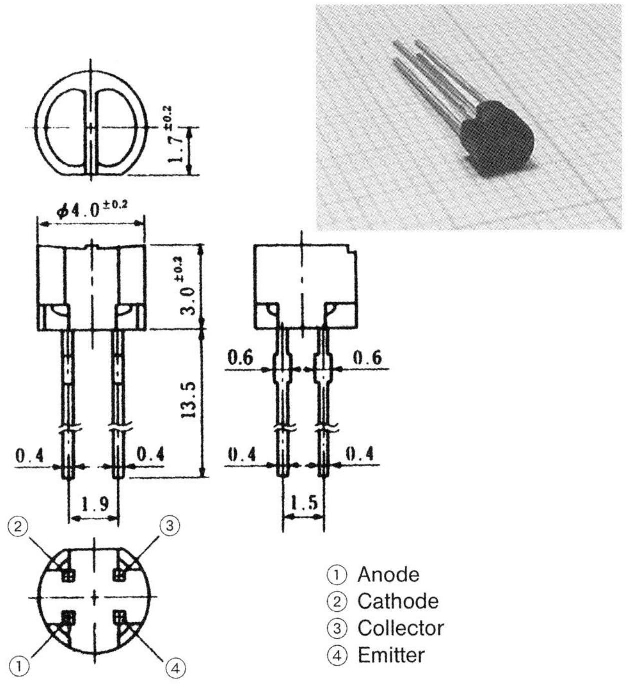


**Figure 15.1** Schematic diagrams of internal connection and outline dimensions of a Sharp GP2L22 photointerrupter (measurements given in millimeters) [18].

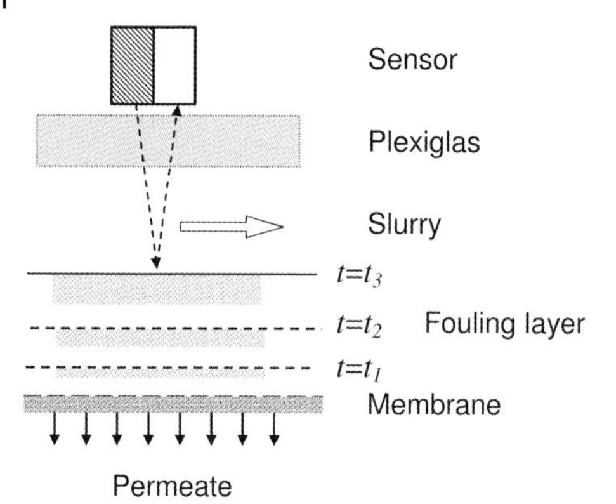


**Figure 15.2** A schematic diagram of cake thickness measurement by a photointerrupter sensor [18].

illustrated in Figure 15.2. In order to verify the validity of this method for measuring cake thickness, Tung et al. [18] applied this method to determine the growth of poly(methyl methacrylate) (PMMA) cake thickness in a crossflow microfiltration (CFMF) system in a range from 10 μm to 5 mm, with a resolution of 10 μm as illustrated in Figure 15.3. For the suspension concentration range tested, 0.05% to 1.0%, the maximum error of thickness measurement was less than ±6.29%. The three-dimensional gray value encoded height profiles of PMMA filter cake formed during a cross-flow microfiltration process offers a better way of discerning topology.

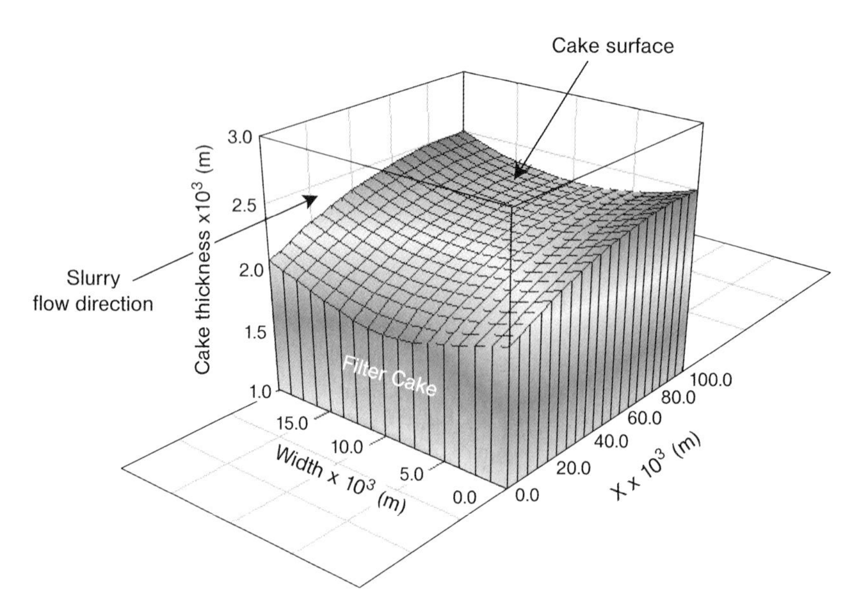


**Figure 15.3** Three-dimensional cake thickness distribution contour: PMMA (0.1 wt%), $\Delta P = 86\,000$ Pa, $U_s = 0.161$ m/s and at $t = 8000$ s [18].

Ouazzani and Bentama [20] further applied this technique to microfiltration of a *Saccharomyces cerevisiae* suspension and found that the response time of the sensor is lower than 1 s and the measurements are stable for a duration in excess of 1 h. Their measurements show that the maximum percentage of deviation from the averaged value from five tests performed under the same conditions is about 4%. Based on the examinations of sensitivity, precision, accuracy and valid range, they concluded that this is a promising technique to measure cake thickness of biological particles. This method has the benefits of low cost, easy installation and accuracy compared to gamma-ray absorption, NMR microimaging and computerized axial tomographic scans (CATSCAN) techniques. The fact that this method can only be applied to a flat plate membrane module is its major limitation.

#### 15.3.1.2 **Acoustic Method**

The second suggested method to measure the thickness of a growing fouling layer for online dynamic analysis uses an ultrasonic time domain reflectometry (UTDR) technique. This method was first developed by Greenberg et al. for measuring membrane compaction and fouling in the RO process [21–23] and was further investigated by Li et al. for observations of membrane fouling and cleaning in MF and UF processes [24, 25]. This technique uses sound waves to measure the location of interfaces in a media and provides information on the physical characteristics of the media through which the waves traverse. Figure 15.4a shows a schematic diagram of a cross-section of a membrane cell with a fouling layer. The ultrasonic transducer was placed on the top plate of the membrane cell and the signals labeled A, B and C represent the reflected echoes at the interfaces of: (a) top plate and feed solution, (b) bottom of fouling layer and membrane surface, and (c) feed solution and fouling layer surface, respectively. The signal labeled B′ is the acoustic difference at the membrane interface after fouling is initiated on the membrane surface. The corresponding time-domain responses of signals A, B, B′ and C are illustrated in Figure 15.4b. The UTDR technique can be applied to measure fouling layer thickness. If the fouling layer is thick enough to be measured by the ultrasonic signal, the thickness can be estimated from a correlation with the difference in arrival time, $\Delta t$, between echos B and C. The thickness of the fouling layer, $L$, can be determined from the following equation:

$$L = \frac{1}{2} c \Delta t \qquad (15.1)$$

where $c$ is the velocity of the ultrasonic wave in the medium. The denser the fouling layer is, the better the reflection as well as the larger the amplitude will be detected. Thus, the detection of the interface echoes allows fouling to be monitored in real time.

Figure 15.5 illustrates the time courses of the thickness of fouling layer and the permeate flux in a cross-flow microfiltration system for treatment of paper mill effluent at an axial velocity of 6.97 cm/s. The thickness obtained by UTDR represents cake and fouling layers. As depicted in this figure, the thickness of the

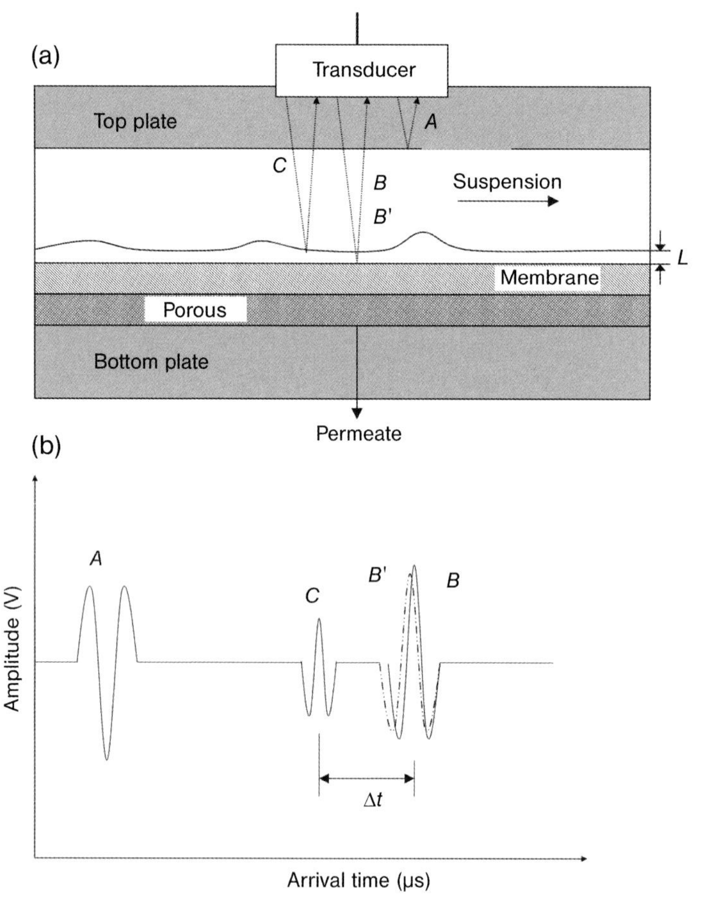


**Figure 15.4** (a) A schematic diagram of a cross-section of a membrane cell with a fouling layer; (b) corresponding time-domain responses. (Redrawn from [25]).

fouling layer increased rapidly at the beginning of fouling because of concentration polarization and cake layer formation. There exists a responding time limitation for UTDR technique that the fouling layer has to be thick enough for the ultrasonic signals to present a new echo at the feed stream and fouling layer interface [24]. Nevertheless, this technique has a diverse applicability for different types of membrane module, for example flat plate, spiral wound and even tubular features, which is its strength.

### 15.3.2
### Dynamic Analysis of Fouling Layer Structure

With the real-time data of fouling layer thickness variation obtained by either of the above-mentioned methods, a dynamic analysis procedure based on mass and

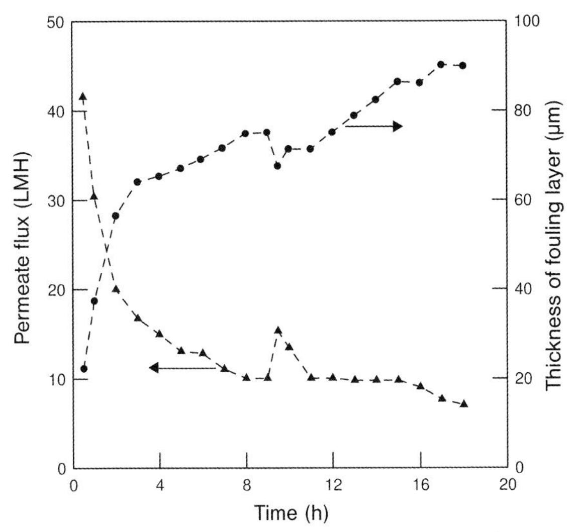


**Figure 15.5** Time courses of fouling layer thickness and permeate flux at crossflow velocity of 6.97 cm/s; stop and restart of the fouling operation at 540 min. (Redrawn from [24]).

force balances of particulate deposition on the membrane surface can be applied to analyze the fouling layer structure in a water treatment membrane process. The local properties in a fouling layer, for instance the porosity distribution, the resistance to fluid flow and the hydraulic pressure distribution, can be estimated online by the proposed dynamic analysis with a preliminary request of the surface porosity of fouling layer and the real-time variation in the fouling layer thickness. The obtained local properties of the fouling layer can be used not only for predicting or monitoring the performance of a membrane filtration system during plant operation but also for designing a membrane filtration process.

#### 15.3.2.1 **Formation of the Surface Fouling Layer**

In a dead-end membrane filtration process, the number of particles instantaneously arriving at the surface of membrane or a formed fouling layer is mainly controlled by the filtration rate and slurry concentration, while the packing structure of the particles depends on their size, shape, physical and chemical properties and so forth. Since the solid compressive loading on a thin layer of the cake surface is limited, the porosity of the surface cake layer, $\varepsilon_i$ (that is $\varepsilon_{1,1}$, $\varepsilon_{2,2}$ and $\varepsilon_{i,i}$ in Figure 15.6), during a membrane filtration process can be assumed to be a constant value and can be preliminarily estimated by a low-head filtration experimental system proposed by Haynes [41].

Thus, at each time increment, the newly formed surface layer has not been compressed and the mass in the surface layer is estimated by the measured increase in fouling layer thickness due to cake growth and the surface porosity of a

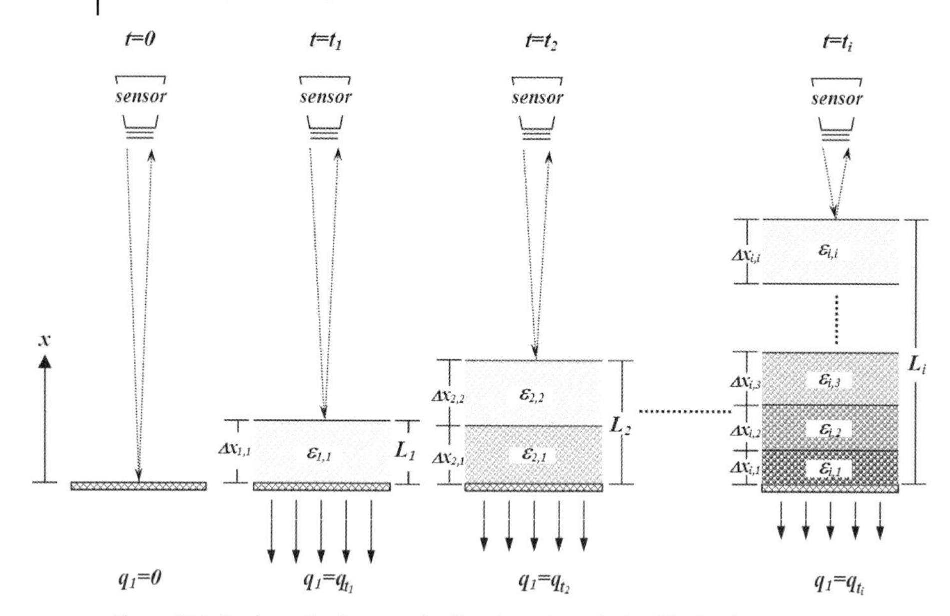


**Figure 15.6** A schematic diagram of online dynamic analysis of fouling layer structure.

fouling layer. Moreover, the mass in each control layer is expressed as:

$$\Delta w = \rho_s(1 - \varepsilon)\Delta x \tag{15.2}$$

where $\Delta w$ is the solid weight in the control fouling layer per unit filtration area, $\rho_s$ is density of the solids and $\Delta x$ is thickness of each control layer. The mass of particle layer formed within a very short period can be estimated by [42]:

$$\Delta w_i(t) = \frac{\rho s}{1 - ms}\left(\frac{dv}{dt} + \frac{sv}{1 - ms}\frac{dm}{dt}\right)\Delta t \tag{15.3}$$

where $v$ is the volume of permeate per unit filtration area, $m$ is the mass ratio of wet to dry cake of the fouling layer and $s$ is the mass fraction of particles in the slurry. At the beginning of a filtration process, $dm/dt = 0$, the permeate flux can be estimated as:

$$\frac{dv}{dt} = q_1 = \frac{1 - ms}{\rho s}\frac{\Delta w_i}{\Delta t} \tag{15.4}$$

where $q_1$ is flow rate at the permeate side.

#### 15.3.2.2 Compression of a Fouling Layer

During a membrane filtration process, the fouling layer is compressed due to frictional drag and the mass of the particulate layer accumulated. For any given control mass, the value of $(1-\varepsilon)\Delta x$ as indicated in the right hand side of Equation (15.2) remains constant during filtration. Thus, the relationship between the thickness of a fouling layer with constant mass and its porosity before and after compression as shown in Figure 15.6 can be expressed as:

$$\frac{\Delta x_{t+\Delta t}}{\Delta x_t} = \frac{(1-\varepsilon_t)}{(1-\varepsilon_{t+\Delta t})} \tag{15.5}$$

Taking a mass balance for a flowing fluid within the controlled mass, the difference between inward and outward permeate flow is equal to the decreased void volume per unit area in a time increment, that is:

$$\left(\frac{\partial q}{\partial x}\right)_t = \left(\frac{\partial \varepsilon}{\partial t}\right)_x \tag{15.6}$$

Equation (15.6) is known as the continuity equation of cake compression [43]. This equation can be employed to describe the relationship between the change of the flow rate of the filtrate and the porosity variation of fouling layer. Moreover, since some flowing fluid is squeezed out toward the permeate side due to compression of the fouling layer, the fluid flow rate of permeate is larger than that at the layer surface. Taking a mass balance for the whole fouling layer, the ratio of the fluid flow rate at the layer surface, $q_i$, to the flow rate of the permeate, $q_1$, can be estimated by [44]:

$$\frac{q_i}{q_1} = \frac{[(1-s)(1-\varepsilon_{av}-Ld\varepsilon_{av}/dL) - s(m_i-1)(1-\varepsilon_i)](1-\varepsilon_{av})}{[(1-ms)(1-\varepsilon_{av}-Ld\varepsilon_{av}/dL)(1-\varepsilon_{av}) - sL(\rho/\rho_s)(d\varepsilon_{av}/dL)]} \tag{15.7}$$

where $L$ is the thickness of the whole fouling layer at each time increment and $m_i$ is the mass ratio of wet to dry cake of in the top layer at the beginning of each time increment.

#### 15.3.2.3 Resistance Estimation of a Fouling Layer

A fouling layer formed and compressed during a membrane filtration process causes a decline in permeate flux due to the increase in filtration resistance from the compressed fouling layer. The Kozeny equation can be used for estimating pressure drop through a fouling layer under a specified filtration rate, that is:

$$\left(\frac{dP_L}{dx}\right) = \frac{q\mu k S_o^2(1-\varepsilon)^2}{\varepsilon^3} \tag{15.8}$$

in which $\mu$ is the viscosity of the fluid, $\varepsilon$ is the porosity of fouling layer, and $k$ is the Kozeny constant. Prior to applying Equation (15.8) to estimate the pressure drop through a differential fouling layer, the value of $k$ should be determined and can be corrected by using the local cake porosity as [45]:

$$k = \frac{2\varepsilon^3}{(1-\varepsilon)\left\{\ln[1/(1-\varepsilon)] - \left[1-(1-\varepsilon)^2\right]/\left[1+(1-\varepsilon)^2\right]\right\}} \tag{15.9}$$

The specific filtration resistance in the fouling layer can be given as:

$$\alpha = kS_o^2 \frac{(1-\varepsilon)}{\rho_s \varepsilon^3} \tag{15.10}$$

where $S_o$ is the effective specific surface area of the particles, which can also be obtained using various measurement methods.

The above-described physical parameters, such as mass of solids in the fouling layer $\Delta w$, porosity of the fouling layer $\varepsilon$, mass fraction of particles in the slurry $s$, mass ratio of wet to dry cake of in the fouling layer $m$, the Kozeny constant $k$, effective specific surface area of the particles $S_o$ and the physical properties of fluid and solids, govern the growth and structure of the fouling layer. Some of them change with time during the membrane filtration process and continuously affect the growth and structure variation of the fouling layer. They are important parameters to be analyzed during the process in order to monitor the membrane filtration progress. The next section describes the procedures to estimate the time course of the variation of those parameters for analyzing fouling layer structure.

#### 15.3.2.4 **Procedures for Analyzing the Fouling Layer Structure During a Membrane Filtration Process**

To perform a dynamic analysis on the formation and compression of a fouling layer during the course of a membrane filtration process, data is needed for: (1) pre-estimated surface porosity of a fouling layer and (2) real-time variation in fouling layer thickness. With these two data and by applying the mass balance and force balance equations given in the preceding sections, the fouling layer structure can be analyzed numerically as follows:

1. At the beginning of a filtration ($t = t_1$), it can be assumed that the fouling layer has not been compressed, thus the porosity of the fouling layer is of the value as the porosity of the surface cake layer, $\varepsilon_i$ (i.e. $\varepsilon_{1,1}$ in Figure 15.6). With the measured fouling layer thickness and the porosity of the surface cake layer, it is possible to calculate the values of: $w_i$ using Equation (15.2), $q_1$ using Equation (15.4), $q_i$ using Equation (15.7), $kS_o^2$ using Equation (15.8) and $\alpha$ using Equation (15.10).

2. After time increase $\Delta t (t = t_2)$, an incremental thin layer of fouling particles has formed on the layer surface. For the newly formed fouling layer, $w_i$, $q_1$ and $q_i$ can be calculated from the online measured value of fouling layer thickness $L$ using Equations (15.2), (15.3) and (15.7). The pressure drop through the newly formed fouling layer can be obtained by Equation (15.8). The Kozeny constant can be corrected by Equation (15.9). Then, the bottom layer is compressed due to the mass of the particulate layer accumulated and frictional drag. The pressure gradient through the bottom fouling layer can be estimated from the difference of the pressure drop through the entire fouling layer and that through the surface top layer. The porosity of the bottom fouling layer $\varepsilon_{2,1}$ can be estimated by solving Equation (15.8); as a result, the average porosity for this time increment can be obtained.

3. The distribution of hydraulic pressure and the average porosity of the previous formed fouling layer can be adopted as the initial values to calculate fouling layer properties for the next time increments ($t > t_2$). An iteration procedure can be employed to estimate the porosity variation of the fouling layer using Equation (15.5) and then to estimate fluid flow rate using Equation (15.6). Substituting these results into Equation (15.8), the local hydraulic pressure can be calculated. The iteration is repeated until the value of the calculated hydraulic pressure in each layer matches the pre-obtained hydraulic pressure distribution.

4. For each time increment ($t > t_2$), the above-proposed iteration procedure is repeated until the calculated total pressure drop through the entire fouling layer is equal to the applying pressure. By substituting the new profile of porosity of the fouling layer into Equation (15.10), the value of local specific filtration resistance for each control mass of the fouling layer can be obtained.

5. Steps 3 and 4 are repeated continuously during a membrane filtration process, thus the variation of the fouling layer properties can be monitored for the entire path of operation.

#### 15.3.2.5 **Case Analysis**

In order to examine the validity of the proposed online dynamic analysis technique, a dead-end microfiltration experimental of cross-linked *S. cerevisiae* suspension was conducted under a constant operating pressure of 61 kPa. The *S. cerevisiae* (density 1150 kg/m$^3$, diameter 4–5 μm) was purchased from Sigma Chemical and suspended in 0.86% NaCl physiological saline to prepare slurry of a concentration of 0.5%. The *S. cerevisiae* particles were then cross-linked with 0.5% glutaraldehyde. *S. cerevisiae* suspension was used for this analysis because it is a microorganism and is more difficult and complicated for a microfiltration process than filtration with an inorganic micron-size particle suspension. The cross-linking treatment is to form deactivated particles to prevent sporulation. The cross-linked *S. cerevisiae* particle is more rigid than an untreated one. Polycarbonate track-etched (TEPC) membrane with a pore size of 0.8 μm (purchased from Millipore) was used as a filter medium in this study. A filter chamber with a filtration area of 0.00196 m$^2$ was used for the experiment. An in situ optical technique with a reflection-type LED photo-interrupter was installed on the filter chamber and adapted to measure the dynamic variation of fouling layer thickness during the membrane filtration experiment [18]. Variations in the fouling layer thickness were recorded with an accuracy of 10 μm by the transverse value of the voltage signal. A time course of fouling layer thickness variation obtained under the preceding described experimental conditions is illustrated in Figure 15.7. The surface porosity was estimated to be 0.56 by means of a low-head filtration system proposed by Haynes [41].

Figure 15.8 illustrates cross-sectional SEM images of cross-linked yeast particle fouling on TEPC membranes under constant applied pressure. The cross-linked *S. cerevisiae* particles were found to be more rigid than non-cross-linked ones. It can be observed from Figure 15.8a that, since the particle size of 4–5 μm is larger than the

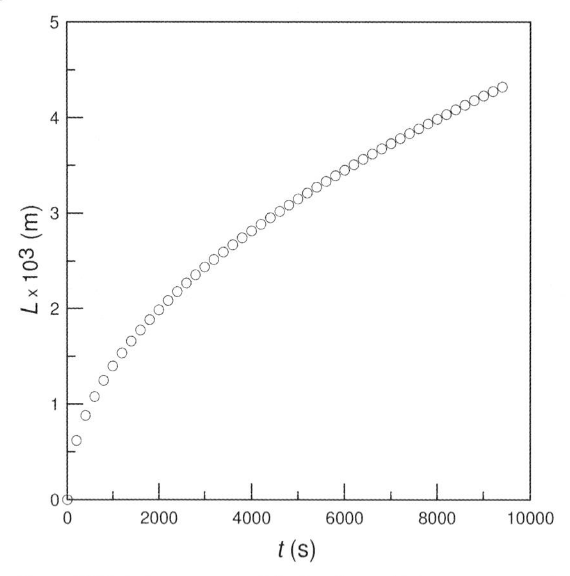


**Figure 15.7** A time course of fouling layer thickness measured under $\Delta P = 61$ kPa for 0.5% cross-linked *S. cerevisiae* suspension.

TEPC membrane pore size (0.8 μm), clogging of membrane pore by yeast particle was not found during the microfiltration. Nevertheless, clogging of membrane pore can be observed when a larger pore size of TEPC membrane was used, as illustrated in Figure 15.8c, d. At the commencement of microfiltration, a loose fouling layer composited by particles was formed on the membrane surface. The surface porosity was estimated to be 0.56 by means of a low-head filtration system proposed by Haynes [41]. As microfiltration continued, the fouling layer was compressed due to frictional drag from the fluid and the mass of foulant consisted of depositing particles as illustrated in Figure 15.8b and hence causing a fast decline in flux. Although to see is to believe, only phenomenological properties can be obtained from these SEM images. It is very difficult to non-destructively obtain local properties in the fouling layer from these SEM images in real time for practical purposes. In order to look for a better fouling control and reduction of irreversible fouling, how to analyze the structure and properties of fouling layer plays a vital role.

The intrinsic properties in the fouling layer can be dynamically and quantitatively analyzed by online dynamic analysis. The materials, their physical parameters and the values used in this simulation are listed in Table 15.4. The important internal properties of a fouling layer include porosity and resistance to filtration. The porosity of the fouling layer is one of the most important quantities for estimating the fouling resistance. Figure 15.9 illustrates the predicted porosity distributions during a course of dead-end microfiltration. From the curves one can notice that the fouling layer is compressed and local porosity in the fouling layer decreases gradually, from a surface porosity of 0.56 to a bottom porosity around 0.51.

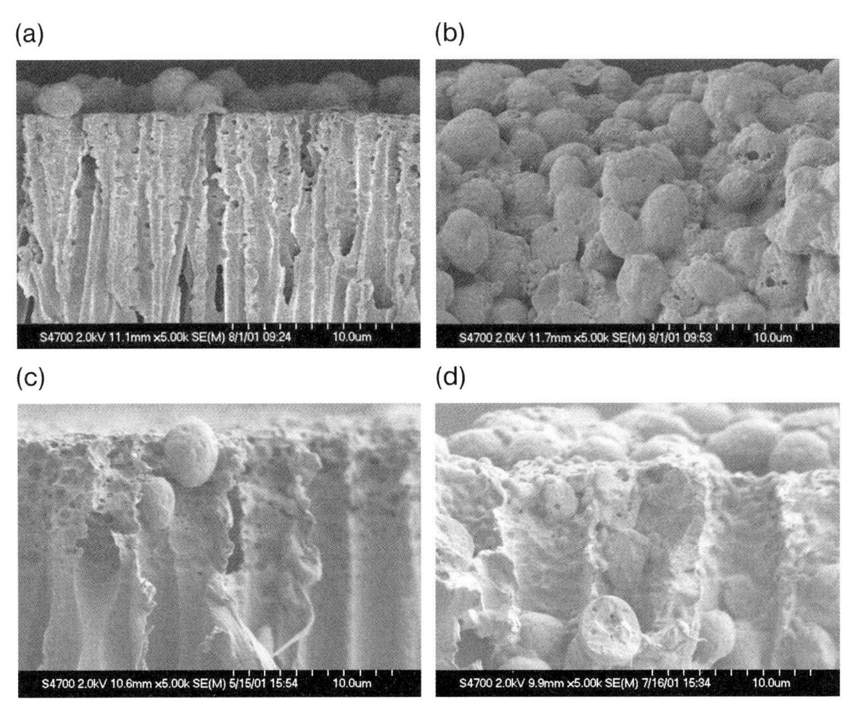


**Figure 15.8** SEM images of cross-linked *S. cerevisiae* particle fouling on various pore sizes of TEPC membranes: (a) using 0.8 μm TEPC membrane at the beginning of a microfiltration, (b) using 0.8 μm TEPC membrane at the surface of a fouling layer, (c) using 3.0 μm TEPC membrane and (d) using 5.0 μm TEPC membrane.

For constant pressure dead-end microfiltration, the values of porosity near the membrane and the surface of fouling layer have almost constant values and a sharp decrease is observed in the main region of the fouling layer. The decrease of porosity within the fouling layer is mainly due to the rearrangement of particles and compression of the fouling layer which are caused by frictional drag from the fluid and the mass of foulant composed of depositing particles.

Fouling layers, in general, are compressible, that is they become more compact as the extent of their compression increases. Solid compressive pressure is responsible for the compression of a fouling layer according to basic filtration theory [46]. In traditional filtration theory, the derivation of the drag equations of filtration for rigid particle slurries assume that particles are in point contact mode and that compression attends instantaneously. Under this assumption, a force balance can be obtained between liquid pressure over the entire cross-section and the solid compressive pressure on the total mass within the porous layer as:

$$P_S + P_L = \Delta P \tag{15.11}$$

**Table 15.4** The materials and the physical parameters and their values used in the simulation.

| Physical parameter | Type/value |
|---|---|
| Particle | *S. cerevisiae* |
| Membrane | 0.8 μm TEPC membrane |
| Density of particle ($\rho_s$) | 1150 kg/m$^3$ |
| Concentration ($s$) | 0.5% |
| Operating pressure ($\Delta P$) | 60 795 Pa |
| Surface porosity ($\varepsilon_o$) | 0.56 |
| Area of filtration (A) | 0.00196 m$^2$ |
| Temperature | 20 °C |

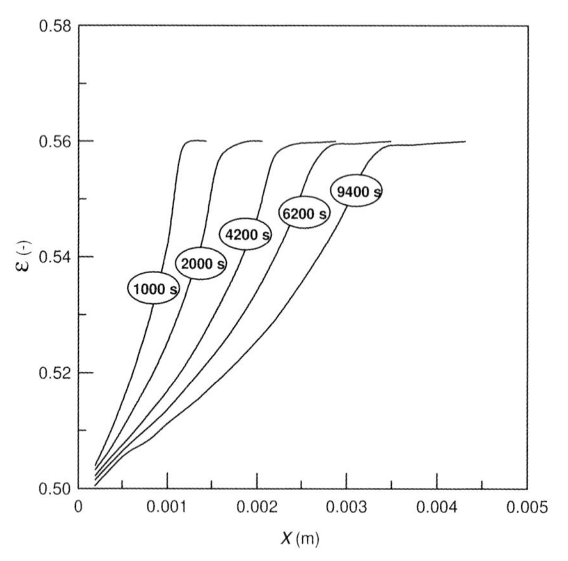


**Figure 15.9** Analyzed porosity distributions in fouling layers during the time course of the dead-end microfiltration under $\Delta P = 61$ kPa for 0.5% cross-linked *S. cerevisiae* suspension.

where $\Delta P$ is the applied pressure, $P_s$ is the solid compressive pressure and $P_L$ is the hydraulic pressure across the fouling layer. The widely used constitutive relationships to correlate the fouling layer property and the variable responsible for fouling layer compression (solid compressive pressure $P_s$) is [47]:

$$\text{either (a): } \alpha = \alpha_o (P_s)^n \text{ or (b): } \alpha = \alpha_o \left(1 + \frac{P_s}{P_o}\right)^n \tag{15.12}$$

where the exponent $n$ is compressibility coefficient. The coefficient can also be estimated by the dynamic analysis during online motoring. Figures 15.10 and 15.11

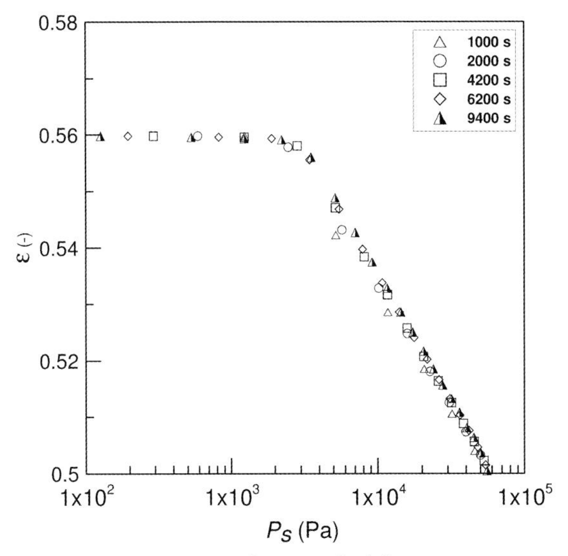


**Figure 15.10** Porosity as a function of solid compressive pressure in the fouling layer formed in the dead-end microfiltration under $\Delta P = 61$ kPa for 0.5% cross-linked *S. cerevisiae* suspension.

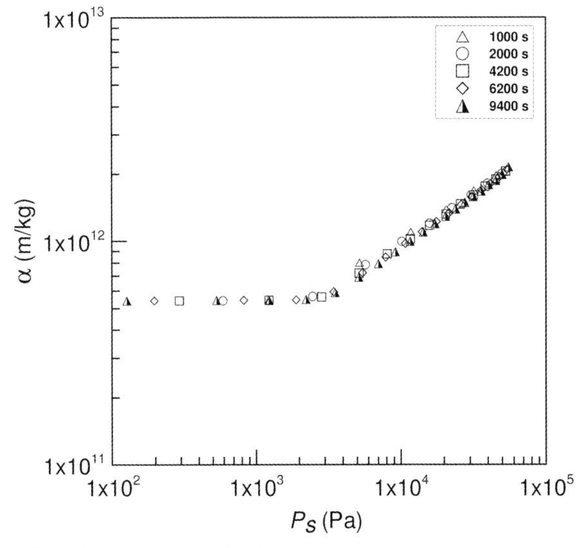


**Figure 15.11** Specific filtration resistance as a function of solid compressive pressure in the fouling layer formed in the dead-end microfiltration under $\Delta P = 61$ kPa for 0.5% cross-linked *S. cerevisiae* suspension.

**Table 15.5** Calculated values of the compressibility parameters.

| Parameter | Calculated value |
|---|---|
| $\Delta P$ | 60 795 Pa |
| $\alpha_o$ | $9.67 \times 10^9$ m/kg |
| $n$ | 0.494 |

illustrate the curves of $\varepsilon$ versus $P_s$ and $\alpha$ versus $P_s$ obtained by dynamic analysis. The calculated values of the compressibility parameters are listed in Table 15.5.

The ratio of the pressure drop over the full fouling layer to the average specific resistance is regarded as being equal to the integral of the differential amount of the local specific resistance:

$$\frac{\Delta P}{\alpha_{av}} = \int_0^{\Delta P} \frac{dP_s}{\alpha} \tag{15.13}$$

Substituting part (a) of Equation (15.12) into Equation (15.13) and integrating and rearranging gives:

$$\alpha_{av} = \alpha_o(1 - n)(\Delta P)^n \tag{15.14}$$

Equation (15.14) is commonly used in filtration for compressible cakes.

This dynamic analysis procedure enables us to obtain these basis data by knowing the surface porosity of a fouling layer and the real-time variation in the fouling layer thickness. In summary, the adopted dynamic procedure proves itself to be useful tool not only for designing a membrane filtration system but also for predicting or monitoring the system performance during plant operation.

### 15.3.3
### Monitoring of Water Quality in a Membrane Filtration Process

In order to measure the particulate fouling potential of feed water, Schippers and Verdouw [48] proposed a modified fouling index (MFI) by using a microfiltration membrane as a quick test of the feed water quality. They further improved the accuracy of the index and developed a new index of MFI-UF by using ultrafiltration membranes to increase the index sensitivity to the presence of colloidal particles for constant pressure operation [49, 50] and constant flux operation [51].

The MFI is based upon cake filtration theory that particles are retained on the membrane surface during filtration. According to the resistance in series model, the reduction in flux due to the presence of cake layer and the additional resistance from the membrane under constant operating filtration can be described as:

$$\frac{dv}{dt} = \frac{\Delta P}{\mu(R_c + R_m)} \tag{15.15}$$

where $\mu$ is the viscosity of slurry and $v$ is the volume of permeate per unit area. The resistance contributed from cake layer under constant pressure filtration can be expressed according to Ruth filtration equation as [46]:

$$R_c = \alpha_{av} \cdot \frac{s\rho_s}{1 - ms} v \tag{15.16}$$

where $\alpha_{av}$ is average specific filtration resistance, $s$ is the fraction particles in slurry, $\rho_s$ is the density of particles and $m$ is the mass ratio of wet to dry cake. In the case of a dilute slurry, the $(1-ms)$ is approximately unit. Substituting Equation (15.16), combining with Equation (15.15) and integrating Equation (15.15) from $t=0$ to $t=t$ at a constant $\Delta P$ results in:

$$\frac{t}{V} = \frac{\mu \cdot \alpha_{av} \cdot s\rho_s}{2\Delta P \cdot A^2} V + \frac{\mu R_m}{\Delta P \cdot A} \tag{15.17}$$

where $V$ is the volume of permeate. According to the definition of MFI defined by Schippers and his co-workers [49–51], the $I$ index is taken to be the production of the specific filtration resistance of the cake layer and the concentration of particles in the feed water, that is:

$$I = \alpha_{av} \cdot s\rho_s \tag{15.18}$$

For a compressible fouling layer, a constitutive relationship of Equation (15.14) is commonly adopted to relate the specific filtration resistance. Substituting Equation (15.14) into Equation (15.17) gives:

$$\frac{t}{V} = \frac{\mu \cdot \alpha_0(1-n)\Delta P^n \cdot s\rho_s}{2\Delta P \cdot A^2} V + \frac{\mu R_m}{\Delta P \cdot A} \tag{15.19}$$

The MFI is defined as the slope of Equation (15.19) to be [49]:

$$\text{MFI} = \frac{\mu \cdot I}{2\Delta P \cdot A^2} = \frac{\mu \cdot \alpha_0(1-n)\Delta P^n \cdot s\rho_s}{2\Delta P \cdot A^2} \tag{15.20}$$

A time course of the MFI represents the filtration resistance variation of a membrane filtration system. The MFI can be determined experimentally from the plot of $t/V$ versus $V$ to monitor the variation of filtration resistance. An example is demonstrated with data obtained during the online dynamic analysis described in the previous section. Figure 15.12 illustrates the $t/V$ versus $V$ plot of the microfiltration of a cross-linked *S. cerevisiae* suspension analyzed in the previous section. An almost linear filtration curve was found in this plot, indicating that a cake filtration mode of filtration behavior occurs during the operation. A time course of MFI during the microfiltration of cross-linked *S. cerevisiae* suspension can be estimated from the slope of the filtration curve shown in Figure 15.12 and is illustrated in Figure 15.13. Examination of the time course variation of the MFI in Figure 15.13 showed that the operation was stable after 1 h, as indicated by an almost constant MFI index value.

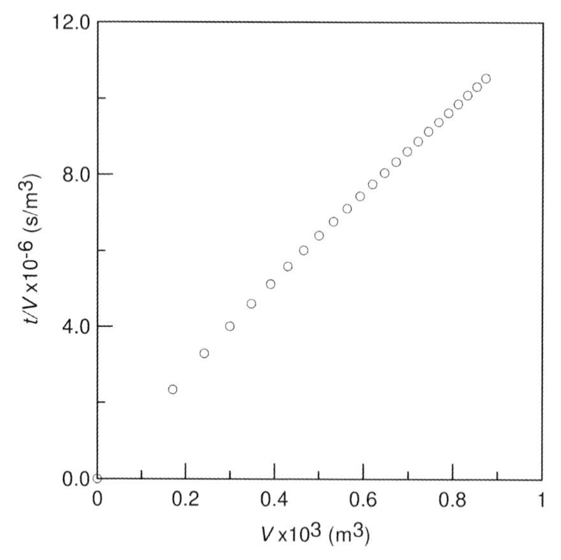


**Figure 15.12** A plot of $t/V$ versus $V$ for microfiltration of 0.5% cross-linked *S. cerevisiae* suspension under $\Delta P = 61$ kPa.

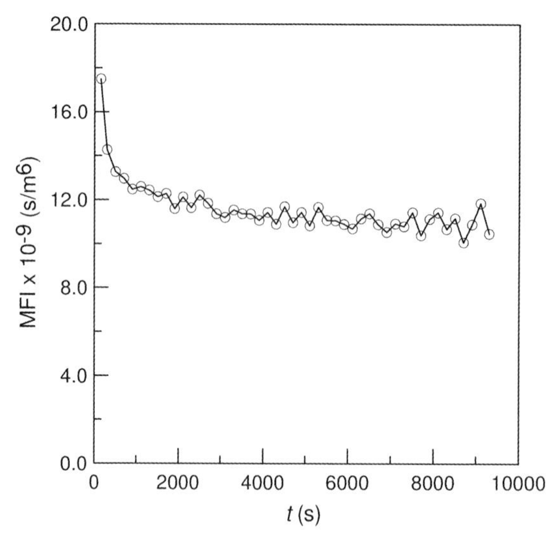


**Figure 15.13** A time course of MFI index estimated during microfiltration of 0.5% cross-linked *S. cerevisiae* suspension under $\Delta P = 61$ kPa by online dynamic analysis.

The MFI can also be used to monitor the particulate fouling in water treatment and serve as the fouling potential of feed water containing particles. The higher fouling potential feed water/suspension has a higher MFI value. A gradual increase in MFI value indicates an increase in filtration resistance which may be caused by compression of the fouling layer or migration of fine particles into the fouling layer. When the MFI value reaches a limiting value, the cleaning process should be followed. An abrupt change in MFI value during operation indicates unexpected failure of the membrane filtration system. For example, a sudden decrease of the MFI value may indicate damages of membrane or membrane module; while a sudden increase of the MFI value may indicate a faulty operation of driving units to cause an abrupt increase of filtration resistance. The MFI can also be applied in the laboratory as a quick test of the feed quality and in long-term testing to monitor the particulate fouling potential of feed water/suspension.

## 15.4
## Conclusions and Future Perspectives

An integrated online monitoring technique that can provide dynamic and real-time information of a fouling process has been developed. An application of this monitoring technique to a water treatment membrane process has been preliminarily demonstrated with a single photosensor. Results of the online dynamic analysis depict that this technique provides process-oriented capabilities of: (1) in situ measurement of fouling layer thickness, (2) dynamic analysis of fouling layer structure and (3) monitoring of membrane fouling potential in a membrane filtration process for water treatment applications. The example reported in this chapter is part of this concept; the integrated online dynamic monitoring technique is being applied to a pilot-scale water treatment unit with a photosensor array to provide real-time 3D profile of fouling layer thickness on a flat plate membrane module. With sensors which allow for scanning of each membrane surface on demand, the integrated online dynamic monitoring technique would be feasible for direct application in a commercial membrane module for water treatment. Moreover, it is a surprising fact that so many robust devices are available for fouling monitoring but they have never moved beyond laboratory status. Significant challenges remain to improve these available monitoring techniques to bridge the gap between laboratory setting and field application.

## Nomenclature

| | |
|---|---|
| $\alpha$ | Specific filtration resistance (m/kg) |
| $\alpha_o$ | Specific filtration resistance of incompressible fouling layer (m/kg) |
| $\varepsilon$ | Porosity of cake (–) |
| $\mu$ | Fluid viscosity (Pa · s) |
| $\rho_s$ | Density of particles (kg/m$^3$) |
| $A$ | Filtration area (m$^2$) |
| $c$ | Velocity of the ultrasonic wave in the medium (m/s) |
| $I$ | Fouling index (s/m$^6$) |
| $k$ | Kozeny constant (–) |
| $L$ | Fouling layer thickness (m) |
| $m$ | Mass ratio of wet to dry cake in fouling layer (–) |
| $n$ | Compressive coefficient (–) |
| $P_0$ | Empirical constant adopted in part (b) of Equation (15.12) (Pa) |
| $P_L$ | Hydraulic pressure (Pa) |
| $P_s$ | Solid compressive pressure (Pa) |
| $\Delta P$ | Applied filtration pressure (Pa) |
| $q$ | Filtration rate (m$^3$/m$^2$s) |
| $q_1$ | Flow rate at the permeate side (m$^3$/m$^2$s) |
| $R_c$ | Filtration resistance of fouling layer (1/m) |
| $R_m$ | Filtration resistance of filter medium (1/m) |
| $s$ | Mass fraction of solids in slurry (–) |
| $S_o$ | Specific surface area of particles (m$^2$/m$^3$) |
| $t$ | Filtration time (s) |
| $U_s$ | Crossflow velocity of suspension in CFMF (m/s) |
| $v$ | Volume of permeate per unit filtration area (m$^3$/m$^2$) |
| $V$ | Volume of permeate (m$^3$) |
| $\Delta w$ | Mass of cake in control layer per unit filtration area (kg/m$^2$) |
| $\Delta w_i$ | Mass of instantaneous fouling layer formation per unit filtration area (kg/m$^2$) |
| $x$ | Distance from the surface of the filter septum to the fouling layer surface (m) |
| $\Delta x$ | Thickness of filter cake deposited in time interval $\Delta t$ (m) |

## Subscripts

| | |
|---|---|
| $av$ | Average value of properties of a fouling layer |
| $i$ | Property of a cake layer at the fouling layer surface, property of a newly formed cake layer on the fouling layer surface |
| $t$ | Property of fouling layer at time $t$ |
| $x$ | Local value of properties of fouling layer at a distance $x$ from the membrane surface |

## References


1 H.C. Flemming, T. Griebe, G. Schaule, *Water Sci. Tech.* **1996**, *34*, 517–524.
2 H.C. Flemming, A. Tamachkiarowa, J. Klahre, J. Schmitt, *Water Sci. Tech.* **1998**, *38*, 291–298.
3 V. Chen, H. Li, A.G. Fane, *J. Membrane Sci.* **2004**, *241*, 23–44.
4 J. Chen, Q. Li, M. Elimelech, *Adv. Colloid Interfac.* **2004**, *107*, 83–108.
5 H.-C. Flemming, *Water Sci. Tech.* **2003**, *47*, 1–8.
6 R.J. Wakeman, *Trans. Inst. Chem. Eng.* **1994**, *72*, 530–540.
7 H. Li, A.G. Fane, H.G.L. Coster, S. Vigneswaran, *J. Membrane Sci.* **1998**, *149*, 83–97.
8 W.D. Mores, R.H. Davis, *J. Membrane Sci.* **2001**, *189*, 217–230.
9 J. Altmann, S. Ripperger, *J. Membrane Sci.* **1997**, *124*, 119–128.
10 M. Mendret, C. Guigui, P. Schmitz, C. Cabassud, *AIChE J.* **2007**, *53*, 2265–2274.
11 M. Hamachi, M. Mietton-Peuchot, *Chem. Eng. Sci.* **1999**, *54*, 4023–4030.
12 M. Hamachi, M. Mietton-Peuchot, *Chem. Eng. Res. Des.* **2001**, *79*, 151–155.
13 A.R. Johnson, *AIChE J.* **1974**, *20*, 966–974.
14 D. Mahlab, N.B. Yoself, G. Belfort, *Desalination* **1978**, *24*, 297–303.
15 M.J. Fernandez-Torres, F. Ruiz-Bevia, J. Fernandez-Sempere, M. Lopez-Leiva, *AIChE J.* **1998**, *44*, 1765–1776.
16 R.M. McDonogh, H. Bauser, N. Stroh, *J. Membrane Sci.* **1995**, *104*, 51–63.
17 A. Tamachkiarow, H.-C. Flemming, *Water Sci. Tech.* **2003**, *47*, 19–24.
18 K.L. Tung, S.J. Wang, W.M. Lu, C.H. Pan, *J. Membrane Sci.* **2001**, *190*, 57–67.
19 W.M. Lu, K.L. Tung, C.H. Pan, K.J. Hwang, *J. Membrane Sci.* **2002**, *198*, 225–243.
20 K. Ouazzani, J. Bentama, *Desalination* **2007**, *206*, 36–41.
21 V.E. Reinsch, A.R. Greenberg, S.S. Kelley, R. Peterson, L.J. Bond, *J. Membrane Sci.* **2000**, *171*, 217–228.
22 A.P. Mairal, A.R. Greenberg, W.B. Krantz, *Desalination* **2007**, *130*, 45–60.
23 A.P. Mairal, A.R. Greenberg, W.B. Krantz, L.J. Bond, *J. Membrane Sci.* **1999**, *159*, 185–196.
24 J.X. Li, R.D. Sanderson, E.P. Jacobs, *J. Membrane Sci.* **2002**, *201*, 17–29.
25 J.X. Li, D.K. Hallbauer, R.D. Sanderson, *J. Membrane Sci.* **2003**, *215*, 33–52.
26 H.G.L. Coster, K.J. Kim, K. Dahlan, J.R. Smith, C.J.D. Fell, *J. Membrane Sci.* **1992**, *66*, 19–26.
27 E.J. Watkins, P.H. Pfromm, *J. Membrane Sci.* **1999**, *162*, 213–218.
28 T.C. Chilcott, M. Chan, L. Gaedt, T. Nantawisarakul, A.G. Fane, H.G.L. Coster, *J. Membrane Sci.* **2002**, *195*, 153–167.
29 L. Gaedt, T.C. Chilcott, M. Chan, T. Nantawisarakul, A.G. Fane, H.G.L. Coster, *J. Membrane Sci.* **2002**, *195*, 169–180.
30 R.H. Davis, D.T. Leighton, *Chem. Eng. Sci.* **1987**, *42*, 275–281.
31 J.M. Pope, S. Yao, A.G. Fane, *J. Membrane Sci.* **1996**, *118*, 247–257.
32 D. Airey, S. Yao, J. Wu, V. Chen, A.G. Fane, *J. Membrane Sci.* **1998**, *145*, 145–158.
33 T.J. Su, J.R. Lu, Z.F. Cui, R.K. Thomas, R.K. Heenan, *Langmuir* **1998**, *14*, 5517–5520.
34 T.J. Su, J.R. Lu, Z.F. Cui, B.J. Bellhouse, R.K. Thomas, R.K. Heenan, *J. Membrane Sci.* **1999**, *163*, 265–275.
35 F. Pignon, A. Magnin, J.M. Piau, B. Cabane, P. Aimar, M. Meireles, *J. Membrane Sci.* **2000**, *174*, 189–204.
36 F. Pignon, A. Alemdar, A. Magnin, T. Narayanan, *Langmuir* **2003**, *19*, 8638–8645.
37 J.S. Vrouwenvelder, J.A.M. van Paassen, L.P. Wessels, A.F. van Dam, S.M. Bakker, *J. Membrane Sci.* **2006**, *281*, 316–324.
38 W.M. Lu, K.J. Hwang, *Sep. Technol.* **1993**, *3*, 122–132.
39 W.M. Lu, Y.P. Huang, K.J. Hwang, *J. Chem. Eng. Jpn.* **1998**, *31*, 969–976.

**40** K.L. Tung, Y.L. Li, K.J. Hwang, W.M. Lu, *Desalination* **2008**, *234*, 99–106.

**41** S. Haynes Jr, *Flow Through Compressible Porous Media Short-time Filtration, Wall Friction in Compression–Permeability Cells*, PhD dissertation, University of Houston, Texas, **1966**.

**42** F.M. Tiller, H.R. Cooper, *AIChE J.* **1960**, *6*, 595–601.

**43** F.M. Tiller, M. Shirato, *AIChE J.* **1964**, *10*, 61–67.

**44** W.M. Lu, *Theoretical and Experimental Analyses of Variable Pressure Filtration and the Effect of Side Wall Friction in Compression–Permeability Cells*, PhD dissertation, University of Houston, Texas, **1968**.

**45** J. Happel, H. Brenner, *Low Reynolds Number Hydrodynamics*, Prentice–Hall, Englewood Cliffs, **1965**.

**46** B.F. Ruth, *Ind. Eng. Chem.* **1935**, *27*, 708–723.

**47** H.P. Grace, *Chem. Eng. Prog.* **1953**, *49*, 303–367.

**48** J.C. Schippers, J. Verdouw, *Desalination* **1980**, *32*, 137–148.

**49** Ś.F.E. Boerlage, M. Kennedy, P.A.C. Bonnè, G. Galjaard, J.C. Schippers, *Desalination* **1997**, *113*, 231–233.

**50** Ś.F.E. Boerlage, M. Kennedy, M.P. Aniye, E.M. Abogrean, G. Galjaard, J.C. Schippers, *Desalination* **1998**, *118*, 131–142.

**51** ŚFE Boerlage, M. Kennedy, Z. Tarawneh, R. de Faber, J.C. Schippers, *Desalination* **2004**, *161*, 103–113.

# Index


*a*
ABS 96
absorption 209ff.
absorption tomography 209
acoustic characterization method 239ff., 242ff., 245, 337
– dead-end ultrafiltration deposit characterization 245
– deposit thickness measurement 247f.
– fouling characterization 243f.
– frequency 244
– general principle 242
– in situ application 247
– principle 247
– signal treatment 247
acoustic signal 244ff.
– clean membrane 245f.
– deposit thickness dependence 244
acridine orange labeling 167
activated carbon 97f., 100
– characteristics 97
– filler 100
– roughness dependence 98
adapted frequency 244f.
adhesion 119f.
adsorption 79, 178, 190f.
– heterogeneous 191
– homogeneous 190
adsorption free energy 191
adsorption-desorption method 79
aeration 307
AFC99 membrane 118f.
aggregation 286
Aimar et al. equation 47
anion adsorption 190
anion exchange membrane 141f., 144
anion permselectivity 196
anodized aluminium (AN) membrane 18f., 22ff., 25ff., 28
– photomicrograph 25, 27
– scanning electron micrograph 18
– videomicrograph 19
anti-Stokes scattering 128ff.
ARA membrane 143
asymmetry 42f.
atomic force microscopy (AFM) 3, 36ff., 44, 47, 78, 82ff., 84, 88f., 91, 95ff., 97ff., 105ff., 111ff., 114f., 117, 122ff.,125, 313f., 323
– application 82, 314
– benefits 125
– cantilever 107, 111
– contact mode 82
– force modulation 83
– height data 44
– image type 44
– imaging principle 105
– interaction detection method 82
– intermittent contact mode 82
– lateral force microscopy 83
– membrane development 122
– micrograph 44
– noncontact mode 82
– operation mode 37, 82
– phase contrast 83, 97, 98
– phase data 44
– phase image 44f.
– possibility range 106
– principle 313
– resolution 313
– scanning mode 313
– schematic 106
– surface pore dimension 111
– tapping mode 82, 89, 91, 95f.
– tip 107
– tipless cantilever 111
– topographical image 44f., 98f.
attenuation length 212ff., 216
autofluorescence 133, 161


*Monitoring and Visualizing Membrane-Based Processes*
Edited by Carme Güell, Montserrat Ferrando, and Francisco López

ISBN: 978-3-527-32006-6

average run length 71
axial resolution 132

***b***

back-transport flux 72
backpressure 29
backpulse 24ff., 27ff., 30
– cleaning efficiency 29
– duration 25ff., 29
– in situ experiment 26
bacteria 308f.
bacterial cell 15
Bancroft rule 285
Beer–Lambert law 209
bentonite 307, 328f.
– filtration 328
– fouling mechanism 328
– temporal fouling deposition change 328
bicontinuous structure 221f.
bidisperse suspension 72
biofilm 267, 308f., 314f.
– cohesive energy 314
– growth 267, 308f.
– thickness 316
biofouling 70, 267, 308
– definition 308
– layer characterization 70
– layer formation 308f.
– schematic illustration 308
biological media 147
biovolume 71
Bode plot 182, 185, 198, 202
bovine serum albumin, *see* BSA
BSA 64ff., 67ff., 156, 166f., 169ff., 172, 243, 306
– fluorophore labeling 166f.
– isoelectric point 192
– label 65
bubble passage effect 317
bubble point method 79
bubble pressure method 78
bubbling 325
bypass flow 333

***c***

cake formation 12ff., 15, 55, 71, 163ff.
cake fouling imaging 160ff.
– ex situ technique 160
– in situ technique 160
cake porosity 70f.
cake structure 160
cake thickness 13, 72, 152, 160, 165, 231f., 233, 236, 335ff.
– growth measurement 233, 335
– photointerrupter sensor measurement 336
– schematic diagram 336
– three-dimensional thickness distribution contour 336
capacitance 181, 198ff., 200f., 204
– salt concentration dependence 198ff., 200
capillary pressure 288, 294, 301
Carman-Kozeny equation 240
cation adsorption 190
cation exchange membrane 145, 192
– intensity map 145
– transport number 194f.
cation permselectivity 196
cell rupture 30
cell-protein fouling 170ff.
– in situ characterization 170ff.
cellulose acetate (CA) membrane 18, 26ff., 29f.
– photomicrograph 27ff.
– scanning electron micrograph 18, 30
– videomicrograph 20
cellulose ester membrane 71
channel 12, 14, 168f.
– height 14
charge 3, 6, 117, 177
chemical adsorption free energy 191
chemical cleaning 67, 69, 169f.
chlorine addition 169
clay suspension 231, 237, 239, 241, 246
cleaning 66ff., 69, 155f., 169
cleaning protocol 66f.
cleaning time 67f.
$CO_2$ 87f., 100
– electronic interaction 87
– permeability 87f., 99
– selectivity 87f., 99
– solubility 87
coalescence 285, 288, 293ff., 301
– rate 301
collagen 155
– SHG property 155
colloid probe 109, 117f., 121f.
– adhesion force 124
– adhesion reduction 122
colloid rejection 109, 120
composite membrane 78, 200, 204
compressibility parameter 348
– calculated values 348
computerized image analysis 83
concentration polarization 322f., 330
conductivity 178f., 201

confocal laser scanning microscopy (CLSM), *see* confocal scanning laser microscopy (CSLM)
confocal microscope 57f.
– main parameter 57f.
– schematic 57f.
confocal microscopy 2, 37f., 131f., 155
– advantage 132
– emission wavelength 155
– single-photon 155
confocal Raman microscopy 131f.
– cell design 140f.
– resolution depth 133
– visualization 138
confocal Raman spectroscopy 127ff.
confocal scanning laser microscopy (CSLM) 37f., 51, 55ff., 58ff., 61ff., 64, 70ff., 73f., 152, 160, 314f., 323
– advantage 56, 315
– application 61f., 73f.
– bio-fouling layer characterization 70
– 3D structural image 314
– drawback 315
– fluorescence imaging 57
– fluorescence mode 57, 73
– fluorescent labeling 59
– fundamental 56ff.
– historical review 62
– image 64
– image analysis 60f.
– limit 73f.
– membrane characterization 62f.
– membrane fouling characterization 63ff.
– online monitoring 72
– principle 37f., 56ff., 314
– resolution 73
– sample preparation 59
– surface porosity measurement 60f.
connected pore 220
connectivity 220f.
constant phase element (CPE) 182
continuity equation of cake compression 341
contrast 214
correlation time 263
covariance 42
creaming 286
critical flux 15ff., 18f., 21f., 24, 31, 120, 123, 165, 226
critical parameter 120
critical Peclet number 124
critical pressure 122
critical velocity 299ff.
cross-check fouling model 167f.
cross-section micrograph 41ff.
– interpretation 41f.
– preparation 41f.
crossflow microfiltration 11f., 14, 20, 156
– schematic DOTM 14
crossflow velocity 13, 15ff., 21f.
cryogenic technique 50
CSLM image 50
current intensity 181
cut-off 5, 45ff.
Cyclopore membrane 114f., 121, 184

***d***

3D characterization 209ff., 219, 314
3D femtosecond imaging (3DFI) 320
2D fluorescence 256, 265ff., 271, 280
– application 280
3D fluorescence 265f.
– transforming into 2D plots 265
2D image 210
3D image 211, 314
3D volumetric reconstruction 69f., 74
Darcy's law 309f.
data matrix 137
dead-end filtration 248, 339, 343, 344, 347f.
– flux curve 248
dead-end ultrafiltration deposit characterization 245
Debye length 119
decay length 119
defect 223ff.
– detection 223f.
degree of asymmetry (DA) 42f.
degree of freedom 129
dendrimer 6
dense membrane 77, 80, 184, 202
– characterization 80
– charged 184
– support 80
deposit 229ff., 233, 236f.
– acoustic wave analysis 230
– cross-section profile 237
– distribution 242
– formation mechanism 238
– growth dynamics 240
– growth kinetics 237, 246
– in situ characterization 230f., 242
– kinetics 237, 249f.
– kinetics related to flow rate variation 249
– local structure 229
– permeability 240
– porosity 229, 250
– thickness 229ff., 233, 236ff., 239ff., 241f., 248f.

– thickness calculation 232f., 236ff., 241
– topography 231
deposition rate 18f.
depth discrimination 132f.
Desal membrane 112ff.
– AFM image 112
– pore size distribution 113
– surface characteristics 113
dextran 47f., 64ff., 67, 69f.
– label 65f.
– molecular weight affecting fouling pattern 65
– permeate flux evolution 65
– retention experiment 47
dielectric constant 178f.
differential scanning calorimeter (DSC) 79f.
diffusion potential 180f., 194ff.
digitization 57f.
direct observation (DO) 316ff., 324
– membrane module 317f.
direct observation of surface of the membrane (DOSM) 17
direct observation through membrane (DOTM) 14ff., 17f., 22, 24, 152, 160, 163, 316f., 324, 332f.
– image 16f.
– model solution 316
– resolution 316
– schematic 14
direct visual observation (DVO) 11ff., 14, 17ff., 20, 22ff., 26, 30f., 316
– image montage 24
– micrograph 23, 27
– schematic 17
Donnan exclusion 194
Donnan potential 180
DOTM, *see* direct observation through membrane
droplet coalescence 293
droplet formation 287ff., 298ff., 301
– mechanism 298f., 301
– microscopic visualization 300
droplet size 288, 291ff., 298
– distribution measurement 291f.
droplet volume density distribution 292ff.

*e*
echo amplitude 243
echo time 246f.
effective decay length 119
electric dipole moment 130
electric force microscopy (EFM) 83
electrical double layer (EDL) 178
electrical measurement 178ff., 188ff.
electrical membrane characterization 177ff.
electrical potential 179, 192
– difference 192
electrical resistance 183, 198ff., 202f.
– electrolyte concentration dependence 198f.
– membrane pore radii/porosity dependence 199
electrochemical microscopy 83
electrochemical shear probe 321, 323f.
electrokinetic charge density 180, 190f.
electromotive force 187
electromotrice force 193
electron microscopy (EM) 2, 35f., 78, 311ff., 323
– overview 311ff.
electroosmotic flow 177f.
emulsification 73, 218, 283ff., 286ff., 289ff.
– advantage 283
– conventional 287ff.
– direct 287ff.
– jetting regime 289, 301
– pore observation 290
– process characterization method 283ff.
– technology 287
emulsifier 283ff., 286, 288, 291ff., 294, 299ff.
– schematic illustration 285
emulsion 283ff., 286ff.
– basic types 284
– double 284, 299
– instability mechanism 285f.
– multiple 284, 299
– oil in water (o/w) 284
– oil in water in oil (o/w/o) 284
– physical stability 284
– preparation 283
– Sauter diameter 293ff.
– schematic illustration 284
– water in oil (w/o) 284
– water in oil in water (w/o/w) 284
endotoxin 63
– blocking 63
environmental scanning electron microscopy (ESEM) 2, 36, 38, 56, 312f.
– acceleration voltage 313
– application 312
– principle 312
– resolution 312f.
– sample conditions 313
– sample preparation 313
excitation 129, 153f.
– energy 129
– multiphoton 154

external fouling 11
extracellular polymeric substance (EPS) 306, 308, 311
– rejection value 311

*f*
factor analysis 137
Fairbother–Mastin approximation 180, 187
$Fe^{2+}$ affinity 280
feed quality quick test 351
feedforward artificial neural network 270f.
– schematic diagram 271
fermentation 156
Ferry-Faxen equation 46
Fick's first law 182
field effect scanning electron microscopy (FESEM) 78, 221, 311f.
– acceleration voltage 313
– application 312
– principle 312
– resolution 82, 312f.
– sample conditions 313
– sample preparation 313
filtration module 232
filtration resistance 342, 348ff.
– variation 349
filtration setup 232
filtration streaming potential (FSP) 178, 186f., 189f.
filtration theory 345, 348
fixed charge concentration 194f.
flat membrane 39ff.
– synthesis 39ff.
Flory parameter 92
Flory–Huggins interaction parameter 93
Flory–Huggins theory 92
flow rate 341
fluorescence 5, 59, 152ff., 155, 161, 163
– 2D, *see* 2D fluorescence
– 3D, *see* 3D fluorescence
– decay, *see also* fluorescence decay 259f.
– decay curve 259
– emission 256, 278
– fingerprint 265, 269
– intensity 59, 258
– labeling, *see also* fluorophore labeling 255
– microscopy 152f.
– polarity dependence 256
– quantum yield 256
– spectra 258, 278
– time-resolved 256, 258, 264
fluorescence anisotropy 256, 261f., 278
– application 263
– calculation 262
– decay 263
– polarizer orientation 262
– time-resolved 263
fluorescence decay 259ff., 263, 275
– complex system 260
– environmental conditions influencing 261
– multi-fluorophore solution 260
fluorescence map 265ff., 268ff., 271
– 3D 266, 268f., 271
– projected surface plot 266
– spectra interpretation 268f.
– spectra substraction 267f., 270
– surface contour plot 266
fluorescence technique 255ff., 266f.
– advance 255ff.
– detection limit 257
– limitation 255ff.
– optical fiber bundle application 266f.
– spectra deconvolution 270
fluorescent foulant 60f.
fluorescent labeling 70
fluorescent marker 5f., 315
fluorescent probe 59, 161
– general requirements 59
fluorometry technique 266f.
– optical fiber bundle application 266f.
fluorophore 59, 161ff., 256, 258, 262, 264
– decay time 258, 261
– intracellular 264
– peak analysis 269
– rotational freedom 262
fluorophore labeling 59, 160ff., 163, 166, 172
– effect 166f.
flux curve 248
flux recovery (FREC) 67f.
foam structure 222, 224
force 114f.
– ionic strength dependence 114f.
force distance curve 109
forward filtration 12, 19f., 25f.
foulant composition analysis 330f.
foulant movement 316
fouling 2, 11, 64, 155f., 167f., 192, 305ff., 317, 325
– affecting factor 306
– bubble passage effect 317
– definition 305
– echo amplitude 243
– empirical model 310
– external 155, 168
– internal 64, 155, 167f.
– mechanism 64, 155, 192, 310
– potential 351

– removal mechanism 318f.
– thickness 70, 322
– thickness quantification 322
– two-stage 155, 168
fouling agent 64
– identifying 64
– locating 64
fouling layer 61, 70, 329, 337f., 342ff., 344ff.
– average porosity 343
– compression 340f., 345ff., 351
– dynamic analysis 329f., 334f., 338, 343, 346f.
– dynamic structure analysis 351
– dynamic thickness variation measurement 343
– formation 339, 342
– intrinsic properties 344
– online measurement 334, 337, 343ff., 349
– physical parameter 342
– porosity 341ff., 344
– porosity distribution 346
– pre-estimated surface porosity 342
– pressure drop 342f.
– real-time thickness variation 342f., 348
– resistance 341, 344
– structure analysis 329f., 338, 342
– structure analysis during filtration 342
– thickness 70, 322, 329, 334, 337ff., 342ff.
– thickness time course 344
– thickness variation 338
fouling monitoring 330ff.
– classification 330f.
– development 330f.
– in situ technique 330
– requirement 330f.
fractal dimension 84, 89, 92
fractional deposition 20f., 124
– quantifying method 21
fractionation 72
free volume 79, 85, 88, 92
Freundlich isotherm 190
full width at half maximum (FWHM) 132, 135

**g**

gadolinium 6
gas permeation 146
gas separation 77f., 80, 85f., 90
gas separation membrane 77f., 80, 85ff., 94
– permeability 85f.
– selectivity 85f.
Gaussian function 135f.
– linear combination 135f.
Gibbs free energy 191
Gibbs function 178
Gibbs-Marangoni effect 285
glass transition temperature 93
global cake specific resistance 240
global porosity 241
global thickness 241
gray level 84, 235
– intensity 235
– maximum determination 235

**h**

Hagen-Poiseuille law 287
Hansen parameter 92
Hansen polar solubility parameter 92
He–Ne laser sensor 231
Helmholtz–Smoluchowski equation 179
hemodialysis module 225f.
Hildebrand solubility parameter 91ff.
hollow fiber 216f., 221f., 224, 226, 317
– defect detection 224
– FESEM image 221
– SRμCT image 217, 221
– tomography image 222
horseradish peroxidase 272f., 277, 279
– structure 277
humic acid 123f.
– fractional deposition 124
– removal 123
hydraulic pressure 343, 346
– distribution 343
– local 343
hydraulic resistance 309
hydrodynamic permeability 189
hydrophilic/lipophilic balance (HLB) value 285

**i**

ideal capillary porous membrane 183
ideal cation exchange membrane 192
IFME software 38f., 41, 43f., 50, 53
image analysis 60f., 83, 234f.
– cake characterization 61
– surface porosity measurement 60f.
imaging depth 58
imaging force 114ff.
imidation 90
impedance 181ff.
– plot 183f., 197f., 200
impedance spectroscopy (IS) 4, 178, 181, 185, 187f., 197f., 199f., 203, 332f.
– fitting 197
– measurement 187f.
in situ fouling assessment 330f.

in situ membrane fouling
characterization 229ff., 329f., 351
inline monitoring 295ff., 299
inner Helmholtz plane (IHP) 179
inorganic filler 96, 100
inorganic fouling 243
integral asymmetric gas separation
membrane 94
– structural characterization 95
Intensity map 142, 145
interaction energy 92f.
internal fouling 11
intrinsic fluorescence 255, 271
– reporter 271
invasive fouling observation method 311ff.
invasive technique 311ff., 323f.
ion exchange membrane 63
– protein binding 63
ion transport number 178, 194ff., 196, 203
ionic permeability 201f.
ionic species distribution visualization 141
ionic strength 114ff., 119, 121f.
IR absorbance 129
isoelectric point 180, 192
isopotential line 116
– ionic strength influence 116

*j*
Jablonski diagram 153
jet 289ff.
jet diameter 289f.
jetting 289, 301
– video imaging 289
jetting regime 289, 301

*k*
Kelvin equation 79
Kozeny constant 341f.
Kozeny equation 341

*l*
β-lactoglobulin 272ff., 275ff.
– fluorescence decay fit 276
– fluorescence emission spectra 274
– normalized steady-state fluorescence
anisotropy variation 277
– permeation 274ff.
– steady-state fluorescence anisotropy 275,
277
– structural alteration 275f.
– time decay 275
– time-resolved fluorescence 275
– TMP-dependent structural alteration 276
Langmuir isotherm 190
Laplace instability 299
Laplace pressure 298
laser 230ff., 320, 323
– application 320
– deflection 234
laser beam excitation 319
laser excitation near infrared region 320
laser sheet at grazing incidence (LSGI)
technique 230, 232, 236f., 249
– capability 236f.
– lateral resolution 236
– limitation 236f.
– thickness measurement accuracy 236
laser triangulometry 231, 332f.
– principle 231
latex 15f., 18, 20, 22, 24
– particle deposition 16, 20
Levenberg–Marquardt method 136f.
light speed 129
line imaging 138
linear attenuation coefficient 210
liquid imaging 114
local thickness 241
Loeb-Sourirajan phase inversion 39f.
long-range electrostatic interaction 119
Lorentzian function 135f.
– linear combination 135f.

*m*
macrovoid 222ff.
– distribution 222f.
magnetic force microscopy (MFM) 83
marker 5f.
– fluorescent 5f.
– radioactive 5
mass transfer 40, 219, 227
Matrimid membrane 94ff.
– active layer thickness 96
– active layer width 96
– dense layer 95f.
– evaporation time 96
– force modulation image 95
– phase contrast image 95
matrix 137
Maxwell law 87
Maxwell theory 86, 100
MBR, *see* membrane bioreactor
membrane, *see also* polymeric membrane
– asymmetric 78, 94, 177, 199, 202
– classification 77f.
– cleaning protocol 66f.
– content visualization 127ff.
– dense, *see also* dense membrane 77, 184
– echo 247

– electrical modification 194
– flat filter 287
– fouling reducer 71
– heteroporous 77
– homoporous 77
– hydrophilicity 6
– integral 94
– micro-engineered 287
– morphological parameter 34f., 38
– morphological parameter quantifying from SEM image 38f.
– morphology 33ff., 36ff., 40ff., 45ff.
– nonporous 177
– performance 45ff.
– permselectivity 196
– porous 177, 183
– selective layer thickness 35
– selectivity 77
– separation characteristics 105
– structural characterization 216ff.
– surface observation 16ff.
– symmetric 78, 177, 202
– synthesis monitoring 146f.
– thickness 35, 210
– tubular 287, 320
membrane bioreactor (MBR) 70ff., 264ff., 267, 305ff., 315
– biofouling layer characterization 70ff., 315
– fouling, *see also* membrane bioreactor fouling 305ff.
– fouling affecting factor 306
– fouling monitoring 305ff., 330
– performance 267
– visualization 305ff.
membrane bioreactor fouling 305ff., 315
– feed nature 306
– foulant 306
– layer visualization 315
– membrane properties 307
– model solution 306
– operating conditions 307
– pore size effect 307
– sludge retention time influence 307
membrane emulsification, *see* emulsification
membrane fouling simulator (MFS) 319, 324, 333
membrane potential (MP) 3, 177f., 180f., 187ff., 189, 192ff., 196f., 201f.
– concentration dependence 194
– measurement 3, 187f.
membrane properties 1, 80
mercury porosimetry 79
methanol crossover 145
micro engineering 286f.
microcapsule 39, 48ff., 52
– cross-section micrograph 50ff.
– CSLM image 52
– ESEM 49f.
– internal porous structure 52
– micrograph 49
– morphological characterization 49
– production set-up image 49
– SEM 49, 52
– size distribution 49
– synthesis 48f.
microchannel 286f., 298, 301
– emulsification, *see also* microchannel emulsification 298f.
– plate 286f.
microchannel emulsification 298f., 301
– droplet formation mechanism 298f.
microelectrode 6
microfiltration membrane 11ff., 114
– cleaning 11ff.
– fouling 11ff.
microRaman confocal spectroscopy 127, 131
microscopic technique 2f., 33ff., 36ff., 80ff.
microsieve 72f.
mixed matrix composite membrane (MMCM) 78, 86f., 98f.
– permeability 87
– phase contrast AFM picture 99
– topographic AFM picture 98f.
mixed matrix membrane 78, 86f., 96ff.
model protein-cell system 156
modified fouling index (MFI) 348ff., 350
– index value 349
– time course 349f.
– value interpretation 351
molecular weight cut-off (MWCO) 46ff., 78, 107, 111ff., 125
– surface pore dimension 111
morphology 33ff., 36ff., 40ff., 45ff.
– parameter 44
multiphoton microscopy (MPM) 2, 151ff., 154ff., 161, 163, 167f., 171f.
– advantage 160, 165, 172
– application 156f.
– filtration circuit 158
– image analysis 158ff.
– imaging time 157
– material 156ff.
– membrane filtration 156f.
– methods 156ff.
– module design 157f.
– principle 153
– schematic setup 159

– side-on image 167f.
– standard experimental procedure 158
– top-down projection 168

***n***

*N*-acetyltryptophanamide (NATA) lifetime 260
Nafion 145f.
– intensity map 145
nanofiltration membrane 108
– pore size distribution 108
– single pore 108
natural fluorescence 255, 271ff., 280f.
– technique 256f., 280f.
net flux 26, 28ff.
– versus backpulse duration 28
net permeate flux measurement 26
nitrate band 143
nitrate ion 143
– concentration profile 144
nomenclature 204f., 250, 302, 352
non-ideal capacitor 182
non-invasive observation method 315
non-invasive technique 229, 244, 315, 323ff., 332f.
– advantage 325
– limitation 325
– overview 332f.
nuclear magnetic resonance (NMR) spectroscopy 79, 322ff., 332f.
nucleic acid 71
Nyquist plot 181f., 185, 197, 198ff.

***o***

objective 57f.
– magnification 58
– numerical aperture (NA) 58
one photon process 153f.
online fluorescence monitoring 5
online fouling layer thickness measurement 334f., 337
– optical method 335ff.
online fouling monitoring 334ff., 343
– dynamic analysis 334f., 343, 349
online measurement 3f., 343
online monitoring 1, 3ff., 55ff., 72f., 343
optical laser sensor 231
– principle 231
optical membrane fouling characterization technique 151ff.
optical method 229ff.
– principle 232
– using laser sheet at grazing incidence (LSGI) 230, 232
optical microscopy 2
order of magnitude verification 240ff.
Ostwald ripening 286
outer Helmholtz plane (OHP) 179
ovalbumin 166f., 169f., 172
– fouling rate 167

***p***

particle deposition 12ff., 15f., 21, 55, 72, 152, 226, 239, 331
– kinetics 331
– observation 14
– wall effect 239
particle image velocimetry (PIV) 6, 322ff.
particle image velocimetry/laser induced fluorescence (PIV/LIV) equipment 6
particle movement 15
particle removal 24
particle size 21f.
particle-specific area calculation 241
Peclet number 124
permeability 45, 85ff., 88, 91, 92, 96f., 100, 219, 240, 309
– active carbon content dependence 100
– solvent influence 92
permeate flux 11, 237f., 305, 309, 339f.
– time course 339
permeation 272ff., 275
permporometry 79
phase contrast agent 214
phase diagram 40
phase inversion 286
phase modulation fluorometry 258
phase segregated membrane 87ff.
phase segregation 88ff.
photo-interrupt sensor 322, 323f.
photodegradation 265
photointerrupter 335f., 343
– principle 335
– schematic diagram 335
photointerrupter sensor 335f.
photomultiplier 57f.
photon 129, 153f., 210, 213
photon energy 213
pinhole size 57f.
pixel 210
Planck constant 129
plane-parallel capacitor expression 201
point-by-point scanning Raman microscopy 138
point-scanning technique 138
point-to-point illumination 139
polarization intensity differential scattering (PIDS) 291, 301

polarized light 261
polyacrylonitrile ultrafiltration membrane 21
polycarbonate (PC) membrane 64ff., 67ff., 166, 183f.
– CSLM image 64
– 3D volumetric reconstruction 69f.
– track-etched (PCTE) 343ff.
polyimide, fluorinated 90
polymeric membrane, *see also* membrane 33, 37, 47, 85f., 90, 133
– autofluorescence 133
– free volume 86, 88
– permeability 85f.
– physicochemical characterization 85
– selectivity 85f.
– separation capability 47
polypeptide 271
polysaccharide 64, 70f.
polystyrene colloid probe 109f.
polysulfone (PSf) 39ff., 43, 47, 50, 183f., 237, 245f.
– SEM 43
– structure 122
– ultrafiltration membrane 183f.
polysulfone-sulfonated poly(ether ether) ketone (PSU/SPEEK) blend membrane 122ff.
– colloid probe adhesion force 124
– colloid probe adhesion reduction 122
– structure 122
pore diameter 34, 43, 46ff., 111ff., 114f., 117, 219ff., 289
pore network 219
pore radii 47
pore regularity 34f., 41
pore size 34, 107f., 115, 219f., 221, 224, 292
pore size distribution 3, 34, 42, 78f., 107f., 112ff., 125
pore statistics study 78
pore surface 60
– fraction 60
pore symmetry 34f., 41f.
pore tortuosity 34
porosity 35, 218ff., 344ff., 347f.
– calculation 218
– solid compressive pressure dependent 346f.
porous membrane 77f.
– application 78
positron annihilation lifetime spectroscopy (PALS) 79
positron annihilation spectroscopy 3
potential 117, 180
premix membrane emulsification 290ff., 293ff., 296ff., 299ff.
– disperse phase fraction 294
– droplet size distribution 292f.
– emulsifier concentration dependence 294
– inline measurement 295ff., 298
– pore size influence 292
– process flux 295f.
– process parameter influence 291
– process principle 290ff.
– repeated processing 292
– schematic illustration 291
– transmembrane pressure difference 292
– volume density 292ff.
pressure 346
pressure drop 287
pressure-driven membrane process 56ff.
Prigogine–Flory–Patterson theory 92
principle component analysis 137f.
process-oriented fouling monitoring system 333f.
– installation 333f.
process-oriented monitoring technique 4f.
profile image 167
projector technique (PT) 316, 323f.
– resolution 316
protein 63, 67, 69f., 108, 110, 155f., 166ff., 169, 272ff., 278
– adsorption 63
– aggregate 171f.
– deposition 168f.
– fluorophore labeling 166f.
– label 65
– microfiltration 64
– MPM image 171
– permeation 272, 280
– purification process 63
– structural alteration 272f.
– unfolding 278, 280
protein fouling 67, 155f., 166ff., 172
– imaging 166ff.
– in situ imaging 166
– two-stage mechanism 166
protein/dextran mixture 64f.
protein/membrane interaction 63
protein-only solution 171f.
protein-polysaccharide solution 64
proton conducting membrane 144
pseudoclearfield equalization 84
pulse fluorometry 258
pulse laser photometry 295ff., 298, 301

*q*
quantum mechanics 128
quencher 256
quenching mechanism 256

*r*
Raman data 133
Raman effect 127ff.
– partial quantum mechanical treatment 128ff.
– theoretical background 127ff.
Raman imaging 127
Raman intensity 132
Raman line assignment 134
Raman microspectrometry 130f.
– confocal setup 131
Raman scattering 129
– anti-Stokes 128ff.
– Stokes 128ff.
Raman spectroscopy 133f., 138ff., 140f., 146f.
– baseline 133f.
– experimental setup 141
– fitting 136
– fluorescence 133
– intensity ratio 134f.
– ionic species distribution visualization 141
– line scanning 138ff.
– membrane synthesis monitoring 146f.
– membrane systems application 140ff.
– nonlinear least squares fitting 135
– plane 140
– quantitative processing 134
– visualization 138ff.
– volume 140
Rayleigh instability 289
Rayleigh scattering 127ff., 130
real-time fouling layer thickness variation 342f.
real-time monitoring 1ff.
region of interest (ROI) 235f.
regularity analysis 42ff.
rejection performance 310f.
relative amplitude variation 244
relative thickness 248f.
relaxation 153, 183, 197, 201
removal 12
resistance 183, 200, 202f., 309, 344
resistance–capacitor (RC) circuit 181f.
resolution 81, 132f., 214f., 225f., 311, 313, 316
reverse filtration 12
reverse osmosis 11f.
RGB merge 168f.
RGB stack 158ff.
rhodamine B 50, 52
Robeson bound 85ff., 88, 93, 100
Robeson chart 100
Robeson diagram 88
rolling motion 22ff.
rotational degree of freedom 129
roughness 84, 96f., 112f., 117
Ruth filtration equation 349

*s*
*S. cerevisiae* 343ff., 346ff., 349f.
– cross-linked 343ff., 346ff., 349f.
– SEM image 345
salt permeability 203f.
sample rotation 215f.
Sauter diameter 293ff., 297ff.
scanning electron microscopy (SEM) 35f., 38, 41, 43, 50, 53, 55f., 78, 81, 95f., 109f., 152, 199, 217, 311ff., 343f.
– acceleration voltage 313
– application 311
– image analysis 38f.
– preparation 81
– principle 311
– resolution 81, 311, 313
– sample conditions 313
– sample preparation 313
scanning probe microscopy (SPM) 77ff., 80
scanning tunneling microscopy (STM) 82
second harmonic generation (SHG) 155
secondary electron 81
secondary fluorescence 59
sedimentation 286
selectivity 11, 85ff., 87, 91, 92, 96, 99
– active carbon content dependence 100
– solvent influence 92
shear force 321
Shirasu porous glass (SPG) membrane 219
silica colloid probe 109, 121
silica sphere 109f.
– BSA-coated 110
– SEM image 109
silicone layer 97f., 132
single cell 111
single photon process 153f.
single-particle theory 21
sludge retention time (SRT) 307
small angle neutron scattering (SANS) 332f.
small angle X-ray scattering (SAXS) 79, 90
Snell-Decartes's law 234
solid compressive pressure 346f.
– porosity dependence 347

soluble microbial product (SMP) 306, 308, 310f.
– rejection value 311
solute retention test 79
solution diffusion potential 194f.
solvent 88ff.
– distribution visualization 144ff.
solvent evaporation 88ff., 93f.
solvent–polymer interaction energy 93
solvent–polymer system 91
spacer 15
specific cake resistance 309f.
specific filtration resistance 342, 348f.
– local 343, 348
– solid compressive pressure dependent 347
spectrofluorometer 257
– schematic diagram 257
SPEEK, *see* sulfonated poly(ether ether) ketone
sponge-like structure 220f.
steady-state fluorescence 256ff., 264, 267, 275, 278
– advantage 264
– limitation 264
– membrane bioreactor monitoring 264ff.
– principle 257f.
– spectra 258, 267, 278
– spectra deconvolution technique 267
– two-dimensional 267
– versus time-resolved fluorescence technique 264
– water treatment process monitoring 264ff.
steady-state fluorescence anisotropy 261f., 275, 278
– schematic representation 262
– variation 279
Stern layer 179
Stern model 179
Stokes scattering 128ff.
Stokes-Einstein equation 263
stream component 118
streaming potential (SP) 3, 117, 177ff., 186ff., 190, 191f., 203f.
– along-the-surface 117
– formation 179
– measurement 186f.
– pH dependence 192
– principle 186f.
– through-the-membrane 117
structural characterization 230
sulfonated EPDM–polypropylene membrane 183
sulfonated poly(ether ether) ketone (SPEEK) 122
surface cake layer 342
surface charge 117, 180
– density 180
surface electrical property 114, 116f.
surface pore diameter distribution 112
surface pore dimension 111
surface porosity 35, 60f., 70, 344ff., 347f.
surface potential 117
surface potential microscopy 83
surface properties 83
surface roughness 35, 118
symmetry group 42
synchrotron radiation microcomputed tomography (SRμCT) 212, 214ff., 217, 227
– advantage 217

***t***

tangential flow filtration 11, 24
tangential streaming potential (TSP) 178, 186, 189f., 204
Teorell–Meyer–Sievers (TMS) theory 180
ternary phase diagram 40
thermoporometry 79
thickener 285
thickness measurement 229f., 233, 322
three-layer model 198
time of flight (TOF) 242, 246f.
total backpulse duration 26
transient phenomena 140
translucent membrane 141
transmembrane pressure (TMP) 290f., 305, 309f.
transmembrane pressure difference 292ff., 295, 301
transmembrane streaming potential 178, 188f.
transmission electron microscopy (TEM) 2, 35f., 38, 50, 53, 55, 78, 81, 90, 312f.
– acceleration voltage 313
– application 312
– preparation 81
– principle 312
– resolution 81, 313
– sample conditions 313
– sample preparation 313
transparent sidewall 12
transport parameter 201
tryptophan 256, 258, 264f., 271ff.
– fluorescence 256, 265, 271ff.
– fluorescence response interpretation 271f.
– relative position change 272
– structural reporter 273

two photon process 153f.
two-layer membrane 183
two-layer model 198
two-stage fouling mechanism 155, 166, 168

*u*
ultrafiltration membrane 108, 183
– pore size distribution 108
– single pore 108
ultrasil 67ff., 170
ultrasonic measurement 242f.
– general principle 242
ultrasonic reflectometry 243
ultrasonic time domain reflectometry (UTDR) 4, 320f., 323f., 332f., 337f.
– minimum thickness 320
ultrasound technique 4

*v*
velocity map 323
– 2D 323
vibrational degree of freedom 129
vibrational energy state 129f.
virtual energy state 129
visualization software 211
voxel 211, 214, 223, 225f.
– size 225f.

*w*
wall effect 239
Warburg impedance 182
wastewater treatment 305, 350f.
water flux 67f., 123
water quality monitoring 348ff.
– quick test 348
water treatment process 264ff., 267, 329ff.
wide angle X-ray diffraction 79
widefield imaging 138

*x*
X-ray microimaging (XMI) 212, 226f.
– resolution 226
X-ray photon spectroscopy 4
X-ray radiography 223
X-ray tomography 4, 209ff., 212ff., 214ff., 225f.
– advantage 226
– application 212, 215ff.
– concentration study 225f.
– data analysis 227
– deposit characterization 225f.
– experiment design 212, 214
– experimental conditions 212, 214ff.
– experimental time 215f.
– fouling study 225f.
– image 210f., 214f.
– limit 212, 225f.
– object size calculation 214
– principle 209ff.
– resolution 214f., 226f.
– sample rotation 215
– voxel size 225

*y*
yeast 15, 18, 20ff., 23f., 26, 28, 30, 111, 156, 160, 162ff., 165, 170, 306, 343f.
– attaching tipless AFM cantilever 111
– autofluorescence 161
– capturing protein aggregate 156
– cell diameter 161
– cell motion 23
– deposition 18, 20f., 23, 164
– direct microscope image 21
– filtration performance 161
– fluorescence absorption 161
– fluorophore labeling 161ff.
– labeling 161ff., 170
– layer predepositing 31
– monolayer 163
– multi-layered cake 25
– multiphoton image 161
– patchy filtration cake enlargement 164
– preformed cake 171
– SEM image 111
– suspension 162ff., 165
– suspension washing effect 161ff.
Yeo's experiment 214

*z*
zeta potential 3, 179f., 187, 188f., 191ff.
– pH dependence 193
– salt concentration dependence 190
zoom magnification 57f.